# Methods in Molecular Biology

*Series Editor*
John M. Walker
School of Life Sciences
University of Hertfordshire
Hatfield, Hertfordshire, AL10 9AB, UK

For further volumes:
http://www.springer.com/series/7651

# Directed Evolution Library Creation

## Methods and Protocols

## Second Edition

Edited by

**Elizabeth M.J. Gillam**

*School of Chemistry and Molecular Biosciences, The University of Queensland, Brisbane, QLD, Australia*

**Janine N. Copp**

*Michael Smith Laboratories, University of British Columbia, Vancouver, BC, Canada*

**David F. Ackerley**

*School of Biological Sciences, Victoria University of Wellington, Wellington, New Zealand*

Humana Press

*Editors*
Elizabeth M.J. Gillam
School of Chemistry and Molecular Biosciences
The University of Queensland
Brisbane, QLD, Australia

Janine N. Copp
Michael Smith Laboratories
University of British Columbia
Vancouver, BC, Canada

David F. Ackerley
School of Biological Sciences
Victoria University of Wellington
Wellington, New Zealand

ISSN 1064-3745 ISSN 1940-6029 (electronic)
ISBN 978-1-4939-1052-6 ISBN 978-1-4939-1053-3 (eBook)
DOI 10.1007/978-1-4939-1053-3
Springer New York Heidelberg Dordrecht London

Library of Congress Control Number: 2014943737



Printed on acid-free paper

Humana Press is a brand of Springer
Springer is part of Springer Science+Business Media (www.springer.com)

# Preface

Since the first edition of this book was published in 2003, directed evolution has become entrenched as a fundamental approach for engineering proteins to enhance activity and explore structure–function relationships. Importantly, improved strategies for directed evolution have also supported the rapid development of the field of synthetic biology over the last decade. All directed evolution strategies mimic Nature's evolutionary algorithm to some degree, via the creation of genetically diverse populations that are subject to some form of selection for advantageous properties. However, an important point of contrast with natural evolution is that the practitioner of directed evolution has the capacity to intervene, and therefore to control the focus and extent of mutagenesis by introducing elements of intelligent design. Since 2003, this has been a dominant trend. There has been a shift towards smaller, more focused libraries, with the recognition that even the most efficient expression and screening systems cannot effectively access more than a small fraction of the protein sequence space that random or recombinatorial mutagenesis methods might generate. In particular, the development of computational methods to guide the exploration of protein sequence space has allowed the scale of library creation and screening to be reduced while concomitantly enriching the pool of robust and functionally useful mutants.

This text is divided into three sections according to the nature of the methods described. The first deals with point mutagenesis including indel-based methods. Random and saturation mutagenesis methods were the first approaches used to introduce genetic diversity into protein libraries and they remain in wide use today. The majority of these techniques are easily implemented, and they can also provide a powerful means by which to identify regions of the protein that are ripe for engineering or to address hypotheses developed from existing structure–function data. Part II covers recombinatorial methods, wherein genetic diversity is sourced from multiple parental genes that are combined via either homology-dependent or homology-independent techniques. Finally, the last section presents a variety of computational methods to guide the design and analysis of mutant libraries.

The excellent and comprehensive first edition of this text, edited by Frances Arnold and George Georgiou, has provided an invaluable model for the preparation of this edition. In soliciting and compiling contributions our aims were threefold: to highlight user-friendly protocols for novel methods that were developed post-2003 and represent a substantial expansion of the protein engineering toolbox; to update the fundamental pre-2003 methods that have stood the test of time, but may themselves have evolved along the way; and to provide information on current computational approaches in a user-friendly format that will allow researchers to make informed choices without needing to comprehend the full technical details of each algorithm. While many chapters in the current edition are presented in a standard protocol format, several represent reviews of available software or groups of related experimental methods. Given the extreme proliferation of library-generation strategies, not all published methods can be included in a volume such as this. Our objective has been to strike a balance between those that have achieved broader usage than simply within the originating laboratory and those that represent a significant innovation with potential to address limitations in the existing engineering toolbox.

A proven strength of directed evolution as an engineering strategy is the ability to effect powerful changes, despite gaps in our knowledge of the interrelation of structure and function in biological molecules. In contrast to rational engineering, the evolutionary process can reveal solutions to biological "problems" without requiring any a priori understanding of how these solutions should work. Directed evolution strategies have, in impressive fashion, accelerated the search for proteins with useful properties. This is apparent from the number and extent of improvements that have been achieved, not only in catalytic activities but also in other properties such as stability. However, if we are to mine the new "evolutionary record" of directed evolution experiments to answer unresolved questions, much more remains to be done. We hope that this book will serve as an authoritative manual for both novice and experienced protein engineers and synthetic biologists, and that its use will enable further technical innovation and the exploitation of directed evolution for a deeper understanding of protein design and function.

*Brisbane, QLD, Australia* — *Elizabeth M.J. Gillam*
*Vancouver, BC, Canada* — *Janine N. Copp*
*Wellington, New Zealand* — *David F. Ackerley*

# Contents

## Contributors

CARLOS G. ACEVEDO-ROCHA • *Organische Synthese, Max-Planck-Institut für Kohlenforschung, Mülheim, Germany; Philipps-Universität Marburg, Fachbereich Chemie, Marburg, Germany*

DAVID F. ACKERLEY • *School of Biological Sciences, Victoria University of Wellington, Wellington, New Zealand*

DAVID L. ALEXANDER • *Department of Biomolecular Engineering SOE2, University of California at Santa Cruz, Santa Cruz, CA, USA*

FRANCES H. ARNOLD • *Division of Chemistry and Chemical Engineering, California Institute of Technology, Pasadena, CA, USA*

JAMES A.J. ARPINO • *School of Biosciences, Cardiff University, Cardiff, UK*

AMY J. BALDWIN • *School of Biosciences, Cardiff University, Cardiff, UK*

JAMES B.Y.H. BEHRENDORFF • *Australian Institute for Bioengineering and Nanotechnology, The University of Queensland, Brisbane, Australia*

JAROSLAV BENDL • *Loschmidt Laboratories, Masaryk University, Brno, Czech Republic*

MIKAEL BODÉN • *School of Chemistry and Molecular Biosciences, The University of Queensland, Brisbane, Australia*

UWE T. BORNSCHEUER • *Institute of Biochemistry, Greifswald University, Greifswald, Germany*

JAN BREZOVSKY • *Loschmidt Laboratories, Masaryk University, Brno, Czech Republic*

MANEL CAMPS • *Microbiology and Environmental Toxicology Department, University of California at Santa Cruz, Santa Cruz, CA, USA*

JANINE N. COPP • *Michael Smith Laboratories, University of British Columbia, Vancouver, BC, Canada*

VANESSA E. COX • *School of Chemistry and Biochemistry, Georgia Institute of Technology, Atlanta, GA, USA*

JIRI DAMBORSKY • *Loschmidt Laboratories, Masaryk University, Brno, Czech Republic*

ASHLEY B. DAUGHERTY • *Department of Chemistry, Emory University, Atlanta, GA, USA*

ALEXANDER DENNIG • *Lehrstuhl für Biotechnologie, RWTH Aachen University, Aachen, Germany*

MATTHEW C. EDMUNDSON • *School of Biosciences, Cardiff University, Cardiff, UK*

RYOTA FUJII • *Synthetic Chemicals Laboratory, Mitsui Chemicals, Inc., Sodegaura, Chiba, Japan*

ERIC A. GAUCHER • *School of Chemistry and Biochemistry, Georgia Institute of Technology, Atlanta, GA, USA*

MONICA L. GERTH • *Department of Biochemistry, University of Otago, Dunedin, New Zealand*

ELIZABETH M.J. GILLAM • *School of Chemistry and Molecular Biosciences, The University of Queensland, Brisbane, Australia*

MOSHE GOLDSMITH • *Department of Biological Chemistry, Weizmann Institute of Science, Rehovot, Israel*

YOSEPHINE GUMULYA • *School of Chemistry and Molecular Biosciences, The University of Queensland, Brisbane, Australia*
PAULINA HANSON-MANFUL • *Institute of Natural Sciences, Massey University, Auckland, New Zealand*
KIYOSHI HAYASHI • *Faculty of Food and Nutritional Sciences, Toyo University, Gunma, Japan*
JAIME HERNANDEZ • *Microbiology and Environmental Toxicology Department, University of California at Santa Cruz, Santa Cruz, CA, USA*
AURELIO HIDALGO • *Center for Molecular Biology, UAM-CSIC, Madrid, Spain*
FLORIAN HOLLFELDER • *Department of Biochemistry, University of Cambridge, Cambridge, UK*
WAYNE A. JOHNSTON • *Institute for Molecular Bioscience, The University of Queensland, Brisbane, Australia*
D. DAFYDD JONES • *School of Biosciences, Cardiff University, Cardiff, UK*
MIRIAM KALTENBACH • *Michael Smith Laboratories, University of British Columbia, Vancouver, Canada*
TSVETAN KARDASHLIEV • *Lehrstuhl für Biotechnologie, RWTH Aachen University, Aachen, Germany*
SABRINA HOEBENREICH • *Philipps-Universität Marburg, Fachbereich Chemie, Junior Research, Group Leader Chemical Biology, Marburg, Germany*
MOTOMITSU KITAOKA • *National Agriculture and Food Research Organization, National Food Research Institute, Ibaraki, Japan*
JOSHUA LILLY • *Microbiology and Environmental Toxicology Department, University of California at Santa Cruz, Santa Cruz, CA, USA*
STEFAN LUTZ • *Department of Chemistry, Emory University, Atlanta, GA, USA*
JAN MARIENHAGEN • *Institut für Bio- und Geowissenschaften, IBG-1: Biotechnologie, Forschungszentrum Jülich GmbH, Jülich, Germany*
YUVAL NOV • *Department of Statistics, University of Haifa, Haifa, Israel*
WAYNE M. PATRICK • *Department of Biochemistry, University of Otago, Dunedin, New Zealand*
ZHEN QIAN • *Du Pont Industrial Biosciences, China Research Center, Shanghai, China*
MANFRED T. REETZ • *Organische Synthese, Max-Planck-Institut für Kohlenforschung, Mülheim, Germany; Philipps-Universität Marburg, Fachbereich Chemie, Marburg, Germany*
DANILO ROCCATANO • *Jacobs University Bremen gGmbH, Bremen, Germany*
LIAT ROCKAH-SHMUEL • *Department of Biological Chemistry, Weizmann Institute of Science, Rehovot, Israel*
JILLIAN ROMSDAHL • *Microbiology and Environmental Toxicology Department, University of California at Santa Cruz, Santa Cruz, CA, USA*
ANNA JOËLLE RUFF • *Lehrstuhl für Biotechnologie, RWTH Aachen University, Aachen, Germany*
ANNA SCHLIESSMANN • *Institute of Biochemistry, Greifswald University, Greifswald, Germany*
ULRICH SCHWANEBERG • *Lehrstuhl für Biotechnologie, RWTH Aachen University, Aachen, Germany*

EVA SEBESTOVA • *Loschmidt Laboratories, Masaryk University, Brno, Czech Republic*
MATTHEW A. SMITH • *Insight Data Science, Palo Alto, CA, USA*
DAN S. TAWFIK • *Department of Biological Chemistry, Weizmann Institute of Science, Rehovot, Israel*
NOBUHIKO TOKURIKI • *Michael Smith Laboratories, University of British Columbia, Vancouver, Canada*
CHRISTOPHER J. TROLL • *Microbiology and Environmental Toxicology Department, University of California at Santa Cruz, Santa Cruz, CA, USA*
RAJNI VERMA • *Department of Chemistry, Wichita State University, Wichita, Kansas, USA*
BENOIT R.M. VILLIERS • *Department of Biochemistry, University of Cambridge, Cambridge, UK*
ELSIE M. WILLIAMS • *School of Biological Sciences, Victoria University of Wellington, Wellington, New Zealand*
TUCK SENG WONG • *ChELSI Institute and Advanced Biomanufacturing Centre, Department of Chemical and Biological Engineering, University of Sheffield, Sheffield, England, UK*
YING YU • *Department of Biochemistry, Emory University School of Medicine, Atlanta, GA, USA*
JULIAN ZAUGG • *School of Chemistry and Molecular Biosciences, The University of Queensland, Brisbane, Australia*

# Part I

## Point Mutagenesis

# Chapter 1

# Error-Prone PCR and Effective Generation of Gene Variant Libraries for Directed Evolution

**Janine N. Copp, Paulina Hanson-Manful, David F. Ackerley, and Wayne M. Patrick**

## Abstract

Any single-enzyme directed evolution strategy has two fundamental requirements: the need to efficiently introduce variation into a gene of interest and the need to create an effective library from those variants. Generation of a maximally diverse gene library is particularly important when employing nontargeted mutagenesis strategies such as error-prone PCR (epPCR), which seek to explore very large areas of sequence space. Here we present comprehensive protocols and tips for using epPCR to generate gene variants that exhibit a relatively balanced spectrum of mutations and for capturing as much diversity as possible through effective cloning of those variants. The detailed library preparation methods that we describe are generally applicable to any directed evolution strategy that uses restriction enzymes to clone gene variants into an expression plasmid.

**Key words** Directed evolution, Random mutagenesis, Error-prone PCR, GeneMorph II, Mutazyme II DNA polymerase, Library generation, Library analysis

## 1 Introduction

Directed evolution at a single gene level is a powerful tool for effecting rapid improvement in a desired protein function. Recent years have seen an increasing trend toward the design of targeted "smart" libraries, which exploit available structural, functional, and/or sequence-derived information to preselect promising residues for mutagenesis. These approaches seek to minimize the number of gene variants that must be screened to achieve a successful outcome [1–3] and are particularly valuable when only a low to moderate throughput screen is available (e.g., $10^2$–$10^4$ variants per day). In contrast, approaches like error-prone PCR (epPCR) provide access to an almost unlimited number of variants, but generally have a lower hit rate and are therefore better suited to high throughput screening or selection strategies. Nevertheless, it is worth noting that epPCR can still yield rapid gains in fitness

Elizabeth M.J. Gillam et al. (eds.), *Directed Evolution Library Creation: Methods and Protocols*, Methods in Molecular Biology, vol. 1179, DOI 10.1007/978-1-4939-1053-3_1, © Springer Science+Business Media New York 2014

for some promiscuous functions; for example, one of us (DFA) was involved in a small-scale epPCR study (screening only 6,000 clones) that generated an evolved chromate reductase variant exhibiting >200-fold improvement in $k_{cat}/K_M$ [4].

The great strengths of error-prone PCR are its simplicity and applicability; anyone who has the capability to amplify and clone a PCR product should be able to generate a high-quality epPCR library. The main limitation stems from the inherent randomness of epPCR; residues distant from the active site are just as likely to be mutated as those within, but while the former are more numerous, the latter are more likely to yield substantial improvements in activity [5]. Thus, relative to more targeted strategies, higher proportions of synonymous, neutral, or deleterious base substitutions are to be expected. Furthermore, even a balanced mutation spectrum at the genetic level yields bias at the protein level. It has been calculated that a single base substitution in a sense codon provides access to only 5.7 of the 19 alternative amino acids on average [6], owing to the degeneracy and structure of the genetic code. In practice epPCR is unlikely to generate multiple base substitutions within the same codon, and this diminishes the amount of sequence space that can be sampled.

While it is important to keep these limitations in mind, each issue also brings potential benefit. The stochastic nature of epPCR can prove advantageous as, unrestricted by preconceived notions about which residues are most likely to yield enhanced variants, entirely unpredictable mutations can be recovered that yield substantial gains in fitness (e.g., mutations of residues distant from the active site that improve substrate access, solubility, or stability [7, 8]). Likewise, although single base substitutions provide access to only a limited pool of alternative amino acids, those residues are slightly more likely to be tolerated than amino acids that are only accessible via multiple base substitutions [9].

In all these considerations, it is implicit that generation of a variant library that exhibits a balanced mutational spectrum is a desirable outcome (this argument is expanded in [10]). The original and most economical methods for performing epPCR are by enhancing the intrinsic error rate of *Taq* polymerase by the addition of $Mn^{2+}$ to the reaction buffer to reduce base pairing specificity and/or by modifying the ratios of dNTPs present [11–13]. However, these methods generate a strong mutation bias. In particular, addition of $Mn^{2+}$ results in an over-representation of mutations at A:T base pairs [14]. Similarly, methods that employ triphosphate derivatives of nucleoside analogues also tend to generate heavily biased libraries [15]. To minimize bias, we advocate use of the GeneMorph® II Random Mutagenesis Kit (Agilent Technologies; manual available at http://www.chem.agilent.com/library/usermanuals/Public/200550.pdf; May 2013). This kit employs a mixture of two proprietary error-prone polymerases that collectively provide a relatively balanced mutational spectrum (albeit still slightly favoring mutations at A:T base pairs). Some fine-tuning of the mutation rate is also feasible by varying the initial concentration of template DNA and the number of amplification cycles. The relatively unbiased mutational spectrum has been independently verified by numerous independent practitioners, e.g., [15], and in our experience the mutation rate is far easier to control than alternative epPCR strategies. Nevertheless, readers who prefer to use a *Taq*/$Mn^{2+}$-based method for epPCR should find it straightforward to adapt the methodology of Fujii et al. (Chapter 2 of this volume) to a linear rather than rolling circle PCR format and to substitute this for the methodology that we describe under Subheading 3.2.

This chapter is based on a consensus of "best practice" approaches for generating an amplified collection of gene variants by epPCR and for restriction enzyme-based cloning of those variants into a plasmid-borne gene library, as employed in two different research laboratories. In this way, it provides a different perspective to a chapter that two of us (P.H.M. and W.M.P.) published in an earlier volume [16]. Wherever we felt that the best strategy was not clear-cut (i.e., wherever we bickered among ourselves!), we have added notes that describe alternative approaches. In this manner, we hope to provide the reader with a comprehensive series of options that can easily be tailored to existing facilities and preferences.

## 2 Materials

### 2.1 Vector Preparation and Analysis

1. Protein expression vector, into which the epPCR product will be cloned (*see* **Notes 1** and **2**).
2. Qiagen Plasmid Midi Kit (Qiagen; Valencia, CA, USA). Equivalent kits from other manufacturers are also suitable. Alternatively, plasmids can be purified using alkaline lysis, to yield DNA at high concentrations [17].
3. Restriction enzymes that enable directional, sticky-ended cloning of the epPCR product. If using two restriction enzymes, as we recommend, then ensure that they are active in a single, compatible buffer (*see* **Note 3**).
4. Restriction enzyme that cuts within the stuffer fragment (*see* **Note 1**) of the expression vector (*see* **Notes 2** and **3**).
5. Agarose gels stained with 1× SYBR Safe DNA gel stain (Invitrogen; Carlsbad, CA, USA).
6. Safe Imager 2.0 Blue-Light Transilluminator (Invitrogen; Carlsbad, CA, USA).
7. Clean razor blades or other implements for excising bands from gels.

8. QiaQuick Gel Extraction Kit (Qiagen; Valencia, CA, USA). Equivalent kits from alternate suppliers are also suitable.
9. DNA Clean & Concentrator Kit (Zymo Research; Irvine, CA, USA). Equivalent kits from other manufacturers are also suitable or DNA can be concentrated using ethanol precipitation [17].
10. Spectrophotometer to measure DNA concentration in small volumes, such as a Nanodrop 1000 (Thermo Scientific; Pittsburgh, PA, USA). If a Nanodrop spectrophotometer is not available, DNA quantification can be performed by agarose gel electrophoresis utilizing a DNA quantification ladder, e.g., the 100 bp DNA ladder from New England Biolabs (Ipswich, MA, USA).

### *2.2 Error-Prone PCR and Preparation of epPCR Insert for Ligation into Vector*

1. Plasmid or genomic DNA containing the gene that is to be amplified by epPCR.
2. Phusion High-Fidelity polymerase (New England Biolabs; Ipswich, MA, USA). This particular brand of DNA polymerase is recommended due to its buffer compatibility with DpnI (https://www.neb.com/tools-and-resources/usage-guidelines/activity-of-restriction-enzymes-in-a-taq-or-phusion-pcr-mix).
3. Oligonucleotide primers for amplification of the target gene (*see* **Note 4**).
4. DNA Clean & Concentrator Kit (*see* Subheading 2.1, **item 9**).
5. Nanodrop spectrophotometer to measure DNA concentration (*see* Subheading 2.1, **item 10**).
6. GeneMorph II Random Mutagenesis Kit (catalog #200550, Agilent; Santa Clara, CA, USA). The kit contains Mutazyme II DNA polymerase (2.5 U/μl), 10× Mutazyme II reaction buffer, a dNTP mix (10 mM of each dNTP) and a 1.1 kb gel quantification standard (20 ng/μl).
7. Thermocycler with a heated lid (e.g., an MJ Mini from Bio-Rad; Hercules, CA, USA).
8. Agarose gels, stained with ethidium bromide at 0.5 μg/ml, and apparatus for agarose gel electrophoresis.
9. Restriction enzymes that facilitate directional, sticky-ended cloning (*see* Subheading 2.1, **item 3**).
10. Restriction enzyme DpnI (New England Biolabs; Ipswich, MA, USA).

### *2.3 Vector Analysis and Test Library Preparation*

1. Digested vector.
2. Digested PCR insert.
3. T4 DNA ligase and ligation buffer. We obtain comparable results using each of the T4 DNA ligases from New England

Biolabs (Ipswich, MA, USA), Fermentas (Thermo Scientific; Pittsburgh, PA, USA) and Enzymatics Inc. (Beverly, MA, USA).

4. Nanodrop spectrophotometer to measure DNA concentration (*see* Subheading 2.1, **item 10**).
5. Aliquots (40 μl) of electrocompetent *E. coli* cells (*see* **Note 5**).
6. Electroporation unit and cuvettes, e.g., Gene Pulser electroporation unit with Pulse Controller (Bio-Rad; Hercules, CA, USA) and Gene Pulser electroporation cuvettes with 0.2 cm electrode gap (Bio-Rad; Hercules, CA, USA).
7. Sterile SOC medium (20 g/L tryptone; 5 g/L yeast extract; 10 mM NaCl; 2.5 mM KCl; 20 mM glucose).
8. LB-agar plates containing the correct antibiotic for selecting plasmid-containing cells.

### 2.4 Analysis of Library Composition

1. Thermocycler with a heated lid (e.g., an MJ Mini from Bio-Rad; Hercules, CA, USA).
2. Primers for amplifying cloned inserts from the epPCR library (*see* **Note 4**).
3. Reagents for a standard PCR screen. While there are many alternative (and equally good) suppliers, we routinely use 2× BioMix Red (Bioline; Alexandria, NSW, Australia). Equivalent kits from other manufacturers are also suitable.
4. Ethidium bromide-stained agarose gels and electrophoresis apparatus (*see* Subheading 2.2, **item 8**).
5. DNA Clean & Concentrator Kit (*see* Subheading 2.1, **item 9**).

### 2.5 Construction and Storage of the Full-Sized Library

1. Digested vector with minimal background level of partially digested DNA.
2. Purified epPCR-amplified insert digested with appropriate restriction enzymes.
3. T4 DNA ligase and ligation buffer (*see* Subheading 2.3, **item 3**).
4. DNA Clean & Concentrator Kit (*see* Subheading 2.1, **item 9**).
5. Nanodrop spectrophotometer to measure DNA concentration (*see* Subheading 2.1, **item 10**).
6. LB-agar plates containing the correct antibiotic for selecting plasmid-containing cells.
7. Electrocompetent *E. coli* cells (*see* **Note 5**).
8. Electroporation cuvettes and apparatus (*see* Subheading 2.3, **item 6**).
9. SOC medium (*see* Subheading 2.3, **item 7**).
10. Supercoiled control plasmid of known concentration (e.g., pUC19, 10 pg/μl; Invitrogen (Carlsbad, CA, USA)).

11. Sterile 50 ml tubes (Falcon or similar).
12. LB medium supplemented with the appropriate antibiotic (~20 ml, total).
13. Refrigerated centrifuge with a rotor that takes 50 ml tubes (*see* Subheading 2.5, **item 11**).
14. Spectrophotometer and cuvettes for measuring cell density ($OD_{600}$), e.g., BioPhotometer and UVettes from Eppendorf (Hamburg, Germany).
15. Sterile glycerol (50 % v/v).
16. Cryogenic vials, suitable for storage at −80 °C (*see* **Note 6**).

## 3 Methods

### *3.1 Vector Preparation and Analysis*

1. Midiprep the plasmid that will be used for cloning and expression of the library using a Qiagen midiprep kit or equivalent (*see* **Notes 1** and **2**).
2. Digest 10 μg of the vector in a 500 μl reaction with 1× restriction enzyme buffer using 25 U of each restriction enzyme. Incubate the reaction at 37 °C overnight (*see* **Note 3**).
3. Add a further 10 U of an enzyme that cuts within the stuffer fragment (*see* **Note 2**). Incubate the reaction for an additional 2 h at 37 °C and then heat inactivate the enzymes according to the manufacturer's guidelines.
4. Run the entire restriction digest on a 0.8 % w/v agarose gel and excise the DNA fragment corresponding to the digested vector. We strongly recommend the use of a blue-light transilluminator and a compatible stain (SYBR Safe) for this step rather than ethidium bromide and a UV transilluminator (*see* **Note** 7). Purify the excised DNA with the QiaQuick Gel Extraction Kit, according to the manufacturer's guidelines. Use 30–50 μl Elution Buffer to elute the DNA from the spin column (*see* **Note 8**).
5. Quantify the purified vector using a Nanodrop spectrophotometer or equivalent.
6. Aliquot the vector and store at −20 °C. We typically aliquot 200–500 ng of the digested vector into multiple microcentrifuge tubes so they can be thawed and used as required. This reduces the number of freeze/thaw cycles and maintains high-quality stocks of digested vector for short-term use.
7. Ligate a test fragment (*see* **Note 9**) with 100 ng of the digested vector. Use a 3:1 molar ratio of digested insert DNA to digested vector in a 10 μl reaction with 1× T4 DNA Ligase buffer and 1 U ligase. Incubate overnight at 16 °C.

8. Prepare a control reaction as above but containing digested vector only (no insert).
9. Thaw three 40 μl aliquots of electrocompetent *E. coli* cells on ice (*see* **Note 5**). Add 1 μl of the test ligation reaction to one of the cell aliquots. Add 1 μl of the control ligation reaction to a second aliquot of cells. Add an appropriate amount of a plasmid control to the third aliquot of cells, to determine their transformation efficiency. We routinely use 10 pg of an uncut plasmid control, containing the same antibiotic resistance cassette as the vector (*see* **Note 10**). Transfer each aliquot to a prechilled electroporation cuvette, electroporate (e.g., 2.5 kV, 200 Ω, and 25 μF in a Gene Pulser unit with Pulse Controller), and immediately add 460 μl of SOC medium to the cuvette. Transfer the cell suspension to a sterilized, capped microcentrifuge test tube (or sterile 15 ml tube). Allow the cells to recover by incubating them for 1 h at 37 °C and 200 rpm. Spread 10, 20, and 50 μl aliquots of the cell suspensions onto selective agar plates. Incubate the plates overnight at 37 °C.
10. Count the number of colonies on each plate. Use the results from the "vector only" plates to calculate the fraction of the test ligation (as represented on the "vector + insert" plates) that contains recircularized or uncut vector. This background must be minimized, to avoid wasting time on futile library screens. If the "vector only" background is >1 % of the "vector + insert" colony count, then we recommend discarding all aliquots of the digested vector and preparing a fresh batch. Use the plasmid control to calculate the transformation efficiency of your electrocompetent cells. The easiest way to increase the size of an epPCR library is to improve the transformation efficiency of the cells (*see* **Notes 5** and **10**).

### 3.2 Template Preparation and Error-Prone PCR

1. Amplify your target gene with Phusion High-Fidelity polymerase. We typically use 20–50 μl reactions following the manufacturer's protocol (*see* **Note 11**).
2. Run 1–2 μl of the PCR reaction on an ethidium bromide-strained gel to confirm appropriate amplification of the target gene.
3. Add 10 U DpnI directly to PCR mix and incubate for 1 h at 37 °C.
4. Heat inactivate the DpnI reaction for 20 min at 65 °C and then purify using the DNA Clean & Concentrator Kit.
5. Quantify the purified DNA using a Nanodrop spectrophotometer.
6. The amount of template and the number of amplification cycles used in the epPCR affects the mutation rate (*see* **Note 12**). Calculate the amount of template DNA and the number of PCR cycles that are required for the desired mutation rate.

7. Prepare the epPCR reagents in a thin-walled, 0.2 ml tube containing 5 μl 10× Mutazyme II reaction buffer, 1 μl dNTP mix (gives 200 μM of each dNTP, final concentration), 2 μl forward primer (from 10 μM stock solution), 2 μl reverse primer (from 10 μM stock solution), 1 μl Mutazyme II DNA polymerase (2.5 U), DNA template, and double-distilled water ($ddH_2O$) to a total volume of 50 μl.
8. Mix the sample and place the tube in the thermocycler. If using a low template concentration, at this point the PCR mix may be split into five separate tubes, each containing 10 μl, to avoid a founder effect from mutations introduced during an early cycle. A negative control (with no template DNA added) should also be prepared to check for contamination (*see* **Note 13**).
9. Run the epPCR program beginning with an initial denaturation step of 1 min at 95 °C, followed by 30 cycles (*see* **Note 12**) of 20 s at 94 °C, 20 s at the optimal annealing temperature for primers (*see* **Note 4**) and 1 min at 72 °C (for a ~1 kb gene; *see* **Note 14**). Complete the epPCR with a 2 min elongation step at 72 °C and then hold at 4 °C (for product storage, if necessary).
10. The total yield of epPCR product is required to calculate the PCR efficiency. Run an aliquot of the epPCR product (typically 2 μl) on an ethidium bromide-stained agarose gel, alongside the 1.1 kb gel standard (*see* Subheading 2.2, **item 6**). Determine the total yield of the epPCR product by comparing the intensity of the epPCR sample with the intensity of the gel standard (*see* **Note 15**).
11. Purify the remainder of the epPCR product using the DNA Clean & Concentrator Kit. Elute the purified DNA from the spin column in 30 μl sterile water heated to 50 °C (*see* **Note 8**).
12. Quantify the purified epPCR product using a Nanodrop spectrophotometer.
13. Digest 5 μg of the epPCR product with 10 U of each of the restriction enzymes in 1× restriction enzyme buffer and a total volume of 200 μl. If your total epPCR yield is lower than 5 μg, use the entire 30 μl elution from Subheading 3.2, **step 11** above. Incubate the reaction at 37 °C for 6–8 h or overnight. Heat inactivate the restriction enzymes, as recommended by the manufacturer. Split the reaction into two 100 μl aliquots and purify using two DNA Clean & Concentrator columns. Elute each of the two aliquots into 10 μl of sterile water prewarmed to 50 °C (*see* **Note 8**).
14. Pool both elution fractions and quantify using a Nanodrop spectrophotometer.

### 3.3 Preparation of a Test Library (See Note 16)

1. Prepare a test ligation containing 1× ligation buffer; 50 ng of vector DNA (*see* Subheading 3.1, **step 6**); T4 DNA ligase (1 U); a 3:1 molar ratio of digested epPCR insert DNA to vector DNA; and water to a final volume of 10 μl. Add the T4 DNA ligase last and mix gently. Prepare a control ligation as above but without the insert DNA.
2. Incubate the ligation reactions at 16 °C for 12–18 h.
3. Use 1 μl of each ligation reaction to transform 40 μl aliquots of electrocompetent *E. coli* cells, as described in Subheading 3.1, **step 9** (*see* **Notes 5** and **10**).
4. Spread aliquots (10 μl and 50 μl) of the two recovery cultures on LB-agar plates. Incubate the plates at 37 °C for 12–18 h.
5. Count the number of colonies on each plate. Use the results from the "vector only" plates to confirm the quality of the library (and compare to previous results from Subheading 3.1, **step 10**).
6. The number of colonies on the "vector + insert" plates also allows the size of the final, scaled-up library to be estimated. The final library is likely to be ~$10^3$ times larger than the total number of colonies on the "vector + insert" test plates (*see* **Note 17**).

### 3.4 Analysis of Library Composition

Amplify the randomly mutagenized gene inserts from 10 to 20 colonies by PCR. We have listed our routine protocol, as a guide. However, many variations are possible; the goal here is merely to generate enough of the amplified product for DNA sequencing (*see* **Note 18**). We typically set up 20 μl PCR reactions in thin-walled 0.2 ml tubes containing 8 μl sterile water, 10 μl 2× BioMix Red (Bioline), 1 μl forward primer (from 10 μM stock solution), and 1 μl reverse primer (from 10 μM stock solution).

1. Use 2 μl pipette tips (or sterile toothpicks) to pick 10–20 colonies at random from the "vector + insert" test plates (Subheading 3.3, **step 6**). Transfer each colony into one of the 20 μl PCR reactions prepared above.
2. Run an appropriate PCR program, such as the one listed in Subheading 3.2, **step 9**. Increasing the initial denaturation step to 3 min is recommended to ensure complete cell lysis.
3. Run a 2 μl aliquot of each PCR product on an agarose gel, to confirm successful amplification.
4. Purify the remainder of each PCR product using the DNA Clean & Concentrator Kit. Elute the purified DNA from each spin column in 20 μl sterile water.
5. Sequence each PCR product. Use the forward and/or reverse primers from the PCR as the sequencing primer(s), as necessary (*see* **Notes 18** and **19**).

**Table 1**
**Mutational spectrum of a *cynT* epPCR library**

| Type(s) of mutations | Frequency | Proportion of total |
|---|---|---|
| *Transitions* | | |
| A→G, T→C | 28 | 22.6 % |
| G→A, C→T | 28 | 22.6 % |
| *Transversions* | | |
| A→T, T→A | 34 | 27.4 % |
| A→C, T→G | 9 | 7.3 % |
| G→C, C→G | 4 | 3.2 % |
| G→T, C→A | 18 | 14.5 % |
| *Insertions and deletions* | | |
| Insertions | 1 | 0.8 % |
| Deletions | 2 | 1.6 % |
| *Summary of bias* | | |
| Transitions/transversions | 0.86 | NA |
| AT→GC/GC→AT | 1 | NA |
| A→N, T→N | 71 | 57.3 % |
| G→N, C→N | 50 | 40.3 % |
| *Mutation rate* | | |
| Mutations/kb | 11.8 | NA |
| Mutations/*cynT* gene[a] | 7.8 | NA |

*NA* not applicable
[a]The cloned *cynT* insert was 657 bp

6. Align the sequence of each PCR product with the known sequence of the unmutated parental gene. Computer programs such as MacVector (Cary, NC, USA) or DNA Baser (Heracle BioSoft; Romania) are useful for this analysis.
7. Tabulate all of the point mutations in the sequenced samples. Also note any insertions, deletions, and stop codons that may have been introduced during the epPCR. The point mutations should be grouped by type (*see* **Note 20**).
8. Use the tabulated data to calculate the overall mutation rate and to assess biases in the mutation spectrum of the epPCR library. There are three key indicators of bias (*see* **Note 20**). The mutation rate and bias measures for an example *cynT* epPCR library are shown in Table 1.
9. The library analysis program PEDEL-AA [19], available online at http://guinevere.otago.ac.nz/stats.html, can be used to predict the utility of the final epPCR library (*see* **Note 21**).

### 3.5 Construction and Storage of Full-Sized Library

1. Prepare a full-sized library "vector+insert" ligation reaction (*see* **Note 22**). The reaction should contain 1× ligation buffer; 500 ng of vector DNA; 10 U T4 DNA ligase; digested insert

DNA (threefold molar excess over vector); and water to a final volume of 100 μl.

2. Prepare a "vector only" control ligation containing 1× ligation buffer; 100 ng of vector DNA; 2 U T4 DNA ligase and water to a final volume of 20 μl.
3. Incubate the ligation reactions at 16 °C for 12–18 h.
4. Transform a 40 μl aliquot of electrocompetent *E. coli* cells with 1 μl of the full-size library ligation (*see* Subheading 3.1, **step 9**). Transform a second 40 μl aliquot of cells with 1 μl of the "vector only" ligation, and a third 40 μl aliquot of cells with an appropriate plasmid to confirm the transformation efficiency of the cells (*see* Subheading 3.1, **step 10**).
5. Purify the remaining full library ligation reaction using the DNA Clean & Concentrator Kit or by drop dialysis [18]. Elute the purified DNA in 50 μl sterile water that has been preheated to 50 °C (*see* **Note 8**) and quantify using a Nanodrop spectrophotometer. Store at −20 °C.
6. After overnight incubation, count the number of colonies on each plate from Subheading 3.5, **step 4** to estimate the final size of the library and to verify that the "vector only" background is <1 % of the total library as predicted from the test ligation (*see* Subheading 3.1, **step 10**).
7. Use the plasmid control to confirm the transformation efficiency of your electrocompetent cells (*see* Subheading 3.1, **step 10**).
8. Use 10–50 ng aliquots of the purified ligation to transform at least ten aliquots (each 40 μl) of electrocompetent *E. coli* cells, as described in Subheading 3.1, **step 9** (*see* **Notes 5** and **23**). If any of the purified ligation reaction products remain, store them at −20 °C. Combine the recovery cultures in a sterile 15 ml tube. The total volume should be 5 ml (ten electroporations; 500 μl per recovery culture) (*see* **Note 24**).
9. Spread the cells on two large square plates (245 mm × 245 mm; spread ~2.5 ml of cell suspension per plate) or, alternatively, spread 250 μl aliquots on 20 standard agar plates.
10. Incubate the plates at 30 °C overnight to avoid the formation of confluent lawns.
11. The plates should each be covered in thousands (or millions) of small colonies. To recover the library from one of the plates, pipette 500 μl (small plates) or 4 ml (large plates) of LB medium (supplemented with the appropriate antibiotic) into the center of the plate. Use a glass spreader to scrape cells off the surface of the plate; pool them in one corner of the plate. Remove the resuspended cells with a P1000 pipettor, and transfer them to a sterile 50 ml tube. Pipette a second aliquot

of LB onto the same plate, and repeat the scraping step. All of the recovered cells should be pooled in the same 50 ml tube.

12. Break up cell clumps and mix well, by pipetting up and down repeatedly with a P1000 pipettor.
13. Pellet the cells by centrifugation at 3,000 × *g* for 15 min at 4 °C. Use a P1000 pipettor to remove the supernatant (*see* **Note 25**).
14. Resuspend the cell pellet in 1 ml of LB medium, plus antibiotic. The pellet is likely to be large; resuspension is likely to involve vigorous pipetting and/or gentle vortexing.
15. Knowing the cell density is likely to be useful for planning downstream screening/selection experiments. Mix 1 μl of the cell suspension with 999 μl of sterile water, and measure the $OD_{600}$ (against a water blank). The final $OD_{600}$ of the resuspended library is usually > 100, corresponding to a cell density of $>2.5 \times 10^{10}$ cells/ml (*see* **Note 26**).
16. Split the library into 100 μl aliquots and transfer each aliquot to a cryogenic vial.
17. Add 50 μl of sterile glycerol (50 % v/v) to each aliquot and mix well by pipetting.
18. Store the aliquots at −80 °C until you are ready to proceed with screening/selection to identify improved variants in the library.

## 4 Notes

1. A high-quality vector preparation is critical for constructing a large library. We have found that it is helpful to make a large-scale preparation of digested vector for use in short-term library construction, as this enables several libraries to be generated and tested within a short time frame, all using a vector preparation of known quality and quantity. However, the prepared vector should ideally be used within 1–2 weeks as we have noted a considerable drop in ligation efficiency after this point, even without repeated freeze/thaw cycles. The protocol can easily be adapted to prepare smaller amounts of vector if required. We find it useful to use a plasmid with a "stuffer" fragment in the cloning cassette. This should be a DNA fragment that is ligated into the vector using the same restriction sites that will be used for cloning the epPCR library. Ideally this fragment should contain a unique internal restriction site not present in your native gene (*see* **Note 2**). Excision of this stuffer fragment from the vector enables the easy monitoring of the restriction enzyme reaction, i.e., a small aliquot of the digestion reaction can be analyzed by gel electrophoresis to

confirm the complete digestion of the vector, determined by the visualization of the stuffer fragment and cut vector as two distinct bands on the agarose gel.

2. Use of a third restriction enzyme that cuts within the stuffer fragment minimizes the number of "vector only" clones in the final library. Where possible, it is ideal if this third restriction enzyme utilizes the same buffer as the other restriction enzymes used to prepare the vector for cloning. New England Biolabs provide an easy-to-use online tool for restriction analyses: http://tools.neb.com/NEBcutter2/. A diagnostic DNA electrophoresis gel would show three distinct DNA bands in this case; the digested vector and two DNA bands showing the separate fragments of the digested stuffer fragment.
3. We recommend the use of NEB High-Fidelity restriction enzymes. This range of enzymes has the same specificity as the native enzyme with reduced star activity and improved buffer compatibility (https://www.neb.com/products/restriction-endonucleases/hf-nicking-master-mix-time-saver-other/high-fidelity-restriction-enzymes/high-fidelity-restriction-endonucleases). The reduced star activity is particularly valuable because it allows for extended incubation times without risk of off-target digestion.
4. Primers should be noncomplementary and ideally have melting temperatures that are within 5 °C of each other. Melting temperatures can be estimated accurately using the OligoAnalyzer tool from Integrated DNA Technologies (http://www.idtdna.com/analyzer/Applications/OligoAnalyzer/) or equivalent online programs, a comprehensive range of which can be found online: http://www.molbiol-tools.ca/PCR.htm. The optimal annealing temperature to use in a PCR is typically 3–5 °C cooler than the lowest primer melting temperature. We typically incorporate a restriction site for directional cloning into each of the primers, and lyophilized primers are resuspended in TE buffer (10 mM Tris–HCl, 1 mM EDTA, pH 8.0) to a concentration of 100 μM. Working stocks (10 μM) are made by tenfold dilution of these master stocks, using sterile water.
5. The choice of *E. coli* strain will depend on the downstream selection or screen that is being employed. In general, the final size of the epPCR library will be directly proportional to the transformation efficiency of the host strain. Therefore, strains with high transformation efficiencies (such as *E. coli* DH5α-E) are preferable. We adapt the method of Sambrook and Russell [17] to prepare electrocompetent cells. Two of us (J.N.C. and D.F.A.) use the modified protocol that follows. In brief, inoculate 400 ml of pre-warmed LB medium in a 2 L flask with 10 ml of an overnight culture (grown from a single *E. coli* colony from a fresh agar plate) and incubate at 37 °C,

200 rpm, until the $OD_{600}$ reaches 0.3. Rapidly cool the culture on ice for 30 min. Transfer the culture to eight prechilled 50 ml Falcon tubes and sediment the cells by centrifugation at 4 °C for 30 min at 1,000 × *g*. Discard the supernatant. Resuspend the cell pellets with 200 ml of prechilled sterile $ddH_2O$ and repeat the centrifugation step above. Discard the supernatant. Resuspend the cell pellets with 100 ml of prechilled 10 % w/v glycerol, combine the cell suspensions into four 50 ml tubes (e.g., Falcon), and repeat the centrifugation step above. Discard the supernatant. Resuspend the cell pellets with 50 ml of prechilled 10 % w/v glycerol, transfer into two 50 ml tubes, and repeat the centrifugation step above. Carefully discard the supernatant, taking care to remove all remaining drops of 10 % glycerol. Resuspend the cell pellet in 200 μl of ice-cold GYT medium (10 % v/v glycerol, 0.125 % w/v yeast extract, 0.25 % w/v tryptone). Measure the $OD_{600}$ of a 1:100 dilution of this resuspension. Calculate the cell concentration (*see* **Note 26**) and dilute the cell suspension to a concentration of ~$2.5 \times 10^{10}$ cells/ml with ice-cold GYT. Transfer 40 μl aliquots of the suspension into sterile microcentrifuge tubes prechilled to −80 °C. Store these aliquots at −80 °C until required.

6. We have found that standard microcentrifuge tubes are also suitable for short-term storage of competent cells and library glycerol stocks.
7. Two of us (P.H.M. and W.M.P.) always purify the digested vector backbone from an agarose gel at this stage, to eliminate the possibility that the unwanted insert fragment interferes with downstream ligation steps. On the other hand, the other two of us (J.N.C. and D.F.A.) have had good results with a faster protocol purifying the digested vector using the DNA Clean & Concentrator Kit (and omitting the gel purification step). If using gel purification, be aware that UV transillumination of ethidium bromide-stained DNA can induce damage, resulting in lower cloning and transformation efficiencies [20]. Even short exposures to UV (<60 s) can have dramatic and deleterious effects. Constructing a large epPCR library (>$10^6$ variants) requires the highest possible quality of DNA. Therefore, we recommend use of SYBR Safe stain and a blue-light transilluminator for preparation of the library vector. In our hands, this results in libraries that are 5–10 times larger than equivalent libraries prepared with ethidium bromide-stained DNA. If a SYBR Safe stain and blue-light transilluminator are not available, a small aliquot (200–300 ng) of the digested vector should be run an agarose gel (without ethidium) next to a DNA ladder and the remainder of the vector digestion should be run on the opposite side of the gel. After electrophoresis,

the lanes corresponding to the small aliquot of vector and the ladder should be separated from the rest of the agarose gel and stained in ethidium bromide (0.5 μg/ml) for 10 min. The location of the DNA fragment corresponding to the digested vector can then be visualized by UV transillumination and should be marked with a clean scalpel. The ethidium-stained agarose gel fragment can then be lined up next to the unstained portion of the agarose gel, and the location of the DNA fragment can be estimated by the scalpel lines and excised for purification.

8. We have routinely found that preheating the Elution Buffer or sterile $ddH_2O$ to approximately 50 °C enhances the yield when using QiaQuick and DNA Clean & Concentrator columns, respectively.

9. We typically amplify the target gene (which will be later used for epPCR) by a standard PCR protocol to serve as a test fragment. Purify the PCR product and digest it overnight with the appropriate restriction enzymes in order to clone the fragment into the prepared vector. Purify and quantify the digested insert prior to ligation.

10. This step gives the researcher an opportunity to test the electrocompetent cell stocks and ensure they are highly efficient and suitable for use in library preparation. Electrocompetent cell aliquots should be discarded, and a fresh batch of electrocompetent cells should be prepared if the efficiency is found to be too low. A good batch of *E. coli* cells should yield $>10^9$ colonies per microgram of the supercoiled control plasmid used in the transformation. As an alternative to homemade cells, we have also obtained excellent transformation efficiencies with premade *E. cloni* 10G Elite electrocompetent cells (Lucigen, Middleton, WI).

11. This step gives the researcher an opportunity to test and optimize PCR conditions prior to Mutazyme amplification. In our experience, the annealing temperatures that are used for Phusion amplification work well for Mutazyme amplification. We have found that use of a PCR purified, DpnI-treated template for Mutazyme amplification reduces the risk of PCR contamination, especially when aiming for a low mutagenesis rate with correspondingly high concentrations of template. If PCR conditions have previously been optimized and researchers are aiming for high mutagenesis rates (with correspondingly low concentrations of template), a plasmid-based template may be used for the epPCR and researchers may directly proceed to Subheading 3.2, **step 7**.

12. The overall mutation frequency depends on the error rate of the polymerase and also the number of times that each template is duplicated in the reaction. If the initial amount of template is

high, it will undergo little duplication in the epPCR. On the other hand, a low amount of template will result in a greater number of duplications and more mutations will be introduced. Low mutation rates may also be achieved by decreasing the number of cycles in the epPCR protocol. This is discussed further in the GeneMorph II Random Mutagenesis Kit manual, available for download from http://www.chem.agilent.com/library/usermanuals/Public/200550.pdf.

We typically aim for a medium-to-high mutation rate, because this generates libraries that contain minimal numbers of "wasted" variants (i.e., non-mutated copies of the template or multiple copies of variants with any one point mutation). This strategy is discussed in more detail elsewhere [10, 19]. Agilent recommends 500–1,000 ng of template DNA for a low mutation rate (0–4.5 mutations/kb); 100–500 ng of template for a medium mutation rate (4.5–9 mutations/kb), and 0.1–100 ng of template for a high mutation rate (9–16 mutations/kb). If you have chosen to use a plasmid template rather than a purified PCR product, it is important to note that the amount of template is not the amount of purified plasmid DNA (*see* GeneMorph II Random Mutagenesis Kit Manual). In general, we find that the resulting mutation rates tend to be slightly lower than those expected from these calculations and therefore use the lowest end of the recommended template concentrations (e.g., we would use 0.1 ng of template for a high mutation rate).

13. We typically use 1/5 of the volume of the experimental reactions for negative controls in order to conserve resources.

14. We routinely use extension times that are calculated at a rate of 1 min/kb. For example, a 30 s extension time is used for a 500 bp product, and a 90 s extension time is used for a 1,500 bp product.

15. Calculating the PCR efficiency parameter allows robust statistical analysis of library composition (*see* Subheading 3.4, **step 9**). When the total product yield and the amount of starting template are known, the number of doublings in the PCR, *d*, can be calculated:

$$d = \frac{\log(\text{Product} / \text{Template})}{\log 2}.$$

The PCR efficiency (i.e., the probability that any particular sequence is duplicated in any one cycle of the PCR, eff) is then given by

$$\text{eff} = 2^{(d/n)} - 1,$$

where $n$ is the number of PCR cycles ($n = 30$ in our protocol). An online tool for calculating eff, given $d$ and $n$, can be found at http://guinevere.otago.ac.nz/cgi-bin/aef/PCReff.pl.

16. Before scaling up to a full-sized library, we find it useful and expedient to construct a test library. This allows the epPCR mutation spectrum to be determined.
17. The size of the final scaled-up library is a rough estimate based on (a) tenfold scale up of the ligation reaction; (b) transformation of ten aliquots of electrocompetent cells; and (c) spreading the entire recovery culture (instead of 10–50 μl aliquots).
18. We recommend using vector primers (e.g., for pET-based vectors, we would use the universal T7 terminator and promoter primers) to improve the quality of sequence at the beginning and end of the gene.
19. Clones from the test library (*see* Subheading 3.3) should be sequenced in order to analyze the mutation rate and the spectrum of mutations that arose in the epPCR. As discussed (*see* **Notes 5** and **6**), some control over the mutation rate is possible. However, in our experience, there is considerable experiment-to-experiment variation in the outcomes of the epPCR process (*see* Subheading 3.2). Therefore, we recommend conducting the analyses described in Subheading 3.4, to avoid wasting time and resources on a scaled-up library that contains little molecular diversity.
20. A library with an unbiased spectrum of mutations will be maximally diverse; that is, it will have the lowest probability of duplicated variants. Therefore, it is more likely to contain at least one improved variant [10]. One indicator of bias is the ratio of transitions (i.e., purine-to-purine and pyrimidine-to-pyrimidine mutations) to transversions (purine-to-pyrimidine and pyrimidine-to-purine mutations). There are four possible transitions and eight possible transversions (listed in Table 1). Therefore, a completely unbiased error-prone polymerase would generate libraries with transition/transversion (Ts/Tv) ratios of 0.5. Provided that the GC content of the gene is ~50 %, the ratio of AT → GC transitions to GC → AT transitions (i.e., AT → GC/GC → AT) in an unbiased epPCR library should be 1. Similarly, the number of mutations at A:T base pairs (A → N, T → N) should also be the same as the number of mutations at G:C base pairs (G → N, C → N). The effects of mutational bias on overall library composition can be assessed by altering the input parameters for PEDEL-AA analysis (*see* **Note 21**).
21. PEDEL-AA has an easy-to-use web interface and takes the following parameters as its inputs:

(a) The sequence of the gene that was randomized

(b) The estimated size of the scaled-up library (*see* Subheading 3.3, **step 6**)

(c) The nucleotide mutation matrix (*see* Subheading 3.4, **step 8**)

(d) The mean number of mutations per gene in the library (*see* Subheading 3.4, **step 9** and Table 1 for an example)

(e) The number of cycles in the epPCR (*see* Subheading 3.2, **step 9**)

(f) The PCR efficiency parameter for the epPCR (*see* **Note 15**)

(g) The mean number of insertions per gene in the library (*see* Subheading 3.4, **step 8**)

(h) The mean number of deletions per gene in the library (*see* Subheading 3.4, **step 8**)

The program output comprises a variety of statistics about the protein variants that are encoded by the epPCR library. Together, these data allow an informed decision to be made about whether to scale up the library (or whether to start over, with different epPCR and ligation conditions).

22. When constructing the full-sized library, the focus should be on scaling everything up by as much as possible. All of the remaining epPCR insert (Subheading 3.2, **step 14**) should be used in the scaled-up ligation, and as many aliquots of electrocompetent *E. coli* as possible should be transformed with the ligated products. The protocol that we describe is a typical example from our laboratory.

23. It is important to keep the DNA concentration low at this point to avoid transformants containing more than one plasmid [21].

24. If your selection or screening method permits, you may immediately proceed to screening/selecting variants from your library:

(a) Add 5 ml of sterile 80 % v/v glycerol to the 5 ml of recovered cells and mix the cells briefly, by inverting the tube 2–3 times. Spread 2 μl, 10 μl, and 50 μl aliquots on LB-agar plates (diluting as necessary to obtain a spreadable volume). Split the remainder of the cells into 1 ml aliquots and store them at –80 °C.

(b) Incubate all of the dilution and control plates at 37 °C for 16 h.

(c) Count the colonies on the plates corresponding to the 2 μl, 10 μl, and 50 μl aliquots and determine the most appropriate dilution to use for your screen/selection (i.e., the dilution that gives you a high number of single colonies but not a confluent lawn).

(d) Thaw separate glycerol aliquots, dilute as necessary, and plate cells on selective media. Electroporate additional 10–50 ng aliquots (*see* Subheading 3.5, **step 8**) of the purified ligation products to achieve full library coverage.

25. If you wish to screen the library in a different *E. coli* strain to the one you have utilized for cloning the library, resuspend the cell pellet from Subheading 3.5, **step 14** in 4.5 ml of LB media, split the suspension into three 1.5 ml aliquots, and purify the library using three plasmid miniprep columns (e.g., Qiagen; Valencia, CA, USA). Quantify using a Nanodrop spectrophotometer and store at −20 °C. Electroporate 10 ng aliquots of the purified library into your preferred screening or selection strain (*see* **Note 23**).

26. For the *E. coli* strain DH5α, we find that $OD_{600} = 1$ corresponds to ~$2.5 \times 10^8$ cells/ml.

## References


1. Lutz S (2010) Beyond directed evolution—semi-rational protein engineering and design. Curr Opin Biotechnol 21(6):734–743
2. Chica RA, Doucet N, Pelletier JN (2005) Semi-rational approaches to engineering enzyme activity: combining the benefits of directed evolution and rational design. Curr Opin Biotechnol 16(4):378–384
3. Lutz S, Patrick WM (2004) Novel methods for directed evolution of enzymes: quality, not quantity. Curr Opin Biotechnol 15(4): 291–297
4. Barak Y, Ackerley DF, Dodge CJ, Banwari L, Alex C, Francis AJ, Matin A (2006) Analysis of novel soluble chromate and uranyl reductases and generation of an improved enzyme by directed evolution. Appl Environ Microbiol 72(11):7074–7082
5. Morley KL, Kazlauskas RJ (2005) Improving enzyme properties: when are closer mutations better? Trends Biotechnol 23(5):231–237
6. Hermes JD, Blacklow SC, Knowles JR (1990) Searching sequence space by definably random mutagenesis: improving the catalytic potency of an enzyme. Proc Natl Acad Sci U S A 87(2): 696–700
7. Horsman GP, Liu AM, Henke E, Bornscheuer UT, Kazlauskas RJ (2003) Mutations in distant residues moderately increase the enantioselectivity of *Pseudomonas fluorescens* esterase towards methyl 3bromo-2-methylpropanoate and ethyl-3-phenylbutyrate. Chemistry 9(9): 1933–1939
8. Kumar S, Chen CS, Waxman DJ, Halpert JR (2005) Directed evolution of mammalian cytochrome P450 2B1: mutations outside of the active site enhance the metabolism of several substrates, including the anticancer prodrugs cyclophosphamide and ifosfamide. J Biol Chem 280(20):19569–19575
9. Wong JT (1980) Role of minimization of chemical distances between amino acids in the evolution of the genetic code. Proc Natl Acad Sci U S A 77(2):1083–1086
10. Patrick WM, Firth AE, Blackburn JM (2003) User-friendly algorithms for estimating completeness and diversity in randomized protein-encoding libraries. Protein Eng 16:451–457
11. Leung DW, Chen E, Goeddel DV (1989) A method for random mutagenesis of a defined DNA segment using a modified polymerase chain reaction. Technique 1:11–15
12. Cadwell RC, Joyce GF (1992) Randomization of genes by PCR mutagenesis. PCR Methods Appl 2(1):28–33
13. Fromant M, Blanquet S, Plateau P (1995) Direct random mutagenesis of gene-sized DNA fragments using polymerase chain reaction. Anal Biochem 224(1):347–353
14. Lin-Goerke JL, Robbins DJ, Burczak JD (1997) PCR-based random mutagenesis using manganese and reduced dNTP concentration. Biotechniques 23(3):409–412
15. Rasila TS, Pajunen MI, Savilahti H (2009) Critical evaluation of random mutagenesis by error-prone polymerase chain reaction protocols,

*Escherichia coli* mutator strain, and hydroxylamine treatment. Anal Biochem 338(1): 71–80

16. Hanson-Manful P, Patrick WM (2013) Construction and analysis of randomized protein-encoding libraries using error-prone PCR. Methods Mol Biol 996:251–267
17. Sambrook J, Russell DW (2001) Molecular cloning: a laboratory manual, 3rd edn. Cold Spring Harbour Laboratory Press, Cold Spring Harbour, NY
18. Saraswat M, Grand RS, Patrick WM (2013) Desalting DNA by drop dialysis increases library size upon transformation. Biosci Biotechnol Biochem 77(2):402–404
19. Firth AE, Patrick WM (2008) GLUE-IT and PEDEL-AA: new programmes for analyzing protein diversity in randomized libraries. Nucleic Acids Res 36:W281–W285
20. Hartman PS (1991) Transillumination can profoundly reduce transformation frequencies. Biotechniques 11:747–748
21. Goldsmith M, Kiss C, Bradbury AR, Tawfik DS (2007) Avoiding and controlling double transformation artifacts. Protein Eng Des Sel 20(7): 315–318

# Chapter 2

# Error-Prone Rolling Circle Amplification Greatly Simplifies Random Mutagenesis

**Ryota Fujii, Motomitsu Kitaoka, and Kiyoshi Hayashi**

## Abstract

We describe a simple and easy protocol to introduce random mutations into plasmid DNA: error-prone rolling circle amplification. A template plasmid is amplified via rolling circle amplification with decreased fidelity in the presence of $MnCl_2$ and is used to transform a host strain resulting in a mutant library with several random point mutations per kilobase through the entire plasmid. The primary advantage of this method is its simplicity. This protocol does not require the design of specific primers or thermal cycling. The reaction mixture can be used for direct transformation of a host strain. This method allows rapid preparation of randomly mutated plasmid libraries, enabling wider application of random mutagenesis.

**Key words** Random mutagenesis, Directed evolution, Point mutation, Rolling circle amplification, Protein engineering

## 1 Introduction

Random mutagenesis followed by high-throughput screening/selection is a powerful tool for improving the properties of proteins, investigating structure–function relationships, and studying the mechanisms of molecular evolution [1–5]. Among the reasons for growing interest in this methodology is the development of new and improved strategies to analyze mutant libraries. Many techniques for high-throughput screening and genetic selection have been devised during the last few decades [6–9], and instruments have been adapted for the rapid analysis of a large number of samples. Spectrophotometers and HPLC can be used to probe 96–1,536 samples at a time using multi-well plate formats.

Protocol simplification has also played an important role in the advances made in random mutagenesis [10]. Currently, the most common approach is error-prone PCR using mutagenesis kits such as GeneMorph® Mutazyme (a mixture of proprietary error-prone polymerases) or *Taq* polymerase PCR amplification with $MnCl_2$ added to a reaction mixture to reduce the fidelity of DNA amplification [11].

Elizabeth M.J. Gillam et al. (eds.), *Directed Evolution Library Creation: Methods and Protocols*, Methods in Molecular Biology, vol. 1179, DOI 10.1007/978-1-4939-1053-3_2, 

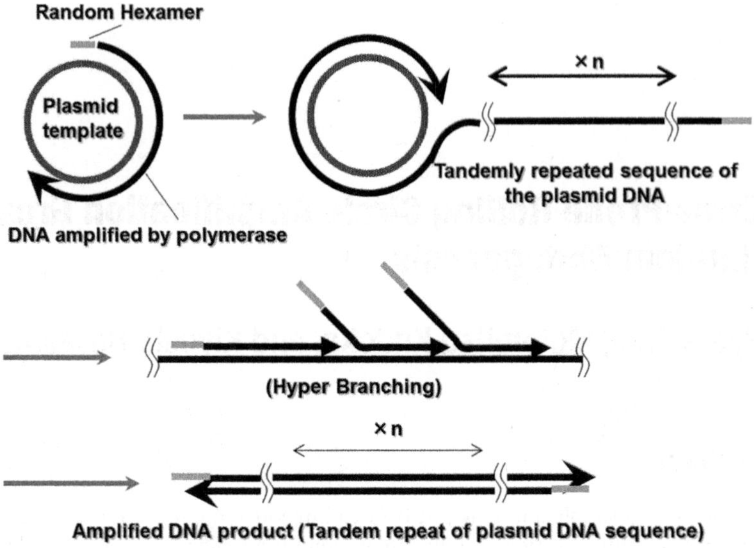


**Fig. 1** Mechanism of rolling circle amplification [14]. Random hexamers (NNNNNN) are hybridized to the circular DNA; the resulting double-stranded segments function as primers in the polymerization reaction conducted by φ29 DNA polymerase, a unique enzyme with very high strand displacement activity [15]. As the "front" of the extending complementary strand of the plasmid encounters double-stranded portions of DNA, the advancing new strand displaces the old one from the template. This extension process covers the entire length of the circular DNA multiple times resulting in the formation of repeated sequences of the template, which are called concatemers. The hexamers also hybridize to these concatemers, which then serve as templates. However, this extension continues only until the terminus of the linear concatemer is reached. The result is the formation of various lengths of double-stranded DNA consisting of repeats of the original template sequence. Reproduced with modification from [16] with the permission of Nature Publishing Group

Another useful random mutagenesis method is the bacterial mutator-strain technique. This method utilizes a bacterial strain in which DNA repair pathways have been blocked, promoting the introduction of a large number of mutations following DNA transformation [12].

Rolling circle replication is a process whereby cells amplify circular DNA molecules such as plasmids and viral DNA [13]. Using the principles of rolling circle replication, rolling circle amplification (RCA) was introduced in the mid-1990s as an isothermal method for amplifying circular DNA (Fig. 1) [17–19]. RCA products are linear DNA duplexes comprising tandem repeats of circular DNA sequences called concatemers.

RCA has several advantages over conventional methods (e.g., PCR) used for amplifying DNA. For example, RCA does not require the design of specific primers because random hexamers

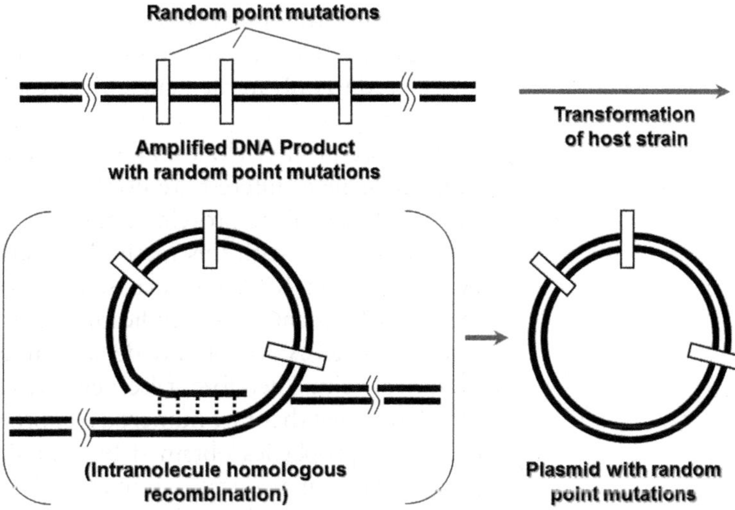


**Fig. 2** Error-prone RCA. Template plasmids are amplified by RCA in the presence of $MnCl_2$. Random point mutations are introduced during amplification because $MnCl_2$ reduces the fidelity of the DNA polymerase. When the RCA product is used for direct transformation of a host strain, it is recircularized by homologous recombination in vivo and a mutant library is thus constructed. Reproduced with modification from [16] with the permission of Nature Publishing Group

can be used as universal primers for any template [14]. Furthermore, RCA does not require optimization of thermal cycling conditions because the amplification reaction proceeds at a constant temperature. All the components required for an RCA reaction are readily available as a kit (TempliPhi DNA Amplification Kit) from GE Healthcare. Moreover, RCA products can be directly transformed into host cells such as *Escherichia coli* or yeast [20, 21]. Although RCA products are linear double-stranded DNA molecules, cells that are transformed with these products will contain plasmid replicas of the original template. This is because the linear RCA products have a tandem repeat structure and can be circularized intramolecularly by the homologous recombination machinery of the transformed strains.

Here we describe error-prone RCA, an extremely simple random mutagenesis protocol that uses RCA (Fig. 2) [16, 20]. Similar to error-prone PCR, this approach involves the addition of $MnCl_2$ to the RCA reaction mixture to promote the introduction of random point mutations into DNA. Subsequently, the host strain is directly transformed with the mutated linear DNA duplexes. These linear molecules will then undergo recircularization in vivo by homologous recombination. The ultimate result of this approach is a library of mutant transformants.

As with the mutator-strain approach to mutagenesis [12], error-prone RCA introduces mutations through the entire plasmid as opposed to only in the gene of interest [16, 20]. In principle, this feature may be a cause for concern because of the possibility of generating false-positive clones. However, in practice, mutations distant from the gene of interest are not likely to result in improvements in the properties of the protein. Thus, false positives are not expected to be a problem with this method. In addition, the transformation efficiency of an RCA product is nearly constant regardless of the mutation frequency [20], indicating that mutations introduced by error-prone RCA do not have deleterious effects on the plasmid replication system. Therefore, false-negative clones should also be quite rare. Given that there are numerous reports of proteins with improved enzymatic properties obtained by mutating the entire length of a plasmid via mutator-strain mutagenesis [22–24], we believe that error-prone RCA will also be useful in protein engineering.

Error-prone RCA can be used in various host organisms. This approach is also simpler than error-prone PCR and bacterial mutator-strain techniques, as there is no need for restriction enzymes, ligases, specific primers, or thermal cycling. Furthermore, because the RCA product is linear, a linearization step is not required before transformation of yeast. This simple method for constructing mutant libraries may facilitate the use of random mutagenesis in a wide variety of applications.

## 2 Materials

1. TempliPhi DNA Amplification Kit from GE Healthcare (Buckinghamshire, UK).
2. Plasmid containing target gene for mutagenesis.
3. Restriction enzymes from Takara (Otsu, Japan).
4. MinElute Reaction Cleanup Kit from Qiagen (Hilden, Germany).
5. MicroPulser electroporator and 0.1 cm electrode cuvettes from Bio-Rad (Hercules, CA, USA).
6. *E. coli* TOP10 from Invitrogen (Carlsbad, CA, USA).
7. Agarose gel apparatus, Mupid-2plus from Advance (Tokyo, Japan).

## 3 Methods

1. Mix 0.5 μL of 50 pg/μL template DNA with 5 μL of sample buffer supplied with the TempliPhi kit (*see* **Notes 1–3**).
2. Heat the mixture at 95 °C for 3 min on a heat block or PCR thermocycler to denature the plasmids. Cool immediately to

room temperature or 4 °C to allow annealing of the hexamer to the single-stranded template.

3. Add a premix of 5 μL reaction buffer and 0.2 μL enzyme mix (both supplied with the TempliPhi kit) to 1 μL of 15 mM $MnCl_2$. Incubate overnight at 30 °C (24 h).
4. Stop the reaction by heating at 65 °C for 10 min. Purify the RCA product using the MinElute Reaction Cleanup Kit (*see* **Note 4**). The concentration of the RCA product can be determined by measuring the absorbance at 260 nm. Formation of the desired RCA product can be confirmed by comparing restriction digests from the RCA product and the template plasmid using appropriate restriction enzyme(s) followed by agarose gel electrophoresis. The band pattern of the digest should be identical to that of the parent plasmid.
5. Transform 40 μL of *E. coli* TOP10 with 1 μL of the purified RCA product (*see* **Note 4**) using the MicroPulser electroporator with a 0.1 cm electrode cuvette at 1.8 kV. Increasing the volume of the RCA product (i.e., increasing the volume of **step 1** and using a larger volume of purified RCA product for electroporation) may increase the number of transformants.
6. Estimate the mutation frequency by sequencing the cloned DNA (*see* **Notes 5–7**).

## 4 Notes

1. We recommend the following standard conditions for error-prone RCA: 2.5 pg template DNA/μL (final concentration), 1.5 mM $MnCl_2$, and a reaction time of 24 h. The expected yield of RCA product is approximately 0.01 μg/μL, and approximately 3 mutations/kb will be introduced under these conditions. This rate corresponds to nearly one amino acid mutation per kilobase of the gene and is thus appropriate for in vitro evolution experiments [10].
2. The plasmid must be fresh supercoiled DNA because nicked plasmids are not suitable for the RCA reaction.
3. The components of the TempliPhi kit can be reproduced according to the methods described by Dean et al. [14]. φ29 DNA polymerase is available from New England Biolabs (Beverly, MA, USA) or Epicentre (Madison, WI, USA). This enzyme has a proofreading 3′–5′ exonuclease activity, which is disadvantageous for random mutagenesis. However, ϕ29 DNA polymerase without proofreading activity is not commercially available at present, although it has been developed [25]. If this enzyme becomes available, the mutation frequency and/or the product yield of this method is likely to improve further.

4. One possible way to increase the library size is to increase the amount of RCA product used during the electroporation step. The RCA reaction mixture may be concentrated and desalted by ethanol precipitation using the following steps: (a) add 30 μL of ethanol; (b) incubate on ice for 30 min; (c) centrifuge at 4 °C, 12,000 × *g* for 20 min and remove supernatant; (d) dry the precipitate using a vacuum centrifuge; (e) dissolve the precipitate in 1 μL of water. This step can be omitted if a sufficient number of mutants can be obtained by direct transformation of the TempliPhi product after purification using the MinElute Reaction Cleanup Kit. Ethanol precipitation will avoid electroporation inefficiency caused by carryover of the buffer salt from the RCA mixture, which increases the conductivity of the solution. Alternatively, the concentration of template DNA can be increased or the concentration of $MnCl_2$ decreased, although both of these approaches may decrease the mutation frequency. Please note that ethanol precipitation is strongly recommended when increasing the volume of electroporated RCA product to 2 μL or more.
5. In our previous work, we obtained 2,200 colonies using 1 μL of the RCA product with pUC19 as the template [20]. The mutation frequency was estimated to be 3.5 ± 1.0/kb. All types of substitution mutations were found in the error-prone RCA variants, although we noted an evident bias. In 61 % of these cases, mutations involved the replacement of T with C and A with G in the complement strand. The transition/transversion ratio was 2.7. Analysis of the plasmids in 174 clones by agarose gel electrophoresis revealed that 25 clones (14 %) contained plasmids that were 2 or 3 times the size of the parent plasmid (multimers).
6. A higher mutation frequency can be achieved by increasing the concentration of $MnCl_2$, decreasing the concentration of the template DNA, or increasing the reaction time of the RCA [16, 20].
7. The transformation efficiency of RCA product may be increased by using a plasmid containing a *loxP* recombination site of bacteriophage Cre recombinase [26].

## Acknowledgments

This study was supported in part by a grant from the Program for Promotion of Basic Research Activities for Innovative Biosciences (PROBRAIN).

## References


1. Bloom JD, Meyer MM, Meinhold P et al (2005) Evolving strategies for enzyme engineering. Curr Opin Struct Biol 15:447–452
2. Jaeger KE, Eggert T (2004) Enantioselective biocatalysis optimized by directed evolution. Curr Opin Biotechnol 15:305–313
3. Arnold FH, Wintrode PL, Miyazaki K et al (2001) How enzymes adapt: lessons from directed evolution. Trends Biochem Sci 26: 100–106
4. Johannes TW, Zhao H (2006) Directed evolution of enzymes and biosynthetic pathways. Curr Opin Microbiol 9:261–267
5. Reetz MT (2006) Directed evolution of enantioselective enzymes as catalysts for organic synthesis. Adv Catal 49:1–69
6. Aharoni A, Griffiths AD, Tawfik DS (2005) High-throughput screens and selections of enzyme-encoding genes. Curr Opin Chem Biol 9:210–216
7. Goddard JP, Reymond JL (2004) Enzyme assays for high-throughput screening. Curr Opin Biotechnol 15:314–322
8. Taylor SV, Kast P, Hilvert D (2001) Investigating and engineering enzymes by genetic selection. Angew Chem Int Ed Engl 40: 3310–3335
9. Lin H, Cornish VW (2002) Screening and selection methods for large-scale analysis of protein function. Angew Chem Int Ed Engl 41:4402–4425
10. Reetz MT, Jaeger KE (1999) Superior biocatalysts by directed evolution. In: Fessner WD, Archelas A, Demirjian DC et al (eds) Biocatalysis—from discovery to application. Springer-Verlag, Berlin, pp 31–57
11. Leung DW, Chen E, Goeddel DW (1989) A method for random mutagenesis of a defined DNA segment using a modified polymerase chain reaction. Techniques 1:11–15
12. Greener A, Callahan M, Jerpseth B (1996) An efficient random mutagenesis technique using an *E. coli* mutator strain. In: Trower MK (ed) In vitro mutagenesis protocols. Humana Press, New Jersey, pp 375–385
13. Kornberg A, Baker T (1992) DNA replication. W.H. Freeman & Company, New York
14. Dean FB, Nelson JR, Giesler TL et al (2001) Rapid amplification of plasmid and phage DNA using Phi 29 DNA polymerase and multiply-primed rolling circle amplification. Genome Res 11:1095–1099
15. Blanco L, Bernad A, Lázaro JM et al (1989) Highly efficient DNA synthesis by phage phi29 DNA polymerase. Symmetrical mode of DNA replication. J Biol Chem 264:8935–8940
16. Fujii R, Kitaoka M, Hayashi K (2006) Error-prone rolling circle amplification: the simplest random mutagenesis protocol. Nat Protoc 1:2493–2497
17. Fire A, Xu SQ (1995) Rolling replication of short DNA circles. Proc Natl Acad Sci U S A 92:4641–4645
18. Liu DY, Daubendiek SL, Zillman MA et al (1996) Rolling circle DNA synthesis: small circular oligonucleotides as efficient templates for DNA polymerases. J Am Chem Soc 118: 1587–1594
19. Lizardi PM, Huang X, Zhu Z et al (1998) Mutation detection and single-molecule counting using isothermal rolling-circle amplification. Nat Genet 19:225–232
20. Fujii R, Kitaoka M, Hayashi K (2004) One-step random mutagenesis by error-prone rolling circle amplification. Nucleic Acids Res 32:e145
21. Ding X, Snyder AK, Shaw R et al (2003) Direct retransformation of yeast with plasmid DNA isolated from single yeast colonies using rolling circle amplification. Biotechniques 35: 774–779
22. Camps M, Naukkarinen J, Johnson BP et al (2003) Targeted gene evolution in Escherichia coli using a highly error-prone DNA polymerase I. Proc Natl Acad Sci U S A 100: 9727–9732
23. Henke E, Bornscheuer UT (1999) Directed evolution of an esterase from Pseudomonas fluorescens. Random mutagenesis by error-prone PCR or a mutator strain and identification of mutants showing enhanced enantioselectivity by a resorufin-based fluorescence assay. Biol Chem 380:1029–1033
24. Bornscheuer UT, Altenbuchner J, Meyer HH (1998) Directed evolution of an esterase for the stereoselective resolution of a key intermediate in the synthesis of epothilones. Biotechnol Bioeng 58:554–559
25. de Vega M, Lazaro JM, Salas M (2000) Phage φ29 DNA polymerase residues involved in the proper stabilisation of the primer-terminus at the 3′-5′ exonuclease active site. J Mol Biol 304:1–9
26. Huovinen T, Julin M, Sanmark H et al (2011) Enhanced error-prone RCA mutagenesis by concatemer resolution. Plasmid 66:47–51

# Chapter 3

# Random Mutagenesis by Error-Prone Pol Plasmid Replication in *Escherichia coli*

**David L. Alexander, Joshua Lilly, Jaime Hernandez, Jillian Romsdahl, Christopher J. Troll, and Manel Camps**

## Abstract

Directed evolution is an approach that mimics natural evolution in the laboratory with the goal of modifying existing enzymatic activities or of generating new ones. The identification of mutants with desired properties involves the generation of genetic diversity coupled with a functional selection or screen. Genetic diversity can be generated using PCR or using in vivo methods such as chemical mutagenesis or error-prone replication of the desired sequence in a mutator strain. In vivo mutagenesis methods facilitate iterative selection because they do not require cloning, but generally produce a low mutation density with mutations not restricted to specific genes or areas within a gene. For this reason, this approach is typically used to generate new biochemical properties when large numbers of mutants can be screened or selected. Here we describe protocols for an advanced in vivo mutagenesis method that is based on error-prone replication of a ColE1 plasmid bearing the gene of interest. Compared to other in vivo mutagenesis methods, this plasmid-targeted approach allows increased mutation loads and facilitates iterative selection approaches. We also describe the mutation spectrum for this mutagenesis methodology in detail, and, using cycle 3 GFP as a target for mutagenesis, we illustrate the phenotypic diversity that can be generated using our method. In sum, error-prone Pol I replication is a mutagenesis method that is ideally suited for the evolution of new biochemical activities when a functional selection is available.

**Key words** Mutagenesis, Pol I, Directed evolution, Genetic adaptation, Mutation spectrum, GFP, Okazaki fragment

## 1 Introduction

Directed evolution is a widely used method for optimization of existing biological activities or for the creation of new ones [1, 2]. This approach involves two basic steps: (a) generation of genetic diversity and (b) identification of mutants with desired properties. Following mutagenesis, individual clones are screened or libraries are put through functional selection to obtain individual mutants with the desired properties.

Elizabeth M.J. Gillam et al. (eds.), *Directed Evolution Library Creation: Methods and Protocols*, Methods in Molecular Biology, vol. 1179, DOI 10.1007/978-1-4939-1053-3_3, 

Approaches based on screening try to maximize the frequency of active mutants with the desired properties by incorporating elements of rational design [1, 3] and by optimizing the mutation spectrum to ensure a balanced representation of mutations and a minimal presence of inactivating mutations. These mutant libraries typically involve random oligonucleotide mutagenesis, thus allowing tight control of the target sites, type, and frequency of mutations.

By contrast, experimental approaches that include functional selections aim at maximizing sequence space exploration. This requires balancing a high mutation density (2–5 mutations/clone) with the preservation of a significant fraction of active mutants. The high mutation density is required to overcome restrictions on available evolutionary trajectories caused by sign epistasis, i.e., the presence of mutations whose effects are neutral, positive, or negative depending on the sequence context [4]. This need to balance genetic diversity with the need to preserve catalytic activity has been addressed by two main approaches: (a) by sequential evolution, i.e., by enriching the library for functional mutants at intermediate steps (although this can create significant bottlenecks) [5], and (b) by shuffling related sequences, orthologs, or paralogs of the gene of interest (this creates hybrid sequences enriched for activity relative to their level of amino acid divergence) [6, 7]. Despite these efforts, the most advantageous mutants can be missed following functional selections in liquid culture due to the phenomenon known as clonal interference, where mutants with modest contributions to fitness compete against each other, preventing the emergence of more infrequent mutants with higher impact on fitness [8]. Biases introduced by clonal interference are increasingly being addressed by deep sequencing of libraries at different stages of selection [9].

Compared to in vitro methods, in vivo mutagenesis approaches are optimal for sequential evolution strategies, as they do not require cloning, thereby greatly facilitating iteration. These methods use mutator strains, i.e., strains that are deficient in one or more mechanisms of replication fidelity [10], or use exposure to mutagens. The genomic instability associated with these nontargeted methods limits the mutation rate (which needs to be tolerated by the host), producing libraries with low mutation densities (~1 point mutation/2–5 kb). In vivo mutagenesis methods are better suited for functional selection strategies, which can identify rare clones from large mutant libraries, because of the limited efficiency for mutagenesis of these methods. In addition, in vivo mutagenesis is not targeted, so mutations outside of the target gene can lead to changes in gene expression. Mutations in regulatory elements such as the promoter of the target gene or the plasmid origin of replication can in turn interfere with selections aimed at optimizing activity through modulation of catalysis. While detrimental in the context of activity, optimization strategies modulating expression can facilitate the evolution of new biochemical activities by enhancing promiscuous activities often present in target enzymes [11, 12]. Thus, in vivo mutagenesis is ideally suited for the evolution of new biochemical activities when a functional selection is available.

Here we present a mutagenesis system that has several advantages over other in vivo mutagenesis approaches. Our method is based on replication of a ColE1 plasmid bearing the gene of interest by an error-prone DNA polymerase I (Pol I). Pol I is a polymerase specialized in ColE1 plasmid replication, although it also plays a role in processing Okazaki primers during lagging-strand synthesis and in small-gap filling during DNA repair. Therefore, error-prone Pol I replication limits mutagenesis to ColE1 plasmid sequence, largely sparing the genome (which is replicated by a different polymerase, Pol III) and allowing a higher mutation load in the target of interest [13].

The error-prone DNA polymerase I that we use (low-fidelity Pol I or "LF-Pol I") bears three mutations, namely, I1709N (in motif A), A759R (in motif B), and D424A (in the proofreading domain) [14, 15]. LF-Pol I is expressed in an *Escherichia coli* strain, JS200, which has a temperature-sensitive allele of Pol I (*polA12*) [16] so that LF-Pol I becomes the predominant Pol I activity at 37 °C. Replication of the ColE1 plasmid-borne target sequence in *polA12* cells under restrictive conditions results in the generation of a random mutant library. Our system also produces mutations in wild-type strains of *E. coli* but at a three- to fivefold lower mutation frequency (data not shown). Mutagenesis is more efficient in saturated cultures as compared with exponential cultures [14]. LF-Pol I mutagenesis is not continuous in culture: mutation rates decrease substantially after the initial culture grown under restrictive conditions reaches saturation; this is true even if this culture is diluted and then subsequently expanded. Therefore, obtaining high mutation densities (>1 mutation/kb) requires multiple iterative rounds of mutagenesis, plasmid recovery, and transformation [13]. The likely explanation for this phenomenon is that the establishment of multi-copy plasmids following transformation requires more cycles of replication than subsequent plasmid maintenance.

This approach is, to our knowledge, the most efficient method of in vivo mutagenesis available, with the added advantage of easy iteration, a relatively balanced spectrum, and very few insertions/deletions. Compared to in vitro mutagenesis methods, the main disadvantages of this approach are lack of ability to restrict mutagenesis to a target gene (with the consequent concern about mutations modulating expression rather than activity) or to a specific area within a target gene, and a partial dependence on *polA* host strains. Error-prone Pol I replication is ideally suited for the evolution of new biochemical activities when coupled with functional selections such as the evolution of extended-spectrum β-lactamase mutants [14] or of two medium-chain-length terminal alkane hydroxylases [17] because this capitalizes on the ability to generate libraries with high complexity and different levels of expression, which is known to favor the evolution of new biological activities.

Here we describe methods for LF-Pol I mutagenesis of a target plasmid bearing a gene sequence of interest. The methods discussed here include protocols for competent cell preparation, for transformation of the target plasmid, for iterative mutagenesis, and for characterization of the resulting libraries. In Subheading 4 of this chapter, we also provide detailed data on the mutation spectrum generated by our method.

## 2 Materials

### 2.1 Cells

1. *E. coli* JS200 strains (*recA718 polA12 (ts) uvrA155 trpE65 lon-11 sulA)*: JS200-WT, i.e., JS200 cells expressing wild-type (wt) Pol I, and JS200-EP, i.e., JS200 expressing error-prone (EP) Pol I. The JS200-EP and JS200-WT strains as well as the LF-Pol I-bearing plasmid, its map, and its sequence are available by request through the Addgene plasmid repository (http://www.addgene.org/).
2. Readout strain, JS200-WT, or (for complementation) a strain lacking the specific activity that is being evolved such as Top10 (Invitrogen, Grand Island, NY, USA).

### 2.2 Media

1. LB agar and LB broth (e.g., Fisher Scientific, Fair Lawn, NJ, USA) prepared according to vendor specifications.
2. 2×YT rich media; 0.016 g/ml Bacto Tryptone, 0.01 g/ml Bacto Yeast Extract, and 0.005 g/ml NaCl suspended in deionized water.

### 2.3 Antibiotics

In our study, antibiotics were purchased from Sigma-Aldrich (Saint Louis, MO, USA).

1. Tetracycline: prepared as a concentrated stock in 50 % ethanol to allow dilution to a final concentration 12.5 μg/ml.
2. Chloramphenicol: prepared as a concentrated stock in 100 % ethanol to allow dilution to a final concentration 35 μg/ml.
3. Carbenicillin: prepared as a concentrated stock in water to allow dilution to a final concentration 100 μg/ml.

### 2.4 Plasmids

1. Pol I Plasmid, i.e., pHSG576 plasmid bearing the sequence of the LF-Pol I gene. This plasmid, which carries a pSC101 (Pol I-independent, ColE1-compatible) origin of replication with chloramphenicol as a resistance marker [18], provides the error-prone polymerase activity.

2. Examples of ColE1 plasmids into which the target gene may be cloned include the pUC, pBR, pLitmus (New England Biolobs, Ipswich, MA, USA), and Topo vectors (Clontech, Mountain View, CA, USA).

**2.5 DNA Purification Kits**

1. Plasmid miniprep kit (e.g., NucleoSpin Plasmid Purification Kit, Macherey-Nagel, Duren, Germany).
2. PCR Clean-Up Kit (e.g., Gel and PCR Clean-Up Kit, Macherey-Nagel, Duren, Germany).

**2.6 Electroporation Equipment**

1. Electroporator (e.g., Electroporator 2510, Eppendorf, New York, NY, USA).
2. 2 mm gap electroporation cuvettes (e.g., from Molecular BioProducts, Santa Clara, CA, USA).

**2.7 Enzymes**

1. Restriction enzymes (e.g., from New England Biolabs, Ipswich, MA, USA).

**2.8 Flow Cytometry**

1. The cytometer used for library characterization in our study was a BD Influx Cytometer, BD Biosciences (San Jose, CA, USA).
2. 8× BioSure Sheath Solution – 1.0 mM disodium EDTA, 1.9 mM potassium phosphate (monobasic), 3.8 mM potassium chloride, 16.6 mM sodium phosphate (dibasic), and 139.0 mM sodium chloride – was purchased from BioSure (Grass Valley, CA, USA) and diluted to 1× in sterile DI water and then used to run samples through the cytometer.

## 3 Methods

**3.1 Preparation of Electrocompetent JS200 Cells (for a Protocol to Make Chemically Competent Cells, See Note 1)**

1. Pick a single *E. coli* JS200 colony transformed with the Pol I plasmid into a flask containing 8 ml of LB plus antibiotic. These colonies are grown on LB plates with appropriate antibiotic selection (for pHSG576, 30 μg/ml chloramphenicol). The culture is grown shaking at 200 rpm overnight at 30 °C (*see* **Note 2**).
2. The next morning, expand the culture by adding the 8 ml overnight culture into a flask containing 400 ml LB with the same antibiotic concentration. Allow this culture to grow at 30 °C while shaking at 200 rpm to an $OD_{600}$ of 0.5–0.7 (ca. 3–4 h).
3. Initiate a glycerol wash by first chilling cells on ice for 20 min. Cells are then pelleted by centrifugation (e.g., in an Eppendorf 5810R, 20 min at 3,220 rcf at 4 °C). Remove supernatant and then resuspend cells in cold 10 % (w/v) glycerol solution using a sterile serological pipette.

4. Transfer resuspended cell solution to a 50 ml conical tube and bring to a final volume of 45 ml in cold 10 % (w/v) glycerol. Pellet cells by centrifugation, remove supernatant, and resuspend cells in cold 10 % (w/v) glycerol. Repeat this step twice so that cells are washed a total of three times in a fresh exchange of 10 % (w/v) glycerol solution to remove all traces of salts. Cells and wash solution need to be kept on wet ice or at 4 °C throughout this process.
5. After final wash, resuspend the cell pellet in ~2 ml of 10 % (w/v) glycerol (approximately twice the pellet volume). Aliquot in volumes for single experimental use to minimize freeze and thaw. Quick-freeze aliquots in dry ice, and store at −80 °C.
6. Cells should be thawed slowly on wet ice for electrotransformation.

### 3.2 Direct-Plating Mutagenesis (for Alternative Liquid Mutagenesis Protocol, See Note 3)

1. Make *E. coli* JS200 EP and WT strains electrocompetent, using the protocol outlined above (permissive conditions, i.e., 30 °C and exponential growth).
2. Transform target plasmid of choice by electroporation into 40 μl of electrocompetent *E. coli* JS200 WT or EP cells. We use an Eppendorf 2510 electroporator and 2 mm gap cuvettes at 1,800 V (*see* **Notes 4** and **5**).
3. Recover cells in 1 ml of LB broth for 40 min at 37 °C with shaking at 200 rpm.
4. Plate cells at a "near lawn" concentration on LB agar plates containing chloramphenicol and the antibiotic selecting for the target plasmid. Plates need to be pre-warmed to 37 °C and maintained at this restrictive temperature during plating (*see* **Notes 6** and **7**).
5. Incubate plates under restrictive conditions, i.e., 37 °C, overnight.

### 3.3 Plasmid Recovery

1. Collect colonies from LB agar plates by washing with 2 ml LB broth. Add 1 ml first, spread with a sterile glass rod, and collect wash into appropriate size tube. Repeat with another 1 ml of LB broth (*see* **Note 8**).
2. Isolate plasmid DNA from the wash using the plasmid miniprep kit to obtain the genetic library (*see* **Note 9**).

### 3.4 Readout/Iteration

1. To make sure both plasmids are present and there is no additional plasmid contamination, digest your isolated plasmid DNA with a restriction enzyme(s) that linearizes both the target plasmid and the pHSG plasmid. Run this digest on an agarose gel and stain to visualize the two bands of appropriate size.

2. To eliminate the Pol plasmid, digest your miniprep with a restriction enzyme that linearizes the Pol plasmid but does not cut the target plasmid (*see* **Note 10**).
3. Clean up the restriction digest using the PCR Clean-Up Kit to remove salts that interfere with subsequent retransformation in the iterative plasmid mutagenesis.
4. For iterative mutagenesis, retransform the restricted target plasmid library preparation into fresh *E. coli* JS200 EP cells and carry out subsequent rounds of mutagenesis. It is estimated that mutations will accumulate at a rate of 0.56 mutations/kbp/cycle.
5. Retransform the isolated restriction digested plasmid library into the readout strain to characterize the mutant phenotypes (*see* **Note 11**).

### 3.5 Sequencing for Mutation Frequency and Mutant Genotype

1. The library is characterized following iterative library generation by direct sequencing of the plasmid-borne target gene in individual colonies. To separate individual plasmids prior to sequencing, the plasmid library needs to be transformed into a readout strain or other bacterial strain not expressing the LF-Pol (*see* **Note 12**).
2. Cultures are plated at a density of ~100 colonies per 100 mm LB agar dish and allowed to grow to 1 mm in diameter.
3. Individual colonies are picked and subjected to rolling circle amplification (RCA) [19]. The product of RCA then serves as a template for single primer extension and dye terminator sequencing of the target gene and regulatory regions of the target plasmid (*see* **Note 13**).
4. Individual mutants coming out of a functional selection are similarly sequenced. In this case, the plasmid DNA is purified prior to sequencing so that mutant plasmids are available for testing. Testing involves retransformation of the sequenced plasmid into naïve readout cells for preliminary phenotypic characterization. Phenotypes need to be confirmed after re-cloning the candidate mutations into a fresh plasmid, as the observed phenotype for a given plasmid could be caused by mutations outside the sequenced area.

### 3.6 Example Application

To characterize our mutagenesis system, we put the plasmid pGFPuv, which bears the "cycle 3" variant of GFP as a reporter [20], through four rounds of mutagenesis as described above. We then characterized the resulting library by transforming the recovered plasmid population into Top10 cells. Figure 1 shows the diversity of fluorescence intensities obtained, both for individual colonies on an LB agar plate (*panel a*) and for individual cells in suspension (*panel b*). For reference, *panel b* also shows the flow cytometry

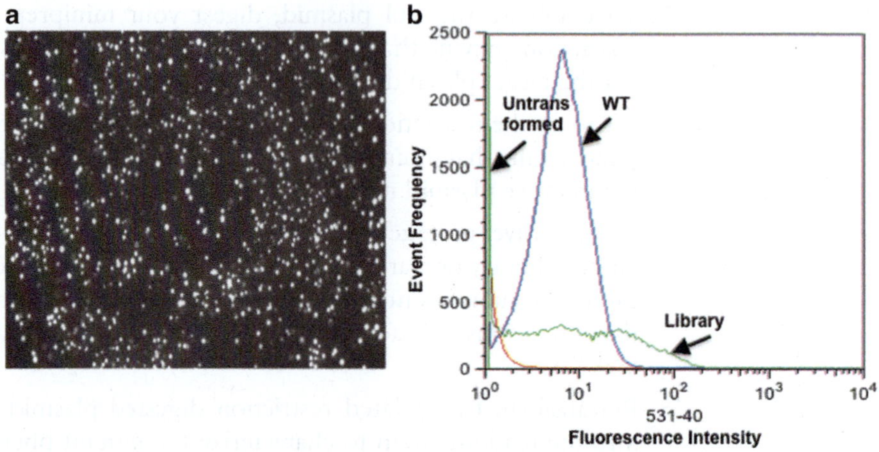


**Fig. 1** Phenotypic diversity. (**a**) Diversity in individual colony fluorescence intensity. *E. coli* Top10 cells were transformed with the pGFPuv library obtained by four rounds of direct-plating mutagenesis and grown under carbenicillin selection (the antibiotic marker for the plasmid). The majority of these colonies represent single plasmid transformations. Colonies were imaged using a UVP bioanalyzer illuminated at 302 nm using a SYBR filter with an emission cutoff between 517 and 570 nm. (**b**) Flow cytometry analysis. Colonies shown in *panel b* were washed with LB and grown to an $OD_{600}$ between 0.7 and 0.9 for optimal GFP fluorescence (*see* **Note 18**). Next, cells were diluted in sheath solution to an event rate of less than 100 cells per second passing through the detector of the BD Influx cytometer. The GFP fluorescence was analyzed using a 531/40 optical filter and excited by a 488 nm laser. This data represents the fluorescence emission of single cells in a cell culture population. In addition to the library, we also show two controls: cells expressing WT pGFPuv plasmid and untransformed cells. These are labeled directly on the figure

distribution of fluorescence intensity for the parental pGFPuv plasmid and for untransformed cells. Note that the fraction of library clones with increased fluorescence is comparable to the fraction exhibiting decreased fluorescence relative to the parental control. Given that for protein-coding sequences gain-of-function mutations generally represent only a small fraction of the total (0.5–1 %, compared to 30–50 % for loss-of-function) [5], the high representation of mutants that are brighter than the wild type in our library is almost certainly attributable to mutations modulating expression.

### *3.7 Mutation Spectrum*

Pol I is a specialized polymerase involved in ColE1 plasmid replication. Pol I initiates ColE1 plasmid replication by extending an RNA primer transcribed from the plasmid *ori* sequence. This extension (corresponding to nascent leading strand) continues until the replication complex with Pol III is loaded, a process generally known as "polymerase switch" (reviewed in [21]). Pol III is a dimeric enzyme containing two core subassemblies (one for each strand) that performs coupled, high-speed replication of the two strands at the replication fork (reviewed in [22]).

We identified an area of 100–150 bp sequence immediately upstream of the polymerase switch that appears to be replicated by Pol I exclusively, providing the mutation spectrum for LF-Pol I in the leading strand in vivo [23] (*see* **Notes 14** and **15**).

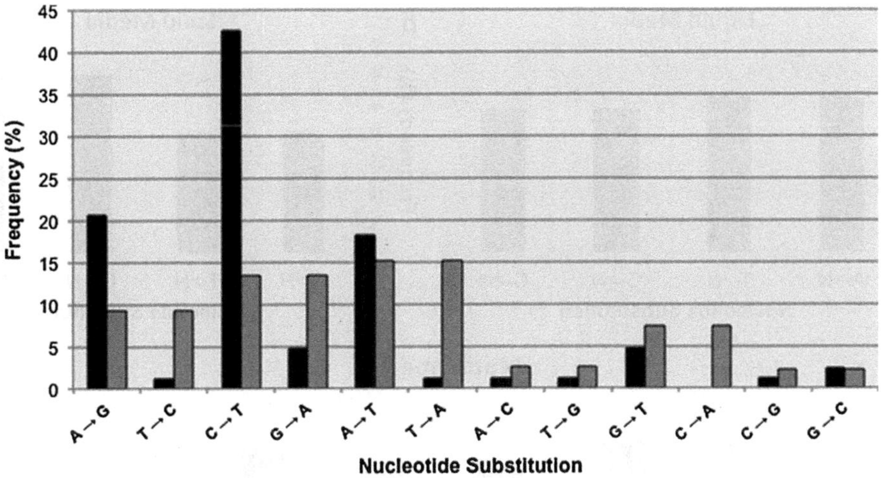


**Fig. 2** Leading-strand mutation spectrum of Pol mutagenesis in vivo. LF-Pol I mutations found in the area immediately upstream of polymerase switch, which appears to correspond to leading-strand synthesis by Pol I, as reported in [23] (*see* **Note 19**), black columns. For reference, this spectrum is compared to the nucleotide substitution spectrum of Mutazyme II, as reported in [24], grey columns

This mutation spectrum, which is shown in Fig. 2, exhibited a dramatic frequency imbalance between complementary pairs, which allowed us to designate the most frequent mutations of the pair (A→G, C→T, A→T, T→G, and G→T) as indicators of leading-strand synthesis and the least frequent (T→C, G→A, T→A, A→C, and C→A) as indicators of lagging-strand synthesis. This analysis showed that beyond the switch (170–250 nt downstream of plasmid replication initiation), Pol I continues but with no apparent strand preference. Double-stranded replication balances out differences in mutation frequency between complementary pairs [23], which explains the remarkably balanced spectrum of LF-Pol I mutagenesis [24] (*see* **Notes 16** and **17**).

When we look at how libraries are generated (solid plate vs. growth in suspension; *see* Subheading 3.2 and **Note 3**), we see differences in mutation spectrum depending on the protocol used: solid media libraries show more C→T mutations (60 %, compared to 40 % in liquid media), whereas libraries generated in liquid culture show more A→G (30 % in liquid media compared to 20 % in solid media) and A→T mutations (20 % in liquid media compared to 10 % in solid media) [23]. In both cases, insertion/deletion mutations are very rare (<1 % of total mutations). Figure 3 summarizes the nucleotide substitution spectrum for double-stranded error-prone Pol I replication for both types of libraries. Both libraries exhibit a well-balanced mutation spectrum. Liquid libraries show the most even distribution of nucleotide substitutions of the two, comparable to that of Mutazyme II in standard error prone PCR protocols, while mutations generated in colonies show a moderate predominance of G and C substitutions.

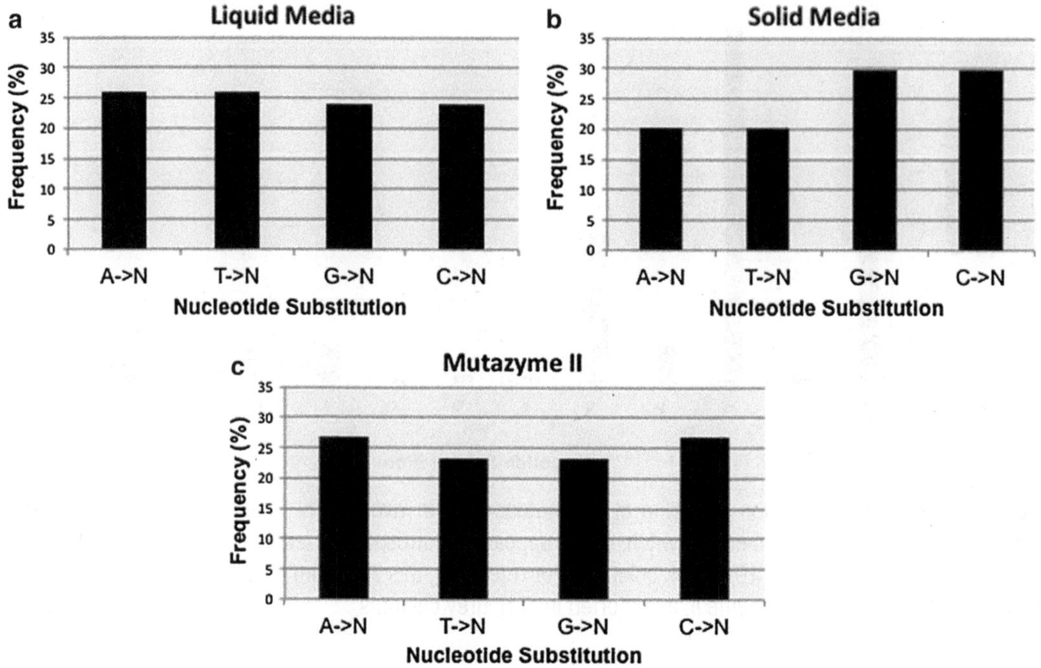


**Fig. 3** Double-strand nucleotide substitution profile of Pol mutagenesis. This figure contains the same mutation spectrum data from Fig. 1 but on both strands of DNA and separating libraries generated on liquid media (*panel a*) from libraries generated in solid plates (*panel b*). For reference, these two spectra are compared to a Mutazyme II nucleotide substitution spectrum, as reported in [24] (*panel c*). The frequency of substitutions (%) for each nucleotide is shown on the *y*-axis, with the nucleotide substitution type (A → N, T → N, G → N, C → N) shown on the *x*-axis. The solid media spectrum includes GFP and ALKBH1 libraries generated on solid plates (*n* = 20 and 40 mutations, respectively), and the liquid mutagenesis spectrum includes GFP and hTK libraries generated in suspension culture (*n* = 33 and 30 mutations, respectively), as reported in [23]

## 4 Notes

1. An alternative here is to prepare chemically competent cells as follows. *E. coli* JS200 cells are grown as described in Subheading 3.1 but washed with a buffered $CaCl_2$ solution: 60 mM $CaCl_2$, 15 % glycerol, 10 mM Hepes (4-(2-hydroxyethyl)-1-piperazineethanesulfonic acid). As per the preparation of electrocompetent cells, cultures are grown to log phase, chilled in wet ice, pelleted by centrifugation, and washed into the $CaCl_2$ solution. Three washes are used to concentrate cells. Pelleted cells are resuspended into approximately twice the pellet volume of cold $CaCl_2$ solution, and aliquots are quick-frozen on dry ice prior to storage at −80 °C.
2. *E. coli* JS200 parental cells are tetracycline resistant. The pHSG576 plasmid is maintained in JS200 cells with chloramphenicol selection.

3. An alternative here is to undertake the mutagenesis in liquid cultures ("liquid mutagenesis"). *E. coli* JS200 cells are transformed with target plasmids as outlined above. The transformants are then recovered for 1 h shaking at 30 °C in 1 ml of LB broth. After the recovery period, cells are plated at 30 °C on LB agar containing 100 μg/ml carbenicillin. Single colonies are then picked and allowed to grow in 4 ml of LB broth at permissive temperature (30 °C) shaking overnight. For mutagenesis, the overnight cultures are diluted to a factor of $1{:}10^3$–$1{:}10^5$ in 4 ml pre-warmed (37 °C) 2×YT media. Cells are grown, shaking at 200 rpm, for 1–3 days to reach saturation or hypersaturation. Plasmids are then isolated using a DNA miniprep kit and sequenced.
4. pGFPuv, a plasmid encoding GFP and available from Clontech (http://www.clontech.com/), is routinely transformed as a control for mutagenesis (*see* **Note 10**).
5. An alternative here is transformation of chemically competent cells. Chemically competent *E. coli* JS200 cells are thawed on wet ice and 100 μl of cells are mixed with up to 1 μg (no more than 10 μl) of the isolated plasmid library. Cells and plasmid are incubated on wet ice for 10 min and then heat-shocked at 42 °C for 2 min. Cells are recovered at 37 °C in 1 ml of LB with shaking at 200 rpm and then plated at a near lawn density on LB plates containing chloramphenicol and the antibiotic selecting for the target plasmid. Transformation of chemically competent *E. coli* JS200 cells is less efficient than electroporation, but this is balanced by the greater amount of DNA that can be used in the transformation, the simplicity of the procedure, and the consistency of the results.
6. A near lawn concentration is defined as distinct but uncountable colonies (>1,000 colonies per 100 mm Petri dish).
7. The dilution of cells plated following transformation is empirically determined for each preparation of electrocompetent cells. Generally 50 μl of cells from the 1 ml culture, where cells are allowed to recover following transformation, will yield a near lawn.
8. Plate washing is transferring the bacterial colonies from the LB plate to LB broth by adding to a small volume (1 ml) of LB broth and "scrubbing" them off the plate with a sterile apparatus, such as a glass or metal plate spreader. Avoid collecting any visible amounts of agar.
9. The wash collected from the LB plate may be too dense to miniprep in its entirety. If this is the case, miniprep the maximum amount recommended by the manufacturer of your miniprep kit (typically, this involves diluting your wash to an $OD_{600}$ of 1.0 and using ~3 ml of the diluted culture for the miniprep).

10. This control can be omitted unless the presence of the EP plasmid interferes with the readout.
11. Quantification of GFP mutagenesis can be carried out in the *E. coli* JS200 strain expressing wild-type Pol I or in Top10 or similar *E. coli* strains supportive of ColE1 plasmid replication. Following one round of LF-Pol mutagenesis, transformation of the recovered pGFPuv library into a readout strain produces ~10 % of colonies with visibly decreased fluorescence on solid plates (under UV light).
12. In the absence of a selection, given that the ColE1 target plasmid is present in multiple copies, mutagenesis in *E. coli* JS200-EP cells would be expected to produce mixed sequences. Retransformation of the library separates individual plasmids, producing a majority of unambiguous sequences, and thus facilitates the characterization of the library.
13. Mutations in the plasmid origin of replication (*ori*) can alter the overall plasmid copy number in individual cells. Rolling circle amplification minimizes the effect of plasmid copy number variation because it uses random hexamers and Phi29 DNA polymerase to produce single-stranded, linear concatenated copies of the circular sequence [19].
14. The source of our mutations is in all likelihood LF-Pol I. This conclusion is based on the high frequency of LF-Pol I mutagenesis, which is 3–4 orders of magnitude above spontaneous mutation levels, and on the fact that the observed mutation frequency in vivo correlates with the fidelity of Pol I in vitro [15].
15. The spectrum of LF-Pol I mutations we see in vivo has in all likelihood been modulated by proofreading mechanisms, notably by mismatch repair (MMR). MMR should not interfere with the overall distribution of mutations, which is the basis for our mutation footprint, but would be expected to have a major impact on the mutation spectrum, selectively suppressing certain base-pair substitutions (transitions, especially T→C mutations).
16. Double-stranded plasmid replication by LF-Pol I compensates for the frequency bias exhibited by LF-Pol I between complementary pairs. For example, A→G is made 20 times more frequently than T→C by the polymerase, but lagging-strand replication produces 20 times more T→C than A→G on the complementary strand, producing a balanced spectrum between complementary pairs. Double-stranded replication of plasmid by Pol I likely explains the balanced mutation spectrum reported for LF-Pol I mutagenesis relative to a variety of other mutagenesis methods [24]. Note, however, that balanced nucleotide spectrum, while generally desirable, does not solve biases inherent to the genetic code, such as the fact that single

nucleotide substitutions only produce a subset of all possible amino acid substitutions and that these mutations tend to be conservative (for a fuller discussion, *see* ref. 25).

17. We found two areas that are predominantly replicated by Pol I on the lagging strand: areas where the RNA fragment priming lagging-strand synthesis is processed by the 5′→3′ exonuclease activity of Pol I [26] and an area associated with replication [23]. These areas are small (approximately 20 and 150 nucleotides, respectively) and therefore comprise a fraction of the total mutations, but have a dramatic effect on the local mutation spectrum, producing a spectrum that is the complementary to that shown in Fig. 2.
18. Cells need to be grown to an $OD_{600}$ of 0.7–0.9 for optimal GFP fluorescence, as GFP has a long half-life and accumulates in the cytoplasm.
19. Mutation hot spots, defined as areas with 6 or more mutations in the same or contiguous positions, were removed as nonrepresentative of baseline LF-Pol I mutations [23].

## Acknowledgments

This work was supported by K08 award CA116429-01A1 of the NCI to M.C. and by R01 award ES019625-01 of NIEHS to M.C. The authors would like to thank Dr. Roel Schaaper for the helpful input on the mutagenic footprint of LF-Pol I.

## References

1. Dalby PA (2011) Strategy and success for the directed evolution of enzymes. Curr Opin Struct Biol 21:473–480
2. Yuan L, Kurek I, English J, Keenan R (2005) Laboratory-directed protein evolution. Microbiol Mol Biol Rev 69:373–392
3. Lutz S (2010) Beyond directed evolution—semi-rational protein engineering and design. Curr Opin Biotechnol 21:734–743
4. Weinreich DM, Watson RA, Chao L (2005) Perspective: sign epistasis and genetic constraint on evolutionary trajectories. Evolution 59:1165–1174
5. Bloom JD, Arnold FH (2009) In the light of directed evolution: pathways of adaptive protein evolution. Proc Natl Acad Sci USA 106 Suppl 1:9995–10000
6. Kagami O, Kikuchi M, Harayama S (2004) Single-stranded DNA family shuffling. Methods Enzymol 388:11–21
7. Zhao H, Zha W (2006) In vitro 'sexual' evolution through the PCR-based staggered extension process (StEP). Nat Protoc 1:1865–1871
8. Kao KC, Sherlock G (2008) Molecular characterization of clonal interference during adaptive evolution in asexual populations of Saccharomyces cerevisiae. Nat Genet 40:1499–1504
9. Araya CL, Fowler DM (2011) Deep mutational scanning: assessing protein function on a massive scale. Trends Biotechnol 29:435–442
10. Nguyen AW, Daugherty PS (2003) Production of randomly mutated plasmid libraries using mutator strains. Methods Mol Biol 231: 39–44
11. Khersonsky O, Tawfik DS (2010) Enzyme promiscuity: a mechanistic and evolutionary perspective. Annu Rev Biochem 79:471–505
12. Soskine M, Tawfik DS (2010) Mutational effects and the evolution of new protein functions. Nat Rev Genet 11:572–582

13. Troll C, Alexander D, Allen J, Marquette J, Camps M (2011) Mutagenesis and functional selection protocols for directed evolution of proteins in E. coli. J Vis Exp 49:e2505
14. Camps M, Naukkarinen J, Johnson BP, Loeb LA (2003) Targeted gene evolution in Escherichia coli using a highly error-prone DNA polymerase I. Proc Natl Acad Sci U S A 100:9727–9732
15. Shinkai A, Loeb LA (2001) In vivo mutagenesis by Escherichia coli DNA polymerase I. Ile(709) in motif A functions in base selection. J Biol Chem 276:46759–46764
16. Uyemura D, Lehman IR (1976) Biochemical characterization of mutant forms of DNA polymerase I from Escherichia coli. I. The polA12 mutation. J Biol Chem 251:4078–4084
17. Koch DJ, Chen MM, van Beilen JB, Arnold FH (2009) In vivo evolution of butane oxidation by terminal alkane hydroxylases AlkB and CYP153A6. Appl Environ Microbiol 75: 337–344
18. Shinkai A, Patel PH, Loeb LA (2001) The conserved active site motif A of Escherichia coli DNA polymerase I is highly mutable. J Biol Chem 276:18836–18842
19. Johne R, Muller H, Rector A, van Ranst M, Stevens H (2009) Rolling-circle amplification of viral DNA genomes using phi29 polymerase. Trends Microbiol 17:205–211
20. Miura H, Inoko H, Inoue I, Tanaka M, Sato M, Ohtsuka M (2011) Simple cloning strategy using GFPuv gene as positive/negative indicator. Anal Biochem 416:237–239
21. Camps M (2010) Modulation of ColE1-like plasmid replication for recombinant gene expression. Recent Pat DNA Gene Seq 4: 58–73
22. McHenry CS (2011) DNA replicases from a bacterial perspective. Annu Rev Biochem 80:403–436
23. Troll CJ, Yoder J, Alexander D, Hernández J, Loh Y, Camps M (2013) Mutagenic footprint of low-fidelity Pol I replication in *E. coli* reveals an extensive interplay between Pol I and Pol III during ColE1 plasmid replication. Curr Genet (Nov 2 epublished ahead of print)
24. Wong TS, Roccatano D, Zacharias M, Schwaneberg U (2006) A statistical analysis of random mutagenesis methods used for directed protein evolution. J Mol Biol 355: 858–871
25. Wong TS, Zhurina D, Schwaneberg U (2006) The diversity challenge in directed protein evolution. Comb Chem High Throughput Screen 9:271–288
26. Allen JM, Simcha DM, Ericson NG, Alexander DL, Marquette JT, Van Biber BP, Troll CJ, Karchin R, Bielas JH, Loeb LA, Camps M (2011) Roles of DNA polymerase I in leading and lagging-strand replication defined by a high-resolution mutation footprint of ColE1 plasmid replication. Nucleic Acids Res 39: 7020–7033

# Chapter 4

# The Sequence Saturation Mutagenesis (SeSaM) Method

**Anna Joëlle Ruff, Tsvetan Kardashliev, Alexander Dennig, and Ulrich Schwaneberg**

## Abstract

Sequence Saturation Mutagenesis (SeSaM) is a random mutagenesis method developed to overcome the limitations of existing error-prone PCR (epPCR) protocols. SeSaM is advantageous with respect to (1) elimination of mutagenic "hot spots", (2) increase in frequency of subsequent nucleotide substitutions, (3) control over the mutational bias through the utilization of universal base analogs, and, consequently, (4) the prospect of generating transversion-enriched mutant libraries. These advanced features lead to chemically diverse mutant libraries on the protein level, essentially making SeSaM a complementary technology to transition biased epPCR mutagenesis methods.

**Key words** Directed evolution, Diversity generation, Protein engineering, epPCR, Random mutagenesis

## 1 Introduction

Directed evolution is a powerful algorithm for tailoring proteins to industrial and medicinal needs and requirements. Error-prone PCR (epPCR)-based methods are the standard for random mutagenesis due to their robustness and simplicity [1]. Despite some impressive success stories [2, 3], these methods are generally limited in their ability to generate highly diverse mutant libraries for the following reasons: (1) the redundancy of the genetic code; (2) commonly low mutagenic frequency in epPCR libraries which disallows subsequent nucleotide exchanges [4]; (3) the innate transitional bias of DNA polymerases used in epPCR (transition mutations typically lead to conservative amino acid substitutions, e.g., a hydrophobic to hydrophobic amino acid exchange) [5, 6]; and (4) mutagenic "hot spots", due to the propensity for polymerases to introduce mutations preferentially in certain DNA sequence contexts [7]. Taken together, these factors significantly limit the chemical diversity of random mutagenesis methods [6, 8]. A good indication for the latter is the observation that on average

Elizabeth M.J. Gillam et al. (eds.), *Directed Evolution Library Creation: Methods and Protocols*, Methods in Molecular Biology, vol. 1179, DOI 10.1007/978-1-4939-1053-3_4, © Springer Science+Business Media New York 2014

only 7 of the 19 possible amino acid substitutions are typically achieved by epPCR [9]. To evaluate the limitations and facilitate the selection of a suitable random mutagenesis method, computational tools such as the mutagenesis assistant program (MAP) [10] and its further improvement MAP3D were developed. MAP3D is implemented on a server (http://map.jacobs-university.de/map3d.html) [11] and can be used as a benchmarking system for random mutagenesis methods on the protein level by comparing amino acid substitution patterns (further information on this program is available in Chapter 19 of this volume, by Verma et al.).

Sequence Saturation Mutagenesis (SeSaM) is a chemo-enzymatic mutagenesis method that aims at eliminating most limitations of current state-of-the-art random mutagenesis techniques. SeSaM is a four-step process that employs selective chemical fragmentation of DNA, phosphorothioate, and universal nucleotide analogs as well as several enzymatic steps in order to introduce nucleotide exchanges with a desirable mutational bias, distribution, and frequency. In step 1 of SeSaM, a pool of single-stranded DNA (ssDNA) fragments with various gene lengths is generated in a PCR that employs a biotin-labeled forward primer and a predefined mixture of standard and cleavable phosphorothioate nucleotides. Phosphorothioester bonds within the resulting PCR products can be cleaved selectively in the presence of iodine, under alkaline conditions and at elevated temperature. The ssDNA fragments are subsequently isolated from non-biotinylated DNA strands using streptavidin-coated magnetic beads. In step 2, the purified ssDNA fragments are "tailed" with a universal base analog of choice in a reaction catalyzed by terminal deoxynucleotidyl transferase (TdT). In step 3, the "tailed" ssDNA fragments are elongated to full gene length. Step 3 amplicons serve as templates in step 4 where the incorporated nucleotide analogs are replaced by standard nucleotides in a final PCR. The two preliminary experiments necessary for SeSaM template generation and determination of the optimal concentration of phosphorothioate nucleotides (Fig. 1, panels 1 and 2) as well as steps 1–4 of SeSaM (Fig. 1, panels 3–6) are represented diagrammatically below.

A SeSaM library can be completed in 2–3 days. Despite the increased workload, SeSaM is a relatively economical and straightforward procedure to perform. More notably, through SeSaM three of the four limitations of most random mutagenesis methods are eliminated, i.e., each codon in a given gene is targeted with equal probability, independently of polymerase bias and with a high probability of consecutive nucleotide exchanges. In SeSaM libraries, up to 30 % of all substitutions are consecutive, which lead to expansion of the obtainable amino acid substitution pattern of epPCR [12]. The significance of SeSaM is proven by several success stories (patents: WO2012119955A1; WO2012028709A3; WO2012017008A1) and reports on evolved enzymes such as proteases [13, 14] and phytase [15].

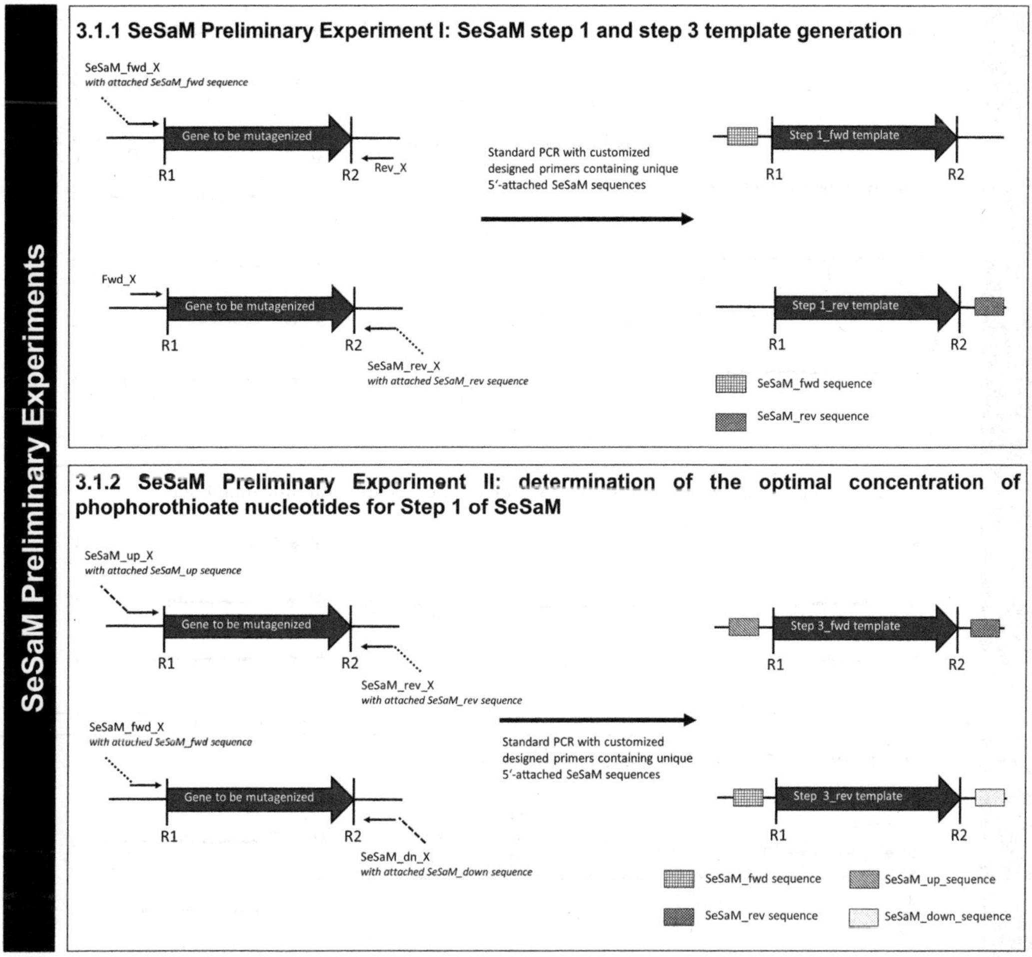


**Fig. 1** Illustration of the SeSaM strategy, consisting of the two preliminary experiments: Preliminary experiment I and II (panels 1 and 2) and the four steps of the SeSaM method (panels 3–6). The *numbering at the top of each panel* refers to Subheading 3 summarized therein

The mutational bias introduced by SeSaM is typically governed by the choice of nucleotide analog employed in step 2. Several universal bases [12, 16, 17] have been used to control the ratio of transition to transversion mutations introduced in SeSaM libraries, with dPTP (6-(2-deoxy-ß-D-ribofuranosyl)-3,4-dihydro-8H-pyrimido-[4,5-C][1,2]oxazin-7-one) being the most frequently used one. dPTP is a pyrimidine base analog used to target adenine (A) and guanine (G) in DNA strands. The substitution of an A- or G-site with dPTP ultimately results in a transversion nucleotide exchange to thymine (T) or cytosine (C) (Fig. 2).

More recently, dRTP (1-ß-D-ribofuranosyl-1,2,4-triazole-3-carboxamide), a purine analog, has been successfully employed in SeSaM to target T- and C-sites in DNA sequences and render

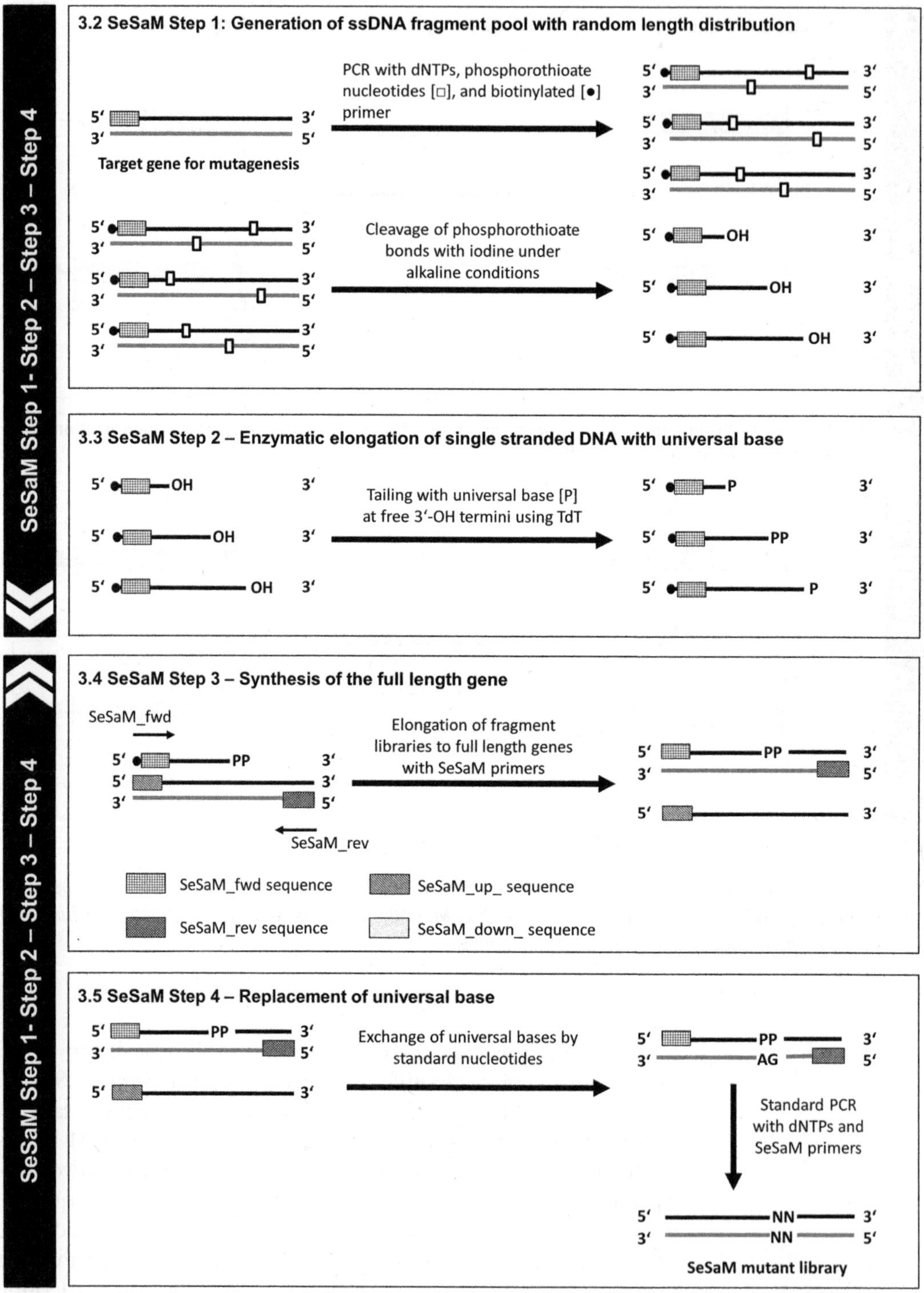


**Fig. 1** (continued)

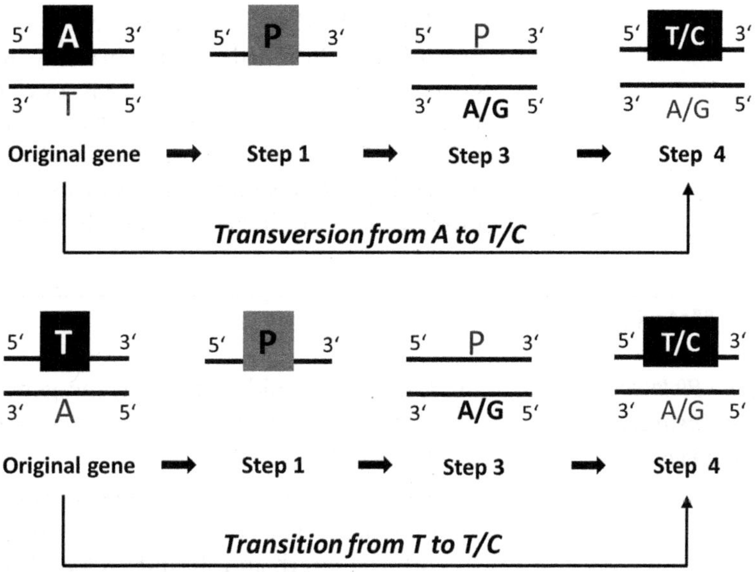


**Fig. 2** The obtainable mutational spectrum when dPTP is introduced at an A- (*top*) or a T- (*bottom*) site. In SeSaM-Tv-classic, dPTP is used to substitute A (or alternatively G, not shown in the figure) in order to obtain transversion-enriched libraries. This nucleotide analog is not used at T-sites (or C-sites) as this would result in either nucleotide preservation or a transition that generates conserved amino acid substitution pattern and could be achieved by epPCR. Step 2 is not illustrated in the figure since this step does not impact the obtainable substitution patterns [17]

exchanges to A and G [17]. Thus, all four nucleotides in a DNA strand can be methodically exchanged by random mutagenesis. Coding and noncoding DNA strands can be independently targeted in step 1 and 2 to generate forward and reverse libraries. These can then be used separately in step 3 or recombined in order to increase the mutational frequency of SeSaM libraries. In step 3, it is also crucial to utilize a DNA polymerase capable of elongation of mismatching primer-template-DNA. Polymerases such as Vent (exo-), 3D1 (a chimera of three genes from the genus *Thermus*), and, more recently, the translesion DNA polymerase Dpo4 have been proven suitable for this application. 3D1 polymerase, for instance, achieves up to 30 % subsequent mutations when dPTP is used in SeSaM and can even "read" through a triple consecutive mismatch in in vitro assays [18] and under application conditions [12]. In the last step of SeSaM, *Taq* polymerase is used in a nested PCR to specifically amplify mutated DNA as well as to exchange the previously introduced universal base analogs with standard nucleotides. The PCR products of step 4 can be directly cloned in an expression vector by restriction cloning, although we recommend ligase-independent cloning methods such as PLICing [19] in order to achieve higher cloning efficiency.

Two SeSaM protocols (SeSaM-Tv-classic, SeSaM-R) employing either dPTP or dRTP are currently in use in our laboratory to generate transversions at all four nucleotides in DNA strands. Here, the SeSaM-Tv-classic protocol using the universal nucleotide dPTP is described in detail. The SeSaM-R protocol [17] employing dRTP slightly differs in step 1 and 2, and the respective modifications are included in Subheading 4.

## 2 Materials

### *2.1 Biological and Chemical Materials*

1. Purified plasmid containing the gene of interest suspended in nuclease-free water or TE buffer: 10 mM Tris–Cl (pH 7.5), 1 mM EDTA.
2. Six PCR primer sets for the preliminary steps and steps 1 to 4 of SeSaM. As a rule, the primers should not exceed 50 nucleotides in length and should preferably be of HPLC purity grade. The melting temperature of the primers should be calculated only for the gene-specific portion excluding the attached synthetic sequences (where applicable).
   (a) Primer set 1 amplifies fragment(s) that contains the region targeted for random mutagenesis not exceeding 1.5 kb in length (as this would negatively affect the efficiency of step 3), preferably including gene-flanking restriction sites. Each primer binding site should be located 20–80 bp up- or downstream of the region to be mutagenized. The gene-specific primer region should have a length of 18–25 bp. Name these primers "fwd_X" and "rev_X", where X stands for the abbreviated name of the gene of interest.
   (b) Primer set 2 comprises the gene-specific forward and reverse primers (i.e., identical to the primer set 1 sequences) but with the following artificial oligos attached at 5′-ends. Name these primers "SeSaM_fwd_X" and "SeSaM_rev_X", respectively.
      - SeSaM_fwd sequence: 5′-CACACTACCGCACTCC GTCG-3′.
      - SeSaM_rev sequence: 5′-GTGTGATGGCGTGAGG CAGC-3′.
   (c) Primer set 3 has gene-specific sequences identical to primers set 1 with the following SeSaM_up or SeSaM_dn sequences attached at the 5′-ends of the forward and reverse primer, respectively. Name the primers "SeSaM_up_X" and "SeSaM_dn_X".
      - SeSaM_up sequence: 5′-CGCCTGTCAC-3′.
      - SeSaM_dn sequence: 5′-GCGGACAGTG-3′.

(d) Primer set 4 comprises solely the following sequences:

- SeSaM_fwd sequence: 5′-CACACTACCGCACTCC GTCG-3′.
- SeSaM_rev sequence: 5′-GTGTGATGGCGTGAGG CAGC-3′.

(e) Primer set 5 comprises the following primer sequences, each of which additionally harbors a biotin tag (Bio-TEG, a biotin attached to a 15-atom mixed polarity triethylene glycol spacer) at 5′-ends. Name these primers "Bio_ SeSaM_fwd" and "Bio_SeSaM_rev":

- Bio_SeSaM_fwd: 5′-[Biotin]CACACTACCGCACTC CGTCG-3′.
- Bio_SeSaM_rev: 5′-[Biotin]GTGTGATGGCGTGAG GCAGC-3′.

(f) Primer 6 is labeled with the fluorescent dye (FITC, fluorescein isothiocyanate) at the 5′-end. This primer is recommended for quality control of SeSaM step 2.

- FITC: 5′-[FITC]GCAATACGCAAACAGTCTCT CT-3′.

3. NucleoSpin Extract II PCR Purification Kit (Macherey-Nagel, Düren, Germany) and the associated buffers NT, NT3, as well as the separately supplied NTC buffer (*see* **Note 1**). This system is highly recommended since the SeSaM protocol has been validated with this particular kit.
4. Nuclease-free water (for PCR amplification and elution of DNA after purification) and deionized water (for buffer preparation).
5. Solution M: 100 mM NaOH in deionized water. Store at 4 °C.
6. Solution R: 0.1 % SDS (w/v) in deionized water. Store at 4 °C.
7. Binding and washing buffer 1 (1× B&W): 10 mM Tris–HCl (pH 7.5), 1 mM EDTA (pH 8.0), 2 M NaCl.
8. Binding and washing buffer 2 (2× B&W): 10 mM Tris–HCl (pH 7.5), 1 mM EDTA (pH 8.0), 1 M NaCl.
9. dNTP buffer (pH 7.5): 1 mM Tris–HCl (pH 7.5), 0.1 mM DTT. Store at −20 °C.
10. Ethanol (99 % purity).
11. Solution C: 20 mM Iodine in ethanol (99 %). Store solution C for not more than 1 week at 4 °C.
12. Oligo buffer: 10 mM Tris–HCl (pH 7.5), 0.1 mM EDTA (pH 8.0). Store at −20 °C.

13. 10× Tris borate EDTA buffer (10× TBE): 0.89 M Tris–HCl (pH 8.0), 0.89 M boric acid, 20 mM EDTA (pH 8.0). Store at room temperature.
14. Tris borate EDTA loading dye (6× TBE-loading dye): 60 % 10× Tris borate EDTA buffer (v/v), 30 % glycerol (v/v), 1 mg/mL (w/v) bromophenol blue. Store at 4 °C.
15. 40 % acrylamide-bis-acrylamide solution. Store at 4 °C.
16. 10 % (w/v) Ammonium persulfate solution (10 % APS), prepared in deionized water. Store at −20 °C.
17. *N*,*N*,*N*′,*N*′-Tetramethylethylenediamine (TEMED, 99 % purity). Store at room temperature.
18. *Taq* DNA polymerase and corresponding buffer (e.g., New England Biolabs, Frankfurt, Germany). Store at −20 °C.
19. High-fidelity (HF) DNA polymerase and corresponding buffer (e.g., Phusion® DNA Polymerase and 5× Phusion® buffer; New England Biolabs, Frankfurt, Germany). Store at −20 °C.
20. Terminal deoxynucleotidyl transferase (TdT) (20 U/μL) and 10× TdT buffer (New England Biolabs, Frankfurt, Germany). Store at −20 °C.
21. 3D1 polymerase (5 U/μL) (*see* **Note 2**) and 10× SuperTaq reaction buffer (HT BioTechnology, Cambridge, UK). Store enzymes at −20 °C.
22. Streptavidin-coated magnetic beads (e.g., M-PVA SAV1, Chemagen, Baesweiler, Germany). Store at 4 °C.
23. 50× TAE buffer: 2 M Tris,1 M acetic acid 50 mM EDTA, with pH adjusted to 8.5 using NaOH. Store at room temperature.
24. 0.8 % (w/v) agarose solution in 1× TAE buffer. Store at 65 °C for up to 2 weeks.
25. 0.001 % (v/v) ethidium bromide prepared in deionized water. Store at room temperature.
26. 6× DNA Loading dye: 67 % (v/v) glycerol, 1 mg/mL bromophenol blue, 10 % 10× TAE buffer. Store at −20 °C.
27. Nucleotide analogs dRTP and dPTPαS as well as phosphorothioated nucleotides (dATPαS, dCTPαS, dTTPαS, dGTPαS) which can be purchased from BIOLOG Life Sciences Institute (Bremen, Germany); standard dNTPs are available from NEB (New England Biolabs, Frankfurt, Germany). Prepare the following dNTP solutions in oligo buffer and store at −20 °C:
    (a) Individual solutions of dATP, dGTP, dCTP, and dTTP; stock concentration of 100 mM.
    (b) dNTP mix: 10 mM each of dATP, dGTP, dTTP, dCTP.

(c) Individual solutions of phosphorothioate nucleotides (dATPαS, dGTPαS, dCTPαS, and dTTPαS); stock concentration of 5 mM.

(d) d(GTP/TTP/CTP)-mix: 10 mM each of dGTP, dTTP, dCTP.

(e) d(ATP/TTP/CTP)-mix: 10 mM each of dATP, dTTP, dCTP.

(f) d(ATP/TTP/GTP)-mix: 10 mM each of dATP, dTTP, dGTP.

(g) d(ATP/GTP/CTP)-mix: 10 mM each of dATP, dGTP, dCTP.

(h) dPTPαS (6-(2-deoxy-β-D-ribofuranosyl)-3,4-dihydro-8H-pyrimido-[4,5-C][1,2]oxazin-7-one); stock concentration of 20 μM (*see* **Note 3**).

(i) dRTP (1-β-D-ribofuranosyl-1,2,4-triazole-3-carboxamide); stock concentration of 3.4 μM.

(j) d(ATP/ATPαS/GTP/TTP/CTP)-mixes: in total 8 mixes with 0 %, 10 %, 15 %, 20 %, 25 %, 30 %, 35 %, and 40 % dATPαS should be prepared and labeled "A0"–"A40", respectively; on these labels, "A" stands for adenine and the number (e.g., 10) signifies the percentage of dATPαS in relation to the total amount of adenine species (dATPαS + dATP); the final concentration for each nucleotide in the mix should be 1 mM (e.g., to prepare 50 μL A10 mix, combine 9 μL dATP (5 mM stock), 1 μL dATPαS (5 mM stock), 5 μL d(GTP/TTP/CTP)-mix (10 mM stock), 35 μL dNTP buffer).

(k) d(GTP/GTPαS/ATP/TTP/CTP)-mixes: in total 8 mixes with 0 %, 10 %, 15 %, 20 %, 25 %, 30 %, 35 %, and 40 % dGTPαS should be prepared and labeled "G0"–"G40", respectively; "G" stands for guanine and the number (e.g., 10) signifies the percentage of dGTPαS in relation to the total amount of guanine species (dGTPαS + dGTP); the final concentration for each nucleotide in the mix should be 1 mM (e.g., to prepare 50 μL G10 mix, combine 9 μL dGTP (5 mM stock), 1 μL dGTPαS (5 mM stock), 5 μL d(ATP/TTP/CTP)-mix (10 mM stock), 35 μL dNTP buffer).

(l) d(CTP/CTPαS/ATP/TTP/GTP)-mixes: in total 8 mixes with 0 %, 10 %, 15 %, 20 %, 25 %, 30 %, 35 %, and 40 % dCTPαS should be prepared and labeled "C0"–"C40", respectively; "C" stands for cytosine and the number (e.g., 10) signifies the percentage of dCTPαS in relation to the total amount of cytosine species (dCTPαS + dCTP); the final concentration for each nucleotide in the mix should be 1 mM (e.g., to prepare 50 μL C10 mix, combine 9 μL dCTP (5 mM stock), 1 μL dCTPαS (5 mM stock), 5 μL d(ATP/TTP/GTP)-mix (10 mM stock), 35 μL dNTP buffer).

(m) d(TTP/TTPαS/ATP/GTP/CTP)-mixes: in total 8 mixes with 0 %, 10 %, 15 %, 20 %, 25 %, 30 %, 35 %, and 40 % dTTPαS should be prepared and labeled "T0"–"T40", respectively; "T" stands for thymine and the number (e.g., 10) signifies the percentage of dTTPαS in relation to the total amount of thymine species (dTTPαS + dTTP); the final concentration for each nucleotide in the mix should be 1 mM (e.g., to prepare 50 μL T10 mix, combine 9 μL dTTP (5 mM stock), 1 μL dTTPαS (5 mM stock), 5 μL d(ATP/GTP/CTP)-mix (10 mM stock), 35 μL dNTP buffer).

### 2.2 Equipment

1. Microcentrifuge (e.g., Eppendorf, Hamburg, Germany) with rotor for 1.5 mL and 2 mL tubes.
2. Thermocycler with gradient and temperature decrement option (e.g., Thermocycler proS, Eppendorf, Hamburg, Germany).
3. Small volume photometer for quantitation of DNA concentrations in sample volumes of 1–2 μL (e.g., NanoDrop photometer, NanoDrop Technologies, Wilmington DE, USA).
4. Agarose gel electrophoresis equipment.
5. Phosphorimager with capability to measure Excitation 494 nm; Emission 518 nm (e.g., FLA-3000 from Fujifilm, Düsseldorf, Germany).
6. Acrylamide gel electrophoresis system (e.g., Mini-Protean Tetra Cell system from Bio-Rad, München, Germany).
7. Recommended: automated gel electrophoresis system (e.g., Experion system and High Sens RNA chip from Bio-Rad, München, Germany).
8. PCR tubes (0.2 mL).
9. Microcentrifuge tubes (1.5 mL or 2 mL).
10. 4 magnets (diameter 20–30 mm).

## 3 Methods

### 3.1 SeSaM Preliminary Experiments

When SeSaM is performed for the first time for a given gene, the following two preliminary experiments should be performed.

#### 3.1.1 SeSaM Preliminary Experiment I: SeSaM Step 1 and Step 3 Template Generation

In the first preliminary experiment, linear DNA templates for step 1 and step 3 of SeSaM must be generated. These templates harbor artificial flanking SeSaM sequences which play an important role in SeSaM steps 1 and 4.

1. Assemble the following pre-mastermix (200 μL) on ice in a 0.2 mL tube. Thaw all nonenzyme components at room temperature and enzymatic components on ice, mix by vortexing, and collect by a brief centrifugation.

| Pre-mastermix (160 μL) | Volume |
|---|---|
| 5× Phusion® polymerase buffer | 40 μL |
| dNTP mix (10 mM) | 4 μL |
| Plasmid template | *X* μL (40 ng/kb of plasmid) (e.g., for a plasmid with a total length of 5 kb, use 2 μL of a 100 ng/μL purified plasmid) |
| Phusion® Polymerase (2 U/μL) (*see* **Note 4**) | 2 μL |
| Nuclease-free $H_2O$ | To a total volume of 160 μL |

2. Mix the pre-mastermix by pipetting.
3. Divide the pre-mastermix into 4× 40 μL aliquots, and then add to each tube a total of 10 μL from the following primer pairs (i.e., 5 μL apiece of forward and reverse primer):
   (a) SeSaM_fwd_X and Rev_X; recommended labeling of the PCR reaction and product—"step 1_fwd_X".
   (b) SeSaM_Rev_X and fwd_X; recommended labeling of the PCR tube reaction and product—"step 1_rev_X".
   (c) SeSaM_fwd_X and SeSaM_dn_X; recommended labeling of the PCR tube reaction and product—"step 3_rev_X".
   (d) SeSaM_rev_X and SeSaM_up_X; recommended labeling of the PCR tube reaction and product—"step 3_fwd_X".
4. If the optimal annealing temperature for these primer sets has been confirmed experimentally, skip **steps 5–7** and proceed to **step 8**.
5. Split each PCR reaction (50 μL) into 4× 12 μL aliquots in fresh PCR tubes. This allows each aliquot to be tested at a different annealing temperature in a gradient PCR in order to identify the most suitable annealing temperature for each primer set.
6. Run PCR samples using a temperature gradient in the range 50–70 °C as outlined below:

   *PCR protocol*—98 °C for 30 s (1 cycle); 98 °C for 15 s, gradient Tm for 30 s, 72 °C for 20 s/kb of the SeSaM template (18 cycles); 72 °C for 5 min (1 cycle); followed by a 4 °C hold to protect the samples.

7. Analyze PCR products on a 0.8 % TAE agarose gel (7.5 V/cm; 40 min). Identify the primer annealing temperature at which the highest amount of specific product is formed and use this annealing temperature for the following PCR reactions.
8. Set up PCR mastermixes (as outlined in **steps 1–3**) whereby a total volume of 200 μL per primer pair is prepared and split into four aliquots of 50 μL before thermocycling using the PCR program with the selected Tm outlined in **step 6**, above. Combine each 4× 50 μL of identical PCR products after the PCR is completed.
9. Confirm that PCR products were successfully obtained by loading a 4 μL aliquot from each reaction on a 0.8 % TAE agarose gel (7.5 V/cm; 40 min) (*see* **Note 5**).
10. Individually column-purify the remainder of each reaction using the NucleoSpin Extract II PCR Purification Kit and determine the concentration of dsDNA spectrophotometrically (e.g., using NanoDrop spectrophotometer). Store the purified samples at −20 °C until needed.

#### *3.1.2 SeSaM Preliminary Experiment II: Determination of the Optimal Concentration of Phosphorothioate Nucleotides for Step 1 of SeSaM*

When SeSaM is performed for the first time for a given gene, it is also crucial to determine the optimal concentration of phosphorothioate nucleotides to use in step 1. This ensures that fragments with uniform size distribution spanning the entire gene sequence are generated in the first step of SeSaM.

1. Prepare a "forward" (fwd) and a "reverse" (rev) mastermix (180 μL each) in 1.5 mL tubes on ice. Thaw all nonenzymatic components at room temperature, mix by vortexing, and collect by brief centrifugation. Keep enzymes on ice.

| Forward (fwd) pre-mastermix (144 μL) | Volume |
|---|---|
| 10× *Taq* DNA polymerase buffer | 18 μL |
| SeSaM_fwd primer (5 μM) | 18 μL |
| Rev_X primer (5 μM) | 18 μL |
| SeSaM_fwd_step 1 template | *X* μL (96 ng/kb of SeSaM template) (e.g., for a gene of 1 kb, a total of 4.8 μL of a 20 ng/μL purified SeSaM_step 1 template is used) |
| *Taq* DNA polymerase (5 U/μL) | 3.6 μL |
| Nuclease-free $H_2O$ | To a total volume of 144 μL |

| Reverse (rev) pre-mastermix (144 μL) | Volume |
|---|---|
| 10× *Taq* DNA polymerase buffer | 18 μL |
| SeSaM_rev primer (5 μM) | 18 μL |
| fwd_X primer (5 μM) | 18 μL |
| SeSaM_rev_step 1 template | *X* μL (96 ng/1 kb SeSaM template) (e.g., for a gene of 1 kb, a total of 4.8 μL of a 20 ng/μL purified SeSaM_step 1 template is used) |
| *Taq* DNA polymerase (5 U/μL) | 3.6 μL |
| Nuclease-free $H_2O$ | To a total volume of 144 μL |

2. Prepare the PCR reactions for the test library on ice using 16× 8 μL aliquots of either reverse or forward pre-mastermix in PCR tubes. Label tubes "AF0"–"AF40" for A-forward test libraries, "AR10"–"AR40" for A-reverse test libraries, "GF0"–"GF40" for G-forward libraries, and "GR10"–"GR40" for G-reverse test libraries. To each tube, add 2 μL of the corresponding phosphorothioate nucleotide premix ("A0"–"A40" or "G0"–"G40", respectively), prepared as described in Subheading 2.1, **item 27** (*see* **Note 6**).
3. Run all samples in a PCR cycler for 15 cycles using the following PCR settings:

   *PCR protocol*—94 °C for 120 s (1 cycle); 94 °C for 30 s, optimal Tm of primers for 30 s, 72 °C for 30 s/kb of SeSaM template (14 cycles); 72 °C for 180 s (1 cycle); followed by a 4 °C hold to protect the samples.
4. After completion of the PCR reaction, add 1.1 μL of freshly prepared solution C (the iodine solution used to cleave phosphorothioester bonds) to each PCR tube. Mix each tube immediately and vigorously by vortexing (10 s).
5. Incubate the samples in a PCR cycler for 30 min at 70 °C. Make sure the lid of the thermal cycler is preheated to 80 °C to prevent sample evaporation. After 30 min the solutions should change color from brownish-orange to transparent.
6. Run the samples (4 μL) on a 0.8 % TAE agarose gel (7.5 V/cm; 40 min). Evaluate the gel images to determine the condition under which the most homogeneous distribution of DNA fragments is achieved, i.e., select the concentration of dNTPαS at which the fragments are homogenously distributed over the entire gene length and only traces of the uncleaved, full-length DNA band are visible (Fig. 3).

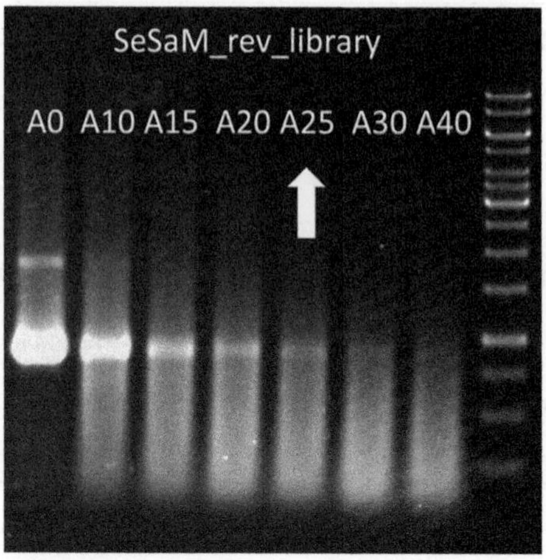


**Fig. 3** An exemplary agarose gel used for evaluation of the optimal phosphorothioate nucleotide concentration. *A0–A40* indicates the fraction of phosphorothioate adenine in relation to standard adenine, expressed as a percentage. Uncleaved PCR product (A0) is loaded in the *leftmost lane*; in the samples prepared with dNTP mix that contains a certain percentage of phosphorothioate nucleotides (*right* of A0), one can observe a decrease in intensity of the full-length band as well as a smearing below that increases with dNTPαS concentration. This smearing signifies DNA cleavage at sites where phosphorothioate nucleotides have been incorporated. The *lane labeled A40* represents a case where the concentration of dNTPαS nucleotides is too high as hardly any uncleaved fragment is visible and the distribution of fragments is uneven (i.e., small fragments are predominant). *Lane A10* shows a case when the concentration of phosphorothioate nucleotides is too low, i.e., uneven smearing with prevailing large fragments and a very high fraction of uncleaved PCR product is observed. A final concentration of phosphorothioate nucleotides of 25 % (A25) for the SeSaM_rev_library would be selected as the optimal condition on this occasion

### 3.2 SeSaM Step 1: Generation of ssDNA Fragment Pool with Random Length Distribution

In step 1 of SeSaM, the incorporation of phosphorothioate nucleotides into PCR products is carried out. The optimal phosphorothioate nucleotide concentration is determined as explained in SeSaM preliminary step II (*see* Subheading 3.1). In SeSaM-Tv classic protocol, A- and G-sites are targeted with the pyrimidine analog dPTP; therefore, dGTPαS or dATPαS are randomly incorporated in the gene sequence (*see* **Note** 7). SeSaM step 1 comprises two parts—the incorporation of the phosphorothioate nucleotides and the iodine cleavage of the PCR products and the isolation of biotinylated ssDNA fragments using streptavidin-coated magnetic beads.

1. Assemble the following "forward" (fwd) and "reverse" (rev) pre-mastermixes (2×423 μL) on ice. Thaw all nonenzyme components at room temperature, mix by a brief vortex, and collect by a brief centrifugation. Keep all enzymatic components on ice.

| **Fwd pre-mastermix (423 µL)** | **Volume** |
|---|---|
| SeSaM template (SeSaM step 1_fwd_X template) | *X*µL (360 ng/kb SeSaM template) (e.g., for a 1 kb gene, use a total of 6 µL of a 60 ng/µL purified SeSaM_fwd_step 1 template) |
| 10× *Taq* DNA polymerase buffer | 45 µL |
| *Taq* DNA polymerase (5 U/µL) | 9 µL |
| SeSaM_fwd_Bio primer (5 µM) | 45 µL |
| rev_X primer (5 µM) | 45 µL |
| Nuclease-free $H_2O$ | To a total volume of 423 µL |

| **Rev pre-mastermix (423 µL)** | **Volume** |
|---|---|
| SeSaM template (SeSaM step 1_rev_X template) | *X*µL 360 ng/1 kb SeSaM template (e.g., for a 1 kb gene, use a total of 6 µL of a 60 ng/µL purified SeSaM_fwd_step 1 template) |
| 10× *Taq* DNA polymerase buffer | 45 µL |
| *Taq* DNA polymerase (5 U/µL) | 9 µL |
| SeSaM_rev_Bio primer (5 µM) | 45 µL |
| fwd_X primer (5 µM) | 45 µL |
| Nuclease-free $H_2O$ | To a total volume of 423 µL |

2. Pipette 188 µL from each mastermix into each of two new PCR tubes and keep the remaining volume of each pre-mastermix (47 µL) in the original PCR tube. Add the respective nucleotide mix according to the pipetting scheme below (*see* **Note 7**).

| **A-fwd library mastermix** | **Volume** |
|---|---|
| *fwd* pre-mastermix | 188 µL |
| d(GTP/TTP/CTP)-mix (10 mM) | 4 µL |
| dATP (5 mM) | *X*µL |
| dATPαS (5 mM) | *Y*µL |
| Total volume | 200 µL |

| A-rev library mastermix | Volume |
|---|---|
| *rev* pre-mastermix | 188 μL |
| d(GTP/TTP/CTP)-mix (10 mM) | 4 μL |
| dATP (5 mM) | *X* μL |
| dATPαS (5 mM) | *Y* μL |
| Total volume | 200 μL |

The total volume of dATP and dATPαS should be 8 μL; *X* represents the volume of dATP and *Y* the volume of dATPαS in the reaction (e.g., to prepare A25 mix (containing 25 % dATPαS), add 2 μL of dATPαS and 6 μL of dATP to the reaction mixture; the percentage of dATPαS used for a given gene should be determined experimentally as outlined in SeSaM preliminary step II (*see* Subheading 3.1)).

| G-fwd library mastermix | Volume |
|---|---|
| *fwd* pre-mastermix | 188 μL |
| d(ATP/TTP/CTP)-mix (10 mM) | 4 μL |
| dGTP (5 mM) | *X* μL |
| dGTPαS (5 mM) | *Y* μL |
| Total volume | 200 μL |

| G-rev library mastermix | Volume |
|---|---|
| *rev* pre-mastermix | 188 μL |
| d(ATP/TTP/CTP)-mix (10 mM) | 4 μL |
| dGTP (5 mM) | *X* μL |
| dGTPαS (5 mM) | *Y* μL |
| Total volume | 200 μL |

The total volume of dGTP and dGTPαS should add up to 8 μL; *X* μL is the volume of dGTP and *Y* μL the volume of dGTPαS used the respective reactions (e.g., to prepare dNTP mix (G25) containing 25 % dGTPαS, we would add 2 μL of dGTPαS and 6 μL of dGTP to the reaction mixture; the percentage of dGTPαS used for a given gene should be determined experimentally as outlined in SeSaM preliminary step II (*see* Subheading 3.1)).

| Positive control mastermix | Volume |
|---|---|
| Pre-mastermix | 47 μL |
| Nuclease-free $H_2O$ | 2 μL |
| dNTP mix (10 mM) | 1 μL |
| Total volume | 50 μL |

3. Split each mastermix (200 μL) into 4× 50 μL aliquots in 0.2 mL PCR tubes and run all samples in a PCR cycler using the following parameters:

   *PCR protocol*—94 °C for 120 s (1 cycle); 94 °C for 30 s, 60 °C for 30 s, 72 °C for 30 s/kb of the SeSaM template-DNA (19 cycles); 72 °C for 3 min (1 cycle); followed by a 4 °C hold to protect the samples.
4. Pool the four identical PCR reactions together and run 4–5 μL from each sample on an 0.8 % TAE agarose gel (7.5 V/cm; 40 min). A specific band corresponding to the length of the gene of interest should be clearly visible on the gel.
5. Column-purify the dsDNA products using NucleoSpin Extract II PCR Purification Kit and NT buffer according to manufacturer's recommendation. Elute DNA in 92 μL elution buffer (NE-buffer or nuclease-free water) and determine dsDNA concentration (*see* **Note 8**).
6. Combine 80 μL of the purified products with 10 μL of 10× *Taq* DNA polymerase buffer in a fresh 0.2 mL tube. Add 10 μL of freshly prepared solution C to each tube, and thoroughly mix by pipetting. Incubate for 30 min at 70 °C in a thermal cycler. During this incubation period, proceed with **steps 7–10**.
7. Resuspend the magnetic beads by a brief vortex, until all beads are in suspension.
8. Transfer 50 μL of the magnetic bead solution into each of four 0.2 mL tubes. Magnetic beads tend to precipitate from solution so make sure they are fully suspended before aliquoting them. Place each tube horizontally on a magnet and wait until the beads have settled down. Discard the liquid while keeping the tube pressed against the magnet.
9. Add 100 μL 2× B&W buffer to each tube and resuspend by pipetting. Place tube on the magnet and wait until the beads have settled down (*see* **Note 9**). Remove the supernatant by pipetting.
10. Add 100 μL 2× B&W buffer to each tube and resuspend by pipetting.
11. Run 4 μL of the cleaved DNA on an 0.8 % TAE agarose gel in order to check the fragmentation pattern of samples after cleavage and before purification (use 7.5 V/cm; 40 min), as explained in SeSaM preliminary step II (*see* Subheading 3.1).

12. Transfer the remainder of each cleaved library to one of the four tubes containing the magnetic beads resuspended in 2× B&W buffer, and mix by pipetting. Incubate each tube for 30 min at room temperature to allow the biotin tag to interact to the streptavidin-coated beads. After the incubation period is over, place each tube on a magnet and wait until the beads have settled down. Remove the supernatant by pipetting.
13. To each tube, add 100 μL solution M (dsDNA melting solution) and resuspend by pipetting. Incubate each tube for 2 min at room temperature. Place each tube on the magnet and wait until the beads have settled down. Remove the supernatant by pipetting.
14. Add 100 μL solution M and resuspend by pipetting. Place tube on a magnet and wait until the beads have settled down. Remove the supernatant by pipetting (it is important to make sure that all traces of M solution are removed).
15. Add 100 μL 1× B&W buffer and resuspend by pipetting. Place tube on a magnet and wait until the beads have settled down. Remove the supernatant by pipetting.
16. Add 75 μL solution R (solution for disruption of streptavidin-biotin interaction) and resuspend by pipetting. Incubate the samples in a PCR cycler for 10 min at 98 °C. Make sure the lid is preheated to 104 °C to avoid sample evaporation. Do not let the tube cool down!
17. Place the hot tube on a magnet, wait until the beads have settled down, and immediately recover the supernatant by pipetting in a fresh 1.5 mL tube. Keep the supernatant.
18. Individually column-purify the recovered ssDNA fragments from each tube with NucleoSpin Extract II PCR Purification Kit. Use NTC buffer (300 μL per 75 μL ssDNA) for binding DNA to the column instead of NT buffer (*see* **Note 1**). Elute DNA in 30 μL NE-buffer or nuclease-free water. Measure the concentration spectrophotometrically using appropriate settings for ssDNA (*see* **Note 10**) and the uniformity of distribution of ssDNA fragments (optional, *see* **Note 11**).

### 3.3 SeSaM Step 2: Enzymatic Elongation of Single-Stranded DNA with Universal Base

In SeSaM step 2, the ssDNA fragments generated in step 1 (SeSaM step 1 products) are elongated with universal base analog dPTPαS (*see* **Note 12**) in a TdT-catalyzed reaction.

1. Prepare five reactions (A-fwd, A-rev, G-fwd, G-rev libraries, and the positive control) in 0.2 mL tubes, on ice, as detailed below. Reaction components should be added in the listed order. Make sure TdT is added last. All nonenzyme components should be thawed at room temperature and all enzymatic components on ice.

| Universal base addition reaction mix | Volume |
|---|---|
| 10× TdT buffer | 2.5 μL |
| 10× $CoCl_2$ (25 mM) | 2.5 μL |
| SeSaM step 1 product | *X* μL (1 pmol of step 1 product) |
| dPTPαS (20 μM) (*see* **Note 3**) | 2 μL |
| TdT (20 U/μL) | 2 μL |
| Nuclease-free $H_2O$ | To a total volume of 25 μL |

| Positive control (25 μL) | Volume |
|---|---|
| 10× TdT buffer | 2.5 μL |
| 10× $CoCl_2$ (25 mM) | 2.5 μL |
| FITC-labeled primer (2 μM) | 2 μL |
| dPTPαS (5 μM) | 2 μL |
| TdT (20 U/μL) | 2 μL |
| Nuclease-free $H_2O$ | 14 μL |

2. Run all samples in a PCR cycler at 37 °C for 120 min, immediately followed by heat inactivation of TdT for 30 min at 75 °C. Make sure the lid of the thermal cycler is preheated to 40 °C during first incubation step and to 80 °C during the inactivation step to prevent evaporation.
3. Keep the positive control aside (SeSaM step 2 positive control; store at −20 °C), and purify the other four samples using the NucleoSpin Extract II PCR Purification Kit and NTC buffer (*see* **Note 1**) according to manufacturer's recommendation. Elute DNA in 25 μL elution buffer or nuclease-free water.
4. Measure the concentration of ssDNA spectrophotometrically and store samples (SeSam step 2 products) at −20 °C until the positive control has been analyzed.
5. Run the positive control of SeSaM step 2 on a 25 % acrylamide gel (*see* **Note 13**) in order to evaluate the resulting universal nucleotide addition pattern. Based on this gel, a conclusion about the number of nucleotide additions can be drawn. Ideally, all FITC primers should be "tailed", whereby fragments elongated with two or three universal base analog additions should be the predominant product (Fig. 4).

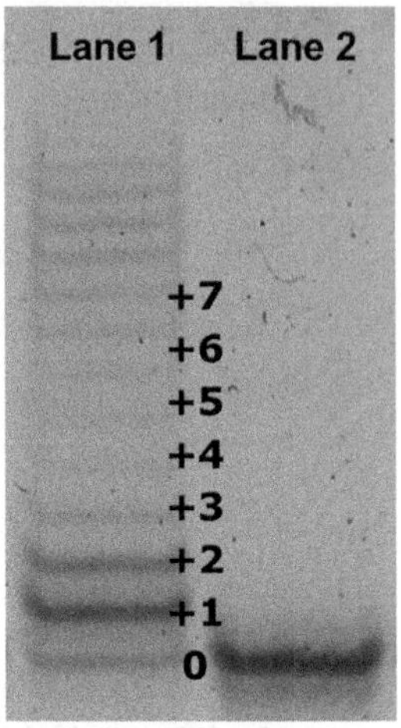


**Fig. 4** Elongation of FITC-labeled primer by TdT separated on acrylamide gel and visualized on phosphorimager (e.g., 494 nm/em. 518 nm). Each stair of the ladder in the lane 1 (*positive control*) represents an addition of one universal base to the FITC-labeled oligonucleotide. In lane 2 (*negative control*) FITC primer that has not been subjected to elongation reaction is loaded as a reference

### 3.4 SeSaM Step 3: Synthesis of the Full-Length Gene

In step 3 of SeSaM, the "tailed" DNA fragments (SeSaM **step 2** products) are elongated to full-length genes. In this step, the SeSaM step 2 products for forward and reverse library preparation can be elongated individually using SeSaM_step3_ templates (generated in SeSaM preliminary experiment II) (optional, *see* **Note 14**).

1. Prepare pre-mastermix for four reactions of 50 μL (two for fwd and two for rev libraries) on ice in 0.2 mL tubes.

| Pre-mastermix (120 μL) | Volume |
|---|---|
| 10× SuperTaq buffer | 20 μL |
| dNTP mix (10 mM) | 5 μL |
| 3D1 polymerase (5 U/μL) (*see* **Note 1**) | 4 μL |
| Nuclease-free $H_2O$ | To a total volume of 120 μL |

2. Split the pre-mastermix into 4× 30 μL aliquots in fresh 0.2 mL tubes.
3. To each aliquot, one of the following combinations of SeSaM_ step 3_templates (*see* Subheading 3.1) and SeSaM step 2 products (*see* Subheading 3.3) should be added; 80 ng from each in a total volume of 50 μL is recommended (e.g., for a 1 kb gene a total of 2 μL of a 40 ng/μL purified SeSaM_step 3_template from preliminary experiment II, 8 μL of a 10 ng/μL purified SeSaM step 2 product, and 10 μL nuclease-free $H_2O$ should be used):

(a) Mastermix + step 3_fwd template + A-Fwd step 2 library
(b) Mastermix + step 3_fwd template + G-Fwd step 2 library
(c) Mastermix + step 3_rev template + A-Rev step 2 library
(d) Mastermix + step 3_rev template + G-Rev step 2 library

4. Run all samples in a PCR cycler for 30 cycles and preheated lid to 105 °C.

   *PCR protocol*: 94 °C for 120 s (1 cycle); 94 °C for 30 s, 52 °C for 1 min, 72 °C for 1 min/kb of the SeSaM template-DNA (29 cycles); followed by a 4 °C hold to protect the samples.

5. Column-purify the elongated ssDNA fragments individually using the NucleoSpin Extract II PCR Purification Kit and NTC buffer according to the manufacturer's recommendation. Elute DNA in 25 μL elution buffer or nuclease-free water.

6. Measure the concentration of dsDNA spectrophotometrically and store samples at −20 °C.

### 3.5 SeSaM Step 4: Replacement of Universal Base

In the last step of SeSaM, *Taq* polymerase is used in a nested PCR to specifically amplify mutated DNA as well as to exchange the previously introduced universal bases with standard nucleotides. The PCR products can be directly used for cloning in an expression vector.

1. Prepare the following pre-mastermix (400 μL) on ice in a 1.5 mL tube:

| Pre-mastermix (360 μL) | Volume |
|---|---|
| 10× *Taq* DNA polymerase buffer | 40 μL |
| SeSaM_fwd primer (5 μM) | 40 μL |
| SeSaM_rev primer (5 μM) | 40 μL |
| dNTP mix (10 mM) | 8 μL |
| *Taq* DNA polymerase (5 U/μL) | 8 μL |
| Nuclease-free $H_2O$ | To a total volume of 360 μL |

2. Divide the pre-mastermix into four aliquots of 90 μL in 0.2 mL PCR tubes.
3. Add 50 ng/kb of the respective SeSaM step 3 products and adjust the final volume to 100 μL with nuclease-free water, if necessary.
4. Split each mastermix into 2× 50 μL in 0.2 mL PCR tubes.
5. Run the reactions in the PCR cycler.

*PCR protocol*: 94 °C for 120 s (1 cycle); 94 °C for 30 s, 60 °C for 30 s, 72 °C for 45 s/kb of the SeSaM template-DNA (20 cycles); 72 °C for 3 min; followed by a 4 °C hold to protect the samples.

6. Combine each set of duplicate aliquots in one tube and run 5 μL of each sample on a 0.8 % agarose gel (7.5 V/cm; 40 min) to evaluate the quality of PCR amplification. PCR clean-up the products using NucleoSpin PCR Clean-Up Kit and NT buffer and elute in 25 μL in elution buffer or nuclease-free water (*see* **Note 15**).
7. Store the completed SeSaM library at −20 °C or proceed to cloning in an expression vector using your preferred methods.

## 4 Notes

1. NTC buffer which is separately supplied by Macherey-Nagel, Düren, Germany, is indispensable for efficient purification of ssDNA in steps 1 and 2 of SeSaM.
2. 3D1 polymerase was expressed and purified to homogeneity according to published protocols [12, 18]. Vent (exo-) DNA polymerase and ThermoPol buffer from New England Biolabs (Frankfurt, Germany) can be used as an alternative enzyme in step 3 of SeSaM. Using the Vent (exo-) instead of the 3D1-polymerase generates fewer consecutive nucleotide exchanges in SeSaM-Tv-classic libraries. Vent (exo-) should be the polymerase of choice in step 3 of SeSaM-R protocol. The reaction conditions with either polymerase remain the same.
3. The rate of incorporation of dPTPαS by TdT is lower relative to that observed for non-phosphorothioate dPTP. This allows more precise control over the addition pattern obtained in SeSaM step 2.
4. High-fidelity polymerases such as Phusion® are recommended in the preliminary step I to ensure error-free synthesis of templates. High-fidelity polymerases should not be used in steps 1–4 of SeSaM.
5. It is important that the SeSaM step 1 template is not contaminated with nonspecific products. If nonspecific products are present, load and separate the PCR product on agarose gel and perform a gel extraction/purification.
6. When performing preliminary experiment II for SeSaM-R protocol, d(CTP/CTPαS/ATP/TTP/GTP)-mixes (C0–C40) or d(TTP/TTPαS/ATP/GTP/CTP)-mixes (T0–T40) should be used to target T- and C-sites. No other modifications in reaction conditions are required.

7. In step 1 of SeSaM-R protocol, instead of preparing an A- and G-library, carry out a T- and C-library (e.g., for preparation of a T-library, replace the d(ATP/TTP/CTP)-mix with d(ATP/GTP/CTP)-mix; similarly dGTP should be replaced by dTTP and dGTPαS by dTTPαS).
8. Concentrations of >100 ng/μL are normally obtained.
9. Prolonged contact between the PCR tubes and the magnet may cause the magnetic beads to firmly stick to the wall of the tube and make resuspension difficult.
10. The concentration of ssDNA fragments after SeSaM step 1 is usually in the range of 10–20 ng/μL.
11. If available, we recommend using a microfluidic electrophoresis device (e.g., Experion Automated Electrophoresis System from Bio-Rad) after DNA purification to confirm and accurately evaluate the size distribution of ssDNA fragments, especially those in the low molecular weight range. The fragments generated should be homogeneously distributed. If the majority of PCR product remains uncleaved for a particular fragment library, repeat SeSaM step 1 by increasing the concentration of dNTPαS by 5 % for the corresponding library. On the other hand, if many short fragments are observed, repeat SeSaM step 1 by decreasing the concentration of dNTPαS by 5 %.
12. In step 2 of SeSaM-R protocol, 0.4 μM dRTP instead of 1.25 μM dPTPαS is required and the universal base addition reaction is carried out for 15 min at 37 °C followed by a heat inactivation of 30 min at 75 °C.
13. To cast a 25 % acrylamide gel, mix in a clean vessel 1.7 mL of deionized $H_2O$, 2 mL of 5× TBE buffer, 6.25 mL of 40 % acrylamide-bis-acrylamide solution, 100 μL of APS, and 8 μL of TEMED. Allow the gel to solidify for at least 15 min. Place the gel in the running chamber filled with fresh 1× TBE buffer and run the gel for 5 min at 100 V prior to applying your samples. In the meantime, mix 5 μL of 6× TBE-loading dye with the 25 μL positive control reaction. Prepare a negative control by mixing together 2 μL FITC primer (2 μM), 8 μL deionized water, and 2 μL of 6× acrylamide loading dye. Load 6 μL of positive control sample and 4 μL of negative control samples per well. Do not use the outermost wells of the gel if possible. Run the gel at 170 V and 300 mA for 2.5 h. Visualize the gel without removing it from the glass casting chamber using phosphorimager/scanner (e.g., 494 nm/em. 518 nm).
14. Forward and reverse SeSaM step 2 products can be mixed together. In the latter case, forward and reverse SeSaM step 2 products will serve as templates for each other; this will result in increase of the mutagenic frequency of SeSaM libraries. If a recombined library, in SeSaM step 3 use equal amounts (80 ng

each) of SeSam step 2 forward and reverse products. No SeSaM_step3_templates should be supplied as forward and reverse products serve to template each other's elongation.

15. If unspecific bands are observed, the specific product bands can be isolated in an agarose gel extraction step.

## References


1. Neylon C (2004) Chemical and biochemical strategies for the randomization of protein encoding DNA sequences: library construction methods for directed evolution. Nucleic Acids Res 32:1448–1459
2. Zhu L, Verma R, Roccatano D, Ni Y, Sun Z-H, Schwaneberg U (2010) A potential antitumor drug (arginine deiminase) reengineered for efficient operation under physiological conditions. Chembiochem 11:2294–2301
3. Glieder A, Farinas ET, Arnold FH (2002) Laboratory evolution of a soluble, self-sufficient, highly active alkane hydroxylase. Nat Biotechnol 20:1135–1139
4. Ruff AJ, Dennig A, Schwaneberg U (2013) To get what we aim for: progress in diversity generation methods. FEBS J 280:2961–2978
5. Wong TS, Zhurina D, Schwaneberg U (2006) The diversity challenge in directed protein evolution. Comb Chem High Throughput Screen 9:271–288
6. Shivange AV, Marienhagen J, Mundhada H, Schenk A, Schwaneberg U (2009) Advances in generating functional diversity for directed protein evolution. Curr Opin Chem Biol 13:19–25
7. Rogozin IB, Pavlov YI (2003) Theoretical analysis of mutation hotspots and their DNA sequence context specificity. Mutat Res 544:65–85
8. Wong TS, Roccatano D, Schwaneberg U (2007) Challenges of the genetic code for exploring sequence space in directed protein evolution. Biocatal Biotransformation 25:229–241
9. Rasila TS, Pajunen MI, Savilahti H (2009) Critical evaluation of random mutagenesis by error-prone polymerase chain reaction protocols, *Escherichia coli* mutator strain, and hydroxylamine treatment. Anal Biochem 388:71–80
10. Wong TS, Roccatano D, Zacharias M, Schwaneberg U (2006) A statistical analysis of random mutagenesis methods used for directed protein evolution. J Mol Biol 355:858–871
11. Verma R, Schwaneberg U, Roccatano D (2012) MAP2.03D: a sequence/structure based server for protein engineering. ACS Synth Biol 1:139–150
12. Mundhada H, Marienhagen J, Scacioc A, Schenk A, Roccatano D, Schwaneberg U (2011) SeSaM-Tv-II generates a protein sequence space that is unobtainable by epPCR. Chembiochem 12:1595–1601
13. Li Z, Roccatano D, Lorenz M, Schwaneberg U (2012) Directed evolution of subtilisin E into a highly active and guanidinium chloride- and sodium dodecylsulfate-tolerant protease. Chembiochem 13:691–699
14. Martinez R, Jakob F, Tu R, Siegert P, Maurer K-H, Schwaneberg U (2012) Increasing activity and thermal resistance of *Bacillus gibsonii* alkaline protease (BgAP) by directed evolution. Biotechnol Bioeng 110:711–720
15. Shivange AV, Serwe A, Dennig A, Roccatano D, Haefner S, Schwaneberg U (2012) Directed evolution of a highly active *Yersinia mollaretii* phytase. Appl Microbiol Biotechnol 95:405–418
16. Wong TS, Roccatano D, Loakes D, Tee KL, Schenk A, Hauer B, Schwaneberg U (2008) Transversion-enriched sequence saturation mutagenesis (SeSaM-Tv+): a random mutagenesis method with consecutive nucleotide exchanges that complements the bias of error-prone PCR. Biotechnol J 3:74–82
17. Ruff AJ, Marienhagen J, Verma R, Roccatano D, Genieser H-G, Niemann P, Shivange AV, Schwaneberg U (2012) dRTP and dPTP a complementary nucleotide couple for the Sequence Saturation Mutagenesis (SeSaM) method. J Mol Catal B Enzym 84:40–47
18. D'Abbadie M, Hofreiter M, Vaisman A, Loakes D, Gasparutto D, Cadet J, Woodgate R, Pääbo S, Holliger P (2007) Molecular breeding of polymerases for amplification of ancient DNA. Nat Biotechnol 25:939–943
19. Blanusa M, Schenk A, Sadeghi H, Marienhagen J, Schwaneberg U (2010) Phosphorothioate-based ligase-independent gene cloning (PLICing): an enzyme-free and sequence-independent cloning method. Anal Biochem 406:141–146

# Chapter 5

# Generation of Effective Libraries by Neutral Drift

**Miriam Kaltenbach and Nobuhiko Tokuriki**

## Abstract

Neutral drift is a recently developed experimental technique used to identify superior starting points for protein engineering. Neutral drift explores accessible sequence space by repeated rounds of mutagenesis and selection to maintain wild-type function. Mutations that are largely neutral for the native function accumulate, and those that are highly detrimental are purged, yielding a library of high diversity and quality. This technique is useful in situations where laboratory evolution is at a dead end, i.e., when the enzyme activity intended for evolution proves recalcitrant to improvements or is too low to be detected.

**Key words** Neutral drift, Directed evolution, Protein engineering, Library creation, Purifying selection, Evolvability, Protein stability

## 1 Introduction

Protein sequences represent only a tiny fraction of accessible sequence space. While large parts of sequence space are probably nonfunctional, the existence of multiple sequences that exhibit the same structure and function demonstrates the range of available sequence space. For example, orthologous proteins can have as little as ≤20 % sequence identity yet perform the same task. A collection of genotypes such as this forms a so-called neutral network [1, 2]. Evolutionary theory predicts that sequences in the network, despite being phenotypically indistinguishable, differ in hidden properties such as protein stability, the level of latent, promiscuous functions, and evolvability. Evolvability, in our definition, is the ability of a sequence to acquire another function or phenotype with only a few mutational steps (Fig. 1a). Identification of more evolvable sequences confers a significant advantage in protein engineering, i.e., a better starting point will facilitate progress of the experiment and increase its success rate.

In traditional protein engineering, a particular starting sequence is primarily selected based on the availability of the gene and the amount of information available from earlier work (i.e., biochemical,

Elizabeth M.J. Gillam et al. (eds.), *Directed Evolution Library Creation: Methods and Protocols*, Methods in Molecular Biology, vol. 1179, DOI 10.1007/978-1-4939-1053-3_5, © Springer Science+Business Media New York 2014

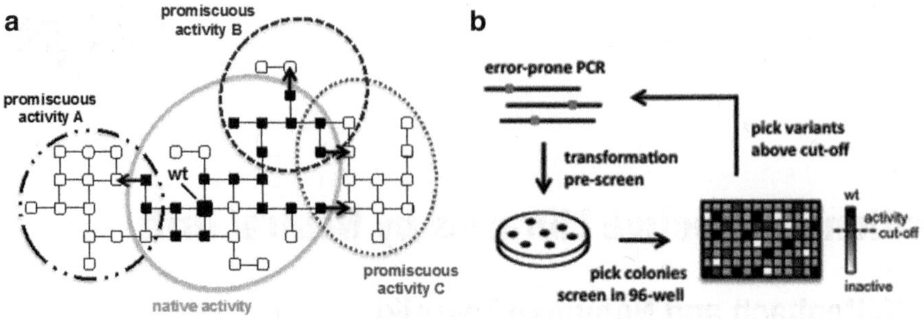


**Fig. 1** Generating neutral drift libraries. (**a**) Concept of neutral drift. Upon purifying selection for the native activity, a population expands across a neutral network of functional sequences. The *white squares* denote all possible sequences belonging to the network, and *black squares* indicate sequences that become populated by neutral drift. Sequences in the intersecting regions exhibit wild-type activity, but also a low level of promiscuous activity A, B, or C. *Arrows* indicate how, upon selection for a promiscuous activity, a sequence exits the intersection region and a different neutral network is accessed. (**b**) Experimental setup. A library is generated by error-prone PCR and transformed into an *E. coli* screening strain. After an optional prescreen, (active) variants are screened in 96-well plates for the wild-type activity. An appropriate cutoff for selection is chosen; all variants above the cutoff are pooled and subsequently used as input for the next selection round

biophysical, or structural characterization). However, in many instances, this sequence might not be the best choice in terms of evolvability. Recent advances in bioinformatics have provided computational tools to predict starting points that are potentially more evolvable than wild-type sequences such as consensus or ancestral sequences [3, 4]. Experimentally, evolvability can be enhanced by exploring sequence space through "neutral drift" using iterative cycles of genetic diversification and purifying selection in the laboratory [5–10]. Unlike traditional directed evolution experiments, the selection pressure is applied to maintain wild-type protein function (rather than enhance it), and mutations accumulate that are largely neutral with respect to this property. However, these mutations can affect evolvability by altering features such as stability and/or promiscuous activities. The strength of this approach is that it generates a diverse pool of structurally and functionally intact sequences. In comparison, conventional random mutagenesis, in particular mutagenesis with high mutation rates, substantially increases the necessary screening effort because the resulting library will largely consist of inactive clones containing deleterious mutations: ~40 % of single amino acid substitutions are highly deleterious, and thus, viable variants in the library ($W$) decline exponentially ($W = 0.6^n$) as the average number of mutations ($n$) increases [11].

The utility of neutral drift for the generation of more evolvable libraries has been demonstrated in several studies (summarized in Table 1). Neutral drift libraries have been shown to contain variants with enhanced promiscuous activities. For example, Bloom et al. showed that after extensive neutral drift (up to 25 rounds), of only a few hundred clones, variants of a cytochrome P450 enzyme exhibited up to four-fold changes in the hydroxylation of five

**Table 1**
**Examples of neutral drift experiments**

| Target protein | Target property[a] | Property under purifying selection[b] | Screening format | Screening throughput | Average number of mutations[c] | No. of rounds | Selection criteria for purifying selection | Findings | References |
|---|---|---|---|---|---|---|---|---|---|
| Cytochrome P450 BM3 | 5 promiscuous hydroxylation substrates<br>Stability | Native monooxygenase activity | 96-well plates | 1–400[d] | 1.4 (1,400 bp) | 15–25[e] | ≥75 % wild-type activity (~40–50 % of analyzed clones) | ≤4-fold change in promiscuous activities<br>Accumulation of stabilizing mutations in polymorphic library | [8, 9] |
| PON1 (recombinant paraoxonase 1) | 5 promiscuous activities (2 phosphotriesters, 2 arylesters, 1 lactone), native activity in the presence of inhibitor | Native lactonase activity and expression level (monitored by fluorescence of GFP-PON fusion protein) | FACS | $10^6$ | 1.9 (1,050 bp) | 3 | 100 variants per round, activity and expression levels > 3 standard deviations above background | Promiscuous activity change ≤ 36-fold, specificity change ≤ 50-fold | [5, 10] |

(continued)

**Table 1**
**(continued)**

| Target protein | Target property[a] | Property under purifying selection[b] | Screening format | Screening throughput | Average number of mutations[c] | No. of rounds | Selection criteria for purifying selection | Findings | References |
|---|---|---|---|---|---|---|---|---|---|
| PON1 (recombinant paraoxonase 1) | 5 promiscuous activities (2 phosphotriesters, 2 arylesters, 1 lactone) | Prescreen for expression level (FACS), followed by screen for expression level and native lactonase activity (96-well plates) | FACS and 96-well plates | $10^6$ | 4.8 (1,050 bp) | 3 | 200–300 variants per round, activity and expression levels > 3 standard deviations above background | Screening of 500 clones from each library for promiscuous activities: 36 improved for HT, 11 for LT library | [5, 10] |
| | | Native lactonase activity and expression level | 96-well plates | 400–1,000[f] | 1 (1,050 bp) | 4 | 200–300 variants per round | Accumulation of stabilizing mutations | |
| TEM-1 β-lactamase | Cefotaxime resistance | Ampicillin resistance of *E. coli* host cells | Agar plates | $10^6$ | 2 (860 bp) | 18 | Survival on high or low concentrations of ampicillin[g], ≥$10^6$ clones maintained throughout, loss of diversity per round ≤ 50 % | Higher frequency of variants conferring increased cefotaxime resistance in high cutoff library<br>Accumulation of stabilizing mutations | [6, 7] |

| | | | | | | | | | |
|---|---|---|---|---|---|---|---|---|---|
| PTE (phosphotriesterase) | Further increase in arylesterase activity | Arylesterase activity | Agar plates and 96-well plates | $10^3$ | 2–3 (1,000 bp) | 4 | ≥30 % activity of the starting point | Accumulation of four mutations that jointly increase activity twofold | [12] |

[a]Property of interest that is to be selected after the neutral drift
[b]Property under selection, e.g., native catalytic activity
[c]Average number of mutations accumulated per round in the unselected libraries. The size of the gene is given in *parentheses*
[d]A monomorphic and a polymorphic library were subjected to neutral drift. In the monomorphic library, only one clone was analyzed after each round. If the clone fulfilled the selection criterion, it was used as starting point for the next round; if not, the previous library was used again. For the polymorphic library, 400 clones were analyzed after each round and all those fulfilling the selection criterion used as input for the next round
[e]15 rounds of drift were performed for the polymorphic library and 25 rounds for the monomorphic library
[f]In the first 3 rounds, analysis of 400 clones was sufficient to find 200–300 active variants; in round 4, it was necessary to analyze 1,000 clones
[g]Two libraries were subjected to neutral drift: a high cutoff library with a high ampicillin concentration (250 μg/mL) and a low cutoff library (12.5 μg/mL)

different promiscuous substrates [9]. Similarly, using serum paraoxonase I (PON1), Amitai et al. demonstrated that 3–4 rounds of neutral drift were sufficient to cause substantial changes in hydrolytic activity (up to 36-fold) and specificity (up to 50-fold) toward five promiscuous substrates [5]. Bershtein et al. demonstrated that neutral drift of TEM-1 β-lactamase on ampicillin led to a number of mutants with increased resistance toward the third-generation cephalosporin antibiotic, cefotaxime [6, 7]. Neutrally drifted P450 [8] and TEM-1 [6, 7] libraries were also found to be enriched in stabilized variants. In the case of TEM-1, many of the accumulated "global suppressor" mutations brought the sequence closer to the family consensus and/or the predicted ancestor. Such back-to-consensus/back-to-ancestor mutations are known to increase protein stability and have been used to generate more evolvable libraries [3, 4]. Neutral drift has also been employed to "rescue" a "dead end" evolved mutant of phosphotriesterase (PTE). In this case, the promiscuous arylesterase function had already been improved by direct selection, but any further improvement conveyed through a single mutation was too small to be detected. After four rounds of drift, the combined effect of four mutations resulted in a detectable, twofold improvement [12].

We propose that neutral drift should be pursued in situations where laboratory evolution is at a dead end, either when the target activity is too low to be detected in a naïve library or when a property has already been enhanced to some extent, but further improvements have become difficult. Generating a neutrally drifted library requires a screening or selection system with a minimum throughput of as little as 100 variants. Neutral drift is a particularly effective approach when a high-throughput screening system (>10,000) is available for the *native* enzyme function, but the desired *promiscuous* function can only be screened in a low-throughput manner.

In the following sections, we will describe a typical neutral drift experiment (Fig. 1b) using the enzyme PTE (phosphotriesterase) as an example. PTE metabolizes the pesticide paraoxon with high efficiency ($k_{cat}/K_m \sim 10^7\ M^{-1}\ s^{-1}$) and also hydrolyzes the arylester 2-naphthyl hexanoate (2NH) at a low level (2NH, $k_{cat}/K_m \sim 10^2\ M^{-1}\ s^{-1}$). PTE has previously been evolved to increase the 2NH hydrolyzing activity [12].

## 2 Materials

### 2.1 Reagents

1. Plasmid DNA containing the target gene, e.g., pET-Strep-PTE vector.
2. Preferred error-prone PCR reagents. For a complete description of the GeneMorph II Mutazyme strategy (Agilent; Santa Clara, CA, USA), refer to Chapter 1 of this volume.
3. Standard cloning reagents, e.g., restriction enzymes and ligase.

4. Plasmid miniprep kit, e.g., QIAprep Spin Kit (Qiagen; Hilden, Germany).
5. *E. coli* strains for cloning (e.g., *E. cloni* 10G (Lucigen; Middleton, WI, USA)) and protein expression (e.g., BL21 (DE3)).
6. LB agar and LB media (Sigma Aldrich; St Louis, MO, USA) containing appropriate antibiotics (100 μg/mL ampicillin for pET-Strep-PTE).
7. Prescreening gel solution: 0.5 % agarose in 50 mM Tris–HCl buffer, pH 7.5, containing 10 μM fluorogenic phosphotriester DEPCyC [13].
8. Lysis buffer: 50 mM Tris–HCl, pH 7.5, containing 0.1 % (w/v) Triton X-100, 200 μM $ZnCl_2$, 100 μg/mL lysozyme, and approximately 1 μL of benzonase (Novagen; Darmstadt, Germany; 25 U/μL) per 100 mL.
9. Paraoxon substrate solution: 50 mM Tris–HCl, pH 7.5, containing 200 μM paraoxon and 0.02 % Triton X-100.
10. 2NH (2-naphthyl hexanoate) substrate solution: 50 mM Tris–HCl, pH 7.5, containing 200 μM 2NH, 200 μM Fast Red (Sigma Aldrich; St Louis, MO, USA), and 0.1 % Triton X-100.

### 2.2 Equipment

1. Thermal cycler.
2. Spectrophotometer for determination of DNA concentration.
3. Thermomixer (Eppendorf; Hamburg, Germany).
4. Electroporator.
5. 37 °C incubator.
6. Gel imager.
7. Incubator/shaker for 96-well plates.
8. Centrifuge for 96-well plates.
9. Microplate reader.

### 2.3 Materials

1. Regular (100 mm diameter) and large (140 mm diameter) Petri dishes.
2. Nitrocellulose membrane (e.g., BioTrace NT Pure Nitrocellulose Transfer Membrane 0.2 μm, PALL Life Sciences; Port Washington, NY, USA).
3. Deep 96-well plates for cell cultures.
4. Transparent 96-well plates for absorbance measurements.

## 3 Methods

The neutral drift procedure is very similar to a traditional directed evolution experiment, with iterative cycles of library generation and phenotypic screening/selection (Fig. 1b). Crucial differences arise in the selection criteria for purifying the "drifted" library between rounds of evolution (*see* Subheading 3.2).

### 3.1 Library Creation

In principle, any procedure can be used for generating a neutral drift library (many of which are described in this book), but because neutral drift requires multiple rounds of mutagenesis and selection/screening, we recommend the use of a fast and simple method. Error-prone PCR has been successfully employed in previous studies and will be described here as our mutagenesis method of choice.

1. Amplify the starting gene by error-prone PCR (*see* **Note 1** and Chapter 1 of this volume).
2. Subclone the gene library into an appropriate expression vector (*see* **Note 2**), transform into an *E. coli* cloning strain (e.g., *E. cloni* 10G), and plate onto LB agar plates (100 mm diameter) supplemented with 100 μg/mL ampicillin.
3. Determine the library size by counting bacterial colonies and estimating the background by comparison to a negative control (transformation of a ligation reaction lacking the insert). If the library size is acceptable, continue to the next step, otherwise the cloning needs to be optimized (for details, refer to Chapter 1 of this volume). We usually obtain a lawn of colonies for the library (>$10^5$ colonies) and a background of <10 % by electroporation transformation.
4. To extract DNA, add 3 mL of LB media onto the transformation agar plate and suspend cells using a spreader. Make sure all colonies are well suspended, then transfer an aliquot to a 1.5 mL centrifuge tube, and purify the resulting gene library using a plasmid miniprep kit (*see* **Note 3**).
5. Determine DNA concentration of the purified library.

### 3.2 Purifying Selection for the Native Activity

The ultimate aim of this experiment is to yield an enzyme which is a better starting point for subsequent evolution to increase 2NH-hydrolyzing activity. Here, we provide a protocol to generate a neutral drift library using an enzymatic assay for PTE activity in crude cell lysate.

It should be noted that any other selection or screening system could be incorporated into a neutral drift experiment. An ideal approach would involve fast, simple high-throughput methods such as genetic selection (antibiotic resistance or compensation of a knockout gene) or an enzymatic screening system using FACS (fluorescence-activated cell sorting).

1. Retransform the purified library from Subheading 3.1, **step 4,** into an expression strain (such as *E. coli* BL21 (DE3)). The amount of DNA should not exceed roughly 1 nmol (4 ng for the ~6,500 bp pET-Strep-PTE plasmid) to avoid transformants containing more than one plasmid [14].
2. Screen or select for wild-type activity. When a genetic selection step (i.e., selective growth on agar) is not available to screen for active clones, it is advisable to include a prescreening step

on agar plates (if possible) to increase throughput and filter out inactive variants, as shown in Subheading 3.2, **steps 3–6**, for PTE. If no prescreen is available, pick colonies randomly from Subheading 3.2, **step 1**, and then continue directly with Subheading 3.2, **step 7**.

3. Spread transformants on large agar plates containing 100 μg/mL ampicillin and incubate overnight at 30 °C.
4. Transfer colonies onto a filter paper; then place paper onto a second plate containing 100 μg/mL ampicillin, 1 mM IPTG, and 200 μM $ZnCl_2$ (to ensure availability of $Zn^{2+}$ ions necessary for PTE activity).
5. Incubate the agar plate overnight at room temperature to express protein (*see* **Note 4**).
6. Place the filter paper, colonies facing up, into an empty Petri dish. Pour 25 mL prescreening gel solution onto the filter, let it cool to solidify, and monitor fluorescent product formation under UV light (e.g., in a gel imager). In the case of the native activity of PTE, fluorescence becomes visible immediately (*see* **Note 5**).
7. Pick (positive) colonies into deep 96-well plates containing 200 μL LB media per well and 100 μg/mL ampicillin, and incubate cultures overnight at 30 °C while shaking at 900 rpm. As a control, pick control cells containing the starting gene and use to inoculate three wells on each plate. If desired, also add a negative control in triplicate.
8. Transfer 25 μL of each pre-culture into deep 96-well plates containing 500 μL of LB media and the appropriate antibiotics.
9. Grow at 30 °C while shaking at 1,050 rpm until the $OD_{600}$ reaches 0.4–0.6, and induce expression by addition of 50 μL LB media containing 10 mM IPTG (approximately 0.9 mM final concentration; any concentration between 0.4 and 1 mM is adequate for PTE expression).
10. After 2 h of protein expression at 30 °C while shaking at 1,050 rpm, measure the $OD_{600}$ for each well before harvesting the cells (*see* **Note 6**).
11. Pellet cells by centrifugation for 5 min at maximum speed (approx. 3,000 × *g* for a swinging bucket rotor), and remove supernatant.
12. Freeze plates for ≥30 min at −80 °C before continuing to the next step (*see* **Note 7**).
13. To lyse cells, add 200 μL lysis buffer per well, vortex to resuspend cells, and incubate for 30 min at room temperature.
14. Clarify lysate by centrifugation for 20 min at maximum speed.
15. Transfer 20 μL of clarified lysate to a clear 96-well plate. If the activity is too high, it might be necessary to dilute the lysate in order to obtain a linear signal.

16. Add 180 μL of paraoxon substrate solution to each well and measure paraoxon hydrolysis by following the change in absorbance at 405 nm for approx. 10 min using a microplate reader, and determine the initial rates of hydrolysis by analyzing the initial linear part of the reaction.
17. Calculate the specific enzymatic activity by normalizing the hydrolysis rate to the number of cells, i.e., dividing the initial rate values by the $OD_{600}$ values (determined in Subheading 3.2, **step 10**).
18. For each variant, calculate the fraction of specific enzymatic activity compared to the starting gene (*see* **Note 8**).

### 3.3 Determining the Selection Cutoff

The crux of neutral drift is not to identify the variant that best fulfils the selection criterion, but to take a mixed population containing a wide range of variants into the next round. The selection cutoff depends on the exact purpose of the experiment. When aiming to increase a promiscuous activity, we propose to use a relaxed cutoff (~30 % of the wild-type activity in the case of PTE) to allow the accumulation of a variety of mutations therefore generating a high-diversity library. Because of trade-offs between different properties, e.g., between function and stability [15] or between native and promiscuous function [16], a cutoff that is too stringent might purge many, if not all, mutations potentially beneficial to the promiscuous activity (*see* Table 1 for details). However, when the cutoff is too permissive, the library will drift toward lower and lower activities and eventually deteriorate [7, 8]. When stability (rather than a promiscuous activity) is the property of interest, we propose to use a more stringent cutoff (as well as a low mutation rate; *see* **Note 1**) to maintain a high level of native activity and promote the accumulation of stabilizing mutations.

1. Chose the selection cutoff.
2. Regrow cell cultures for all variants displaying activities above the selection cutoff in deep 96-well plates containing 200 μL LB media per well and 100 μg/mL ampicillin while shaking at 900 rpm and 30 °C overnight.
3. Pool cells and extract DNA as described in Subheading 3.1, **step 4**.
4. Use this plasmid mixture as a starting point for a new round of evolution and repeat the mutagenesis and screening process from Subheading 3.1, **step 1**.

### 3.4 Screening and Selection for the Desired Property

No standard recipe exists to determine when a library will be sufficiently drifted. In general, the accumulation of several mutations will be necessary to generate sufficient genetic diversity (*see* Table 1). We propose monitoring the accumulation of mutations by sequencing several of the selected variants every few rounds [2–5] and in conjunction, testing for the desired enzyme activity.

1. After several rounds, measure the property that is the ultimate protein engineering target (*see* **Note 9**). For example, determine 2NH activity in lysate (as described in Subheading 3.2, **steps 7–17,** with minor modifications; *see* **Note 10**).
2. Once improved variants are identified and/or between rounds of error-prone PCR, DNA recombination (DNA shuffling [17] or StEP PCR [18]) can be performed to remove destabilizing mutations that did not improve function and to combine favorable ones. Stabilizing mutations can be assessed by sequencing variants to identify mutations that have become enriched (i.e., occur many times) within the library (*see* **Note 11**).
3. Improved variants can serve as starting points for subsequent rounds of directed evolution with direct selection for the desired trait, e.g., 2NH activity.

## 4 Notes

1. We propose a medium mutation rate (2–3 nucleotide substitutions/gene) for systems based on 96-well plates (<1,000 variants throughput). For high-throughput systems, higher mutation rates can be applied, e.g., libraries containing ~5 substitutions/gene have been successfully used for FACS; *see* Table 1. Mutation rates that are too high will result in a high proportion of nonfunctional variants. If protein stability is the ultimate selection target, rather than increases in a promiscuous activity, we propose to keep the mutation rate low (1 nucleotide substitution/gene) to avoid the possibility that the effect of stabilizing mutations is masked by other destabilizing mutations.
2. In the case of PTE, we use a pET vector that confers ampicillin resistance and contains an N-terminal strep-tag (pET-Strep) for protein purification.
3. Use a 1 mL aliquot of the resuspended cells for plasmid extraction to prevent overloading of the plasmid miniprep spin column, which can cause poor extraction yields and impurities in the eluted sample.
4. We generally use mild expression conditions to promote the folding of variants with reduced stability caused by the acquisition of slightly deleterious mutations. Alternatively, proteins can be expressed for 1 h at 37 °C to save time and/or promote the acquisition of compensatory, stabilizing mutations [19].

5. To improve the signal for lower activities, cells can be lysed on the nitrocellulose membrane prior to the activity assay by alternating three times between 10 min storage at −20 °C and 37 °C.
6. For $OD_{600}$ measurement, we find it convenient to take out 50 μL of cells from each well and to add them to clear 96-well plates filled with 150 μL of LB per well. After the measurement, cells are discarded to prevent experimental errors and contamination.
7. Make sure that the cells are completely frozen; otherwise, cell lysis will not be uniform, causing inconsistencies in the activity assay. If a −20 °C freezer is used, freeze plates for at least several hours.
8. For high-throughput systems such as genetic selections or FACS, it is neither possible nor necessary to accurately calculate the relative activities, and variants above a certain level of activity (Subheading 3.2, **step 2**) can be pooled directly (Subheading 3.2, **step 3**). Nevertheless, the number of surviving variants and overall library distribution should be monitored to judge the development of the drift (and prevent loss of diversity).
9. Because of the high quality and high diversity of the drifted library, it should be sufficient to screen only a small number of clones to find a hit, i.e., for improvements in 2NH activity, we only screen several hundred variants in a 96-well format.
10. For the activity assay, use 2NH substrate solution, which contains Fast Red. Upon hydrolytic release, the naphtholate leaving group forms a red precipitate with Fast Red that can be detected at 500 nm.
11. Stabilizing mutations can be identified by sequencing many variants (>20) of the final, neutrally drifted population. Mutations that are enriched (i.e., appear in many variants) are likely to be stabilizing [6, 8]. If only a few potentially stabilizing mutations are identified, confirm their effect by creating single point mutants. Alternatively, or if multiple mutations are enriched, perform DNA recombination and screen/select again to eliminate destabilizing mutations and/or combine stabilizing mutations.

## Acknowledgments

This work was supported by the Natural Sciences and Engineering Research Council of Canada (NSERC) and Canada Institutes of Health Research (CIHR). N.T. is a Michael Smith Foundation for Health Research Scholar.

## References


1. Hartl DL, Dykhuizen DE, Dean AM (1985) Limits of adaptation: the evolution of selective neutrality. Genetics 111(3):655–674
2. Huynen MA, Stadler PF, Fontana W (1996) Smoothness within ruggedness: the role of neutrality in adaptation. Proc Natl Acad Sci U S A 93(1):397–401
3. Dalby PA (2011) Strategy and success for the directed evolution of enzymes. Curr Opin Struct Biol 21(4):473–480
4. Goldsmith M, Tawfik DS (2012) Directed enzyme evolution: beyond the low-hanging fruit. Curr Opin Struct Biol 22(4):406–412
5. Amitai G, Gupta RD, Tawfik DS (2007) Latent evolutionary potentials under the neutral mutational drift of an enzyme. HFSP J 1(1):67–78
6. Bershtein S, Goldin K, Tawfik D (2008) Intense neutral drifts yield robust and evolvable consensus proteins. J Mol Biol 379(5):1029–1044
7. Bershtein S, Tawfik DS (2008) Ohno's model revisited: measuring the frequency of potentially adaptive mutations under various mutational drifts. Mol Biol Evol 25(11):2311–2318
8. Bloom JD et al (2007) Evolution favors protein mutational robustness in sufficiently large populations. BMC Biol 5:29
9. Bloom JD, Romero PA, Lu Z, Arnold FH (2007) Neutral genetic drift can alter promiscuous protein functions, potentially aiding functional evolution. Biol Direct 2:17
10. Gupta RD, Tawfik DS (2008) Directed enzyme evolution via small and effective neutral drift libraries. Nat Methods 5(11):939–942
11. Tokuriki N, Tawfik DS (2009) Stability effects of mutations and protein evolvability. Curr Opin Struct Biol 19(5):596–604
12. Tokuriki N et al (2012) Diminishing returns and tradeoffs constrain the laboratory optimization of an enzyme. Nat Commun 3:1257
13. Aharoni A, Amitai G, Bernath K, Magdassi S, Tawfik DS (2005) High-throughput screening of enzyme libraries: thiolactonases evolved by fluorescence-activated sorting of single cells in emulsion compartments. Chem Biol 12(12):1281–1289
14. Goldsmith M, Kiss C, Bradbury AR, Tawfik DS (2007) Avoiding and controlling double transformation artifacts. Protein Eng Des Sel 20(7):315–318
15. Tokuriki N, Stricher F, Serrano L, Tawfik DS (2008) How protein stability and new functions trade off. PLoS Comput Biol 4(2):e1000002
16. Aharoni A et al (2005) The 'evolvability' of promiscuous protein functions. Nat Genet 37(1):73–76
17. Stemmer WP (1994) DNA shuffling by random fragmentation and reassembly: in vitro recombination for molecular evolution. Proc Natl Acad Sci U S A 91(22):10747–10751
18. Zhao H, Giver L, Shao Z, Affholter JA, Arnold FH (1998) Molecular evolution by staggered extension process (StEP) in vitro recombination. Nat Biotechnol 16(3):258–261
19. Tokuriki N, Tawfik DS (2009) Chaperonin overexpression promotes genetic variation and enzyme evolution. Nature 459(7247):668–673

# Chapter 6

# Site-Saturation Mutagenesis by Overlap Extension PCR

**Elsie M. Williams, Janine N. Copp, and David F. Ackerley**

## Abstract

Site-saturation mutagenesis is a proven strategy for generating high-quality variant gene libraries of a defined size. Variation is introduced via incorporation of degenerate base combinations at specific codon locations, giving rise to a precise series of amino acid substitutions in the encoded protein. Here we describe a simple and efficient overlap PCR protocol for the introduction of degenerate bases at either single or multiple codon locations. The resulting libraries can then be directly screened for improved protein function as either an independent directed evolution study or an adjunct to random mutagenesis strategies (such as error-prone PCR) that are, in isolation, unlikely to access the full repertoire of possible amino acid substitutions at any given position.

**Key words** Saturation mutagenesis, Random mutagenesis, Smart library, Overlap PCR, Degenerate codons

## 1 Introduction

Site-saturation mutagenesis (SSM) is a simple, semi-rational approach to protein engineering that occupies a useful middle ground between fully rational site-directed mutagenesis strategies (which typically examine one to ten variants) and non-rational mutagenesis strategies (typically $10^6$–$10^{12}$ variants per library). In SSM, the semi-rational designation derives from the fact that target residues are identified in a specific, knowledge-guided manner, but are then mutated randomly, in either an individual or combinatorial fashion. Target residues are generally chosen on the basis of structural, functional, and/or sequence-derived information, and different methods for deriving this information are surveyed in detail elsewhere in this volume (in particular Chapters 7 and 20). Irrespective of how these target residues are chosen, the crux of this approach is that focusing on just one, or a few, residue positions per round of mutagenesis enables a comprehensive spectrum of mutations to be considered at each position without the requirement for an ultrahigh-throughput screening capability.

Elizabeth M.J. Gillam et al. (eds.), *Directed Evolution Library Creation: Methods and Protocols*, Methods in Molecular Biology, vol. 1179, DOI 10.1007/978-1-4939-1053-3_6, © Springer Science+Business Media New York 2014

The potential benefits of employing such targeted "smart" libraries have been detailed in several excellent reviews (e.g., [1, 2]).

Depending on the throughput capability of the available screen, target residues may be mutated in either a sequential or a combinatorial fashion. The "saturation" aspect of SSM generally refers to randomization of the codons that specify each target residue, such that any of the 20 standard proteinogenic amino acids might be incorporated at that position. However, rather than full randomization at a genetic level via NNN degeneracy, a reduced codon set such as NNK or NNS is generally employed (where N is any nucleotide, K is either G or T, and S is either C or G); these still provide access to all 20 amino acids, but with a more even distribution and a lower probability of generating a stop codon than NNN [3]. Moreover, in situations where the screen is potentially limiting, either due to a low-throughput analytical technique or because several residues are being targeted in parallel, partial saturation of each codon can still prove highly effective. The codon set NDT is particularly valuable in this context as it reduces the degeneracy to only 12 possible codons that each specifies a different amino acid, while still providing a balanced representation of large and small, polar and nonpolar, aliphatic and aromatic, and negatively and positively charged side chains [4]. Finally, we refer readers who are concerned that their screen may not suffice for complete interrogation of their library to the recent statistical analyses of Yuval Nov ([5] and Chapter 18 of this volume), which neatly illustrate the diminishing rewards that accrue with successive rounds of screening. In brief, achieving only partial library coverage may not uncover the best variant, but will likely still yield substantial increases in fitness over wild type.

While the primary focus of this book is on directed evolution, it should be noted that SSM also provides a powerful tool for deconvoluting structure-function relationships, in that it allows the mutational tolerance at specific residue positions to be probed readily. SSM also provides a useful complement to imperfect random mutagenesis approaches such as error-prone PCR, where multiple base substitutions are unlikely to occur within the same codon, leading to a limited repertoire of potential substitutions at each residue location. This repertoire can subsequently be expanded by using saturation mutagenesis to mutate residues that yielded hits in the initial screens, essentially "fine-tuning" the lead mutants [2]. In our own SSM studies, we have frequently observed that the most active single-residue variants we recovered required base substitutions at two codon positions relative to wild type and thus would almost certainly not have been realized by epPCR alone (e.g., [6]).

Our preferred method for generating SSM libraries is via overlap extension PCR. In essence, this employs overlapping

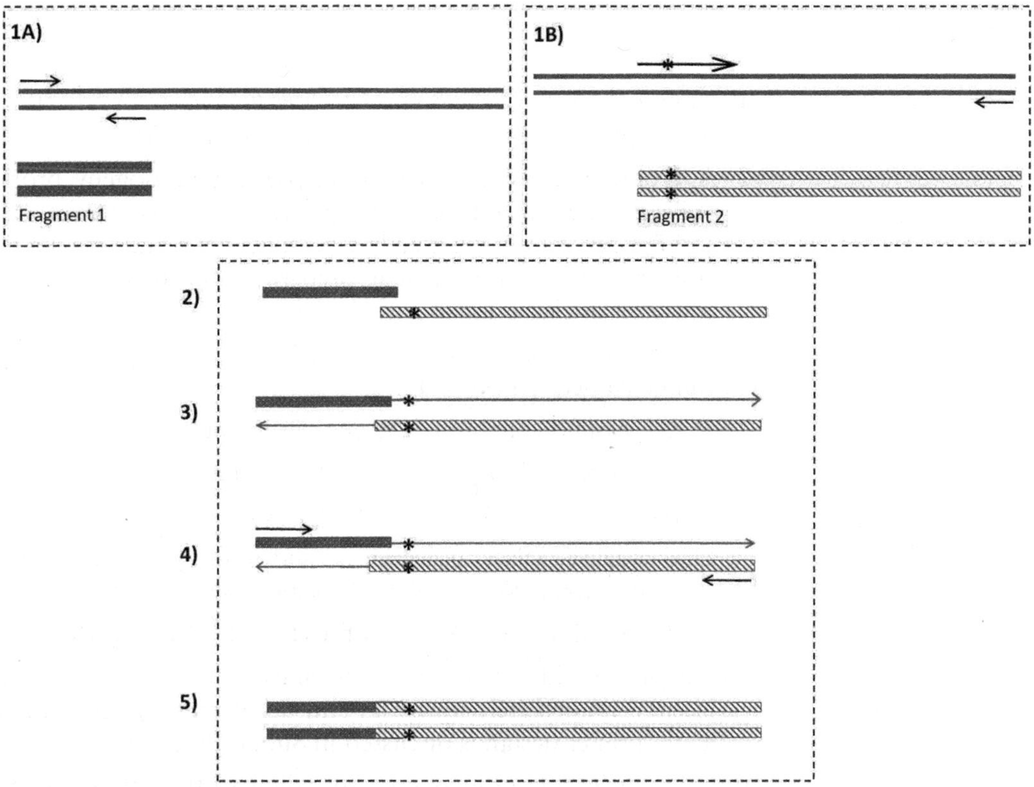


**Fig. 1** Schematic diagram of an overlap extension PCR to generate a site-saturation mutagenesis gene library at a single codon site. (*1*) Two gene fragments are amplified in separate PCR reactions (*a*) and (*b*) using the wild-type gene as the PCR template, one internal primer and one complementary flanking primer in each reaction. The second fragment contains a degenerate codon specified by the internal primer, as marked by an *asterisk*. (*2*) The two individually amplified fragments, which have overlapping sequence due to the specific design of the internal primers, are mixed together, denatured, and subsequently allowed to anneal. (*3*) A primerless PCR reaction is performed, reconstituting a full length gene from the overlapping fragments. (*4*) After 10–15 PCR cycles, gene-specific flanking primers are added to enable amplification of the full length gene. (*5*) The full length gene product should now contain the specified codon degeneracy as well as unique restriction sites for cloning at the 5′ and 3′ ends

internal gene primers that include degenerate base combinations at the target codons to amplify short gene fragments, which are then recombined to reconstitute a full length gene (e.g., Fig. 1). In our hands, this approach has proven more reliable and efficient when generating larger libraries (i.e., greater than $10^2$ variants) than either Quikchange [7, 8] or megaprimer PCR [9] strategies. Our protocol is readily adaptable for SSM of multiple target residues in parallel. To date the greatest number of residues we have targeted in parallel is eight (on five distinct gene fragments); however, we are confident that the method could easily accommodate greater variation.

## 2 Materials

1. Competent cells of the desired *Escherichia coli* strain (selection of competent cells is discussed in Subheading 3.5).
2. Appropriate growth media for *E. coli*, e.g., Lysogeny broth (LB): per liter of LB broth mix 5 g yeast extract, 10 g tryptone, and 5 g NaCl. Autoclave at 121 °C for 25 min and let cool to ambient temperature. LB agar: prepare LB broth as above, but add 15 g of agar prior to autoclaving.
3. *E. coli* expression plasmid into which the mutant gene library will be cloned (*see* **Note 1**).
4. *E. coli* plasmid containing the gene to be mutated.
5. Appropriate antibiotics for plasmid selection.
6. Sterile distilled $H_2O$ (e.g., Milli-Q (MQ)).
7. High-fidelity DNA polymerase and corresponding buffer (e.g., Phusion®; New England Biolabs, Ipswich, MA, USA).
8. dNTP solution (10 mM each of dATP, dTTP, dCTP, dGTP).
9. Oligonucleotide primers that anneal to flanking regions of the plasmid immediately upstream and downstream of the target gene (primer design is discussed in Subheading 3.1).
10. Oligonucleotide primers corresponding to the part(s) of the gene to be mutated, designed to contain degenerate codons to enable appropriate site saturation at target locations (primer design is discussed in Subheading 3.1).
11. PCR clean and concentrate kit (e.g., DNA Clean & Concentrator kit; Zymo Research, Irvine, CA, USA).
12. Plasmid Miniprep kit (e.g., Zyppy™ Plasmid Miniprep kit; Zymo Research, Irvine, CA, USA).
13. Plasmid Midiprep kit (e.g., Qiagen Plasmid Midi Kit; Qiagen, Valencia, CA, USA).
14. Appropriate restriction enzymes to enable directional, sticky-ended cloning of the variant gene product. These should both be active in a single, compatible buffer (*see* **Notes 2** and **3**).
15. Restriction enzyme that cuts uniquely within the "stuffer" fragment of the expression vector (*see* **Notes 2**).
16. DNA ligase and corresponding buffer (e.g., T4 DNA ligase; Invitrogen, Carlsbad, CA, USA).
17. Thermocycler with a heated lid (e.g., an MJ Mini; BioRad, Hercules, CA, USA).
18. Agarose gels pre-stained with ethidium bromide at 0.5 μg/ml, and apparatus for agarose gel electrophoresis.
19. Agarose gels pre-stained with 1× SYBR Safe DNA gel stain (Invitrogen, Carlsbad, CA, USA).

20. Blue-Light Transilluminator (e.g., Safe Imager 2.0 Blue-Light Transilluminator; Invitrogen, Carlsbad, CA, USA).
21. Gel extraction kit (e.g., PureLink™ Quickgel extraction kit; Invitrogen Carlsbad, CA, USA).
22. Clean razor blades for excising DNA fragments from agarose gels.
23. Spectrophotometer to measure DNA concentration in small volumes (e.g., Nanodrop 1000; Thermo Scientific, Pittsburgh, PA, USA). If a Nanodrop spectrophotometer is not available, DNA quantification can be performed by agarose gel electrophoresis utilizing a DNA quantification ladder (e.g., the 100 bp DNA Ladder from New England Biolabs, Ipswich, MA, USA).
24. Reagents for a standard PCR screen; any relatively cheap, easy to use, and robust PCR mix can be used, and high-fidelity amplification is not required (e.g., BioMix Red; Bioline, Alexandria, NSW, Australia).

## 3 Methods

Described herein is a simple and flexible method for constructing gene libraries with saturated residues at single or multiple sites using overlap extension PCR. We have used this method to create very small libraries (32 possible gene variants), in which only one codon was randomized, as well as larger libraries (ca. 2,000–3,000,000 gene variants), in which multiple codons were randomized.

The method consists of five distinct phases: (1) identification of target codons and design of primers to introduce variation at those codons, (2) amplification of the gene as $n+1$ distinct fragments (where $n$ is the number of gene sites at which a codon (or multiple adjacent or near-adjacent codons) will be targeted by saturation mutagenesis), (3) annealing of individual fragments by overlap extension PCR to generate full length gene constructs, (4) cloning of the full length gene constructs into the desired expression plasmid to create a variant gene library, and (5) transformation of the variant gene library into an *E. coli* expression strain for screening or selection of improved gene variants.

### 3.1 Primer Design

1. Identify the codon(s) to be targeted for site-saturation mutagenesis (*see* **Note 4**) and design degenerate codon sets for appropriate randomization of these sites. Calculate the number of individual gene variants the collective degeneracy will yield and consequently the size of the variant library that will need to be screened to achieve the desired level of library coverage (*see* **Note 5**).

**Partial wild type gene sequence surrounding desired mutation site** 5'-CGCTACTGCCAATACAGCCCATCCAGCACCAACTCCCAGCCGTGGCAT-3'

**Fwd primer** 5'-**CAATACAGCCCATCCAGC**NNKAACTCCCAGCCGTGG-3'

**Rev primer** 5'-**GCTGGATGGGCTGTATTG**-3'

**Fig. 2** Example of internal primers designed to carry out full saturation mutagenesis of a target codon. The target codon and the degenerate codon set (in this case, NNK) in the Fwd primer are *underlined*. The regions of primer overlap are indicated in *bold*

2. For each target codon design a pair of overlapping internal primers. If two or more target codons are in close proximity (i.e., within 21 base pairs of each other), they can be randomized using one pair of primers (*see* **Notes 6** and **7**). For each primer pair, the forward (Fwd) internal primer should comprise three parts: (a) a 15–20 nucleotide (nt) section identical to the gene sequence immediately upstream of the target codon(s) (e.g., Fig. 2, Fwd primer, bold sequence), (b) the randomized target codon(s) (e.g., Fig. 2, Fwd primer, underlined sequence), and (c) a 15–20 nt section identical to the gene region immediately downstream of the target codon(s) (e.g., Fig. 2, plain font). The reverse (Rev) internal primer should be the reverse complement of the first 15–20 nt of the "Fwd" primer (e.g., Fig. 2, Rev primer, bold sequence; *see* **Note 8**).
3. Flanking primers that will amplify the full length gene are also required (*see* **Notes 9** and **10**). Each primer should contain a unique restriction site at its 5′ end to enable directional in-frame cloning into the expression plasmid. To increase restriction enzyme digestion efficiency, a four nucleotide G/C clamp should be included at the 5′ end of the primer immediately adjacent to the restriction site (*see* **Note 11**).

### 3.2 Amplification of Individual Gene Fragments

1. Set up a separate reaction for the amplification of each fragment in individual thin-walled PCR tubes (e.g., three distinct reactions to amplify the three fragments required for site-saturation mutagenesis at two distantly located target codons, as illustrated in Fig. 3).

   Each reaction should contain 0.1–10 ng plasmid template DNA, 1× polymerase buffer, 200 μM (final concentration) of each dNTP, 0.5 μM (final concentration) of each appropriate primer, and 0.2 U high-fidelity DNA polymerase (*see* **Note 12**), and sterile MQ to a total volume of 10 μL.
2. A negative control reaction (omitting the plasmid template but otherwise identical) should be set up and run in parallel for each primer pair.
3. Proceed with the following PCR protocol: an initial denaturization of 98 °C for 1 min, followed by 30 repeating cycles of

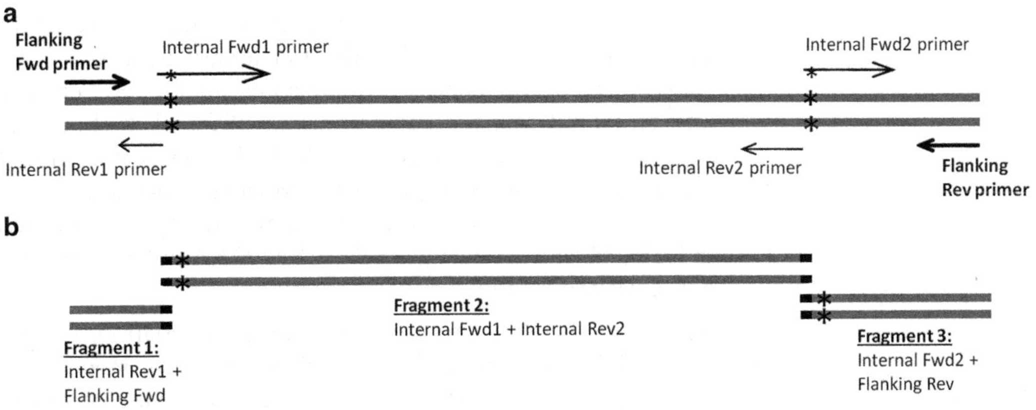


**Fig. 3** Schematic illustration of the primer (**a**) and fragment (**b**) combinations required for target codon mutagenesis at two distant locations in the gene. Overlapping fragment regions are shown in *black*, primers are shown by *directional arrows*, and target site for mutagenesis is indicated by an *asterisk*. As noted in Subheading 3.1, **step 2**, distinct primer pairs should be used to achieve randomization of codons that are separated by more than 21 bp in the linear gene sequence

98 °C for 10 s, 56 °C for 20 s (*see* **Note 13**), and 72 °C for 10 s (*see* **Note 13**). Complete the PCR with a final elongation step of 72 °C for 2 min, and then hold samples at 4 °C for storage (if necessary).

4. Confirm successful and specific amplification of gene fragments of the predicted size by analysis of 1 μL of each PCR product via agarose gel electrophoresis (*see* **Notes 14–16**).
5. Purify the remaining PCR reaction using a DNA Clean and Concentrate column; elute in 10 μL pre-warmed (~50 °C) MQ (*see* **Note 17**).
6. Quantify DNA, preferably using a Nanodrop spectrophotometer. In our hands, the resulting DNA concentration is typically between 10 and 30 ng/μL. Store at −20 °C until required.

### *3.3 Generation of Full Length Mutant Gene Library Using Overlap Extension PCR*

1. Set up the overlap extension primerless PCR reaction with the following components: 1× polymerase buffer, 200 μM (final concentration) of each dNTP, 0.6 U high-fidelity DNA polymerase, each of the gene fragments generated in Subheading 3.2, combined in an equal molar ratio (*see* **Note 18**) and sterile MQ to 30 μL.
2. Set up a negative PCR control that does not contain any amplified gene fragments (*see* **Note 19**).
3. Run the overlap extension PCR reaction using the following protocol: an initial denaturization step of 98 °C for 2 min, followed by 15 cycles of 98 °C for 10 s, 50 °C for 30 s (*see* **Note 20**), and 72 °C for 15 s (*see* **Note 21**). Then hold samples at 4 °C.

4. Prepare an additional PCR mix which includes the two flanking primers that will amplify the full length gene (e.g., the Flanking Fwd and Flanking Rev primers in Fig. 3). Prepare sufficient PCR mix to enable addition of 20 μL to both the negative control and the overlap PCR. The PCR mix should contain 1× polymerase buffer. 200 μM (final concentration) of each dNTP, upstream and downstream flanking primers each at 1.25 μM (final concentration), 0.8 U DNA polymerase, and sterile MQ to 40 μL.
5. Add 20 μL PCR mix to the paused 30 μL overlap PCR reaction and negative control from Subheading 3.3, **step 3** and resume the PCR by proceeding with the following protocol: an initial denaturization step of 98 °C for 2 min, followed by 15 repeating cycles of 98 °C for 10 s, 56 °C for 20 s (*see* **Notes 10** and **13**), and 72 °C for 30 s (*see* **Note 13**). Complete the PCR with a final elongation step of 72 °C for 2 min.
6. Confirm successful assembly and amplification of the full length gene by analysis of 3–5 μL of PCR product via ethidium-stained agarose gel electrophoresis. Run all of the negative control. There should be no observable DNA in the lane corresponding to the negative control and one bright clear band the size of the reconstituted full length gene in the PCR sample. If the overlap PCR shows either no observable fragment of the correct size or else multiple randomly sized DNA bands and/or smears as well as the desired fragment, the PCR protocol will require further optimization (*see* **Notes 22** and **23**).
7. Purify the remaining overlap extension PCR reaction using the DNA Clean & Concentrator kit, and elute in 20 μL prewarmed MQ (*see* **Note 17**).
8. Quantify the purified PCR product; we generally achieve an overall DNA yield of 0.75–2 μg. Store at −20 °C until required.
9. At this stage, we recommend direct sequencing of an aliquot of the PCR product to confirm randomization of nucleotides (nts) at the targeted sites (*see* **Note 24**). The sequencing chromatogram should be visually inspected to confirm randomization at these sites. At the randomized nt site(s), the sequencing chromatogram should show multiple different colored peaks, comprising a peak for each nt allowed by the redundancy codon.

### *3.4 Cloning of Site-Saturation Mutagenesis Gene Library into Expression Plasmid*

1. Midiprep the plasmid that will be used for cloning and expression of the library using a Qiagen midiprep kit or equivalent (*see* **Notes 1** and **2**).
2. Digest 10 μg of the plasmid vector in a 500 μL reaction containing 1× restriction enzyme buffer and 25 U of each restriction enzyme. Incubate the reaction at 37 °C overnight (*see* **Note 3**).

3. Add a further 10 U of an enzyme that cuts uniquely within the stuffer fragment (*see* **Note 2**). Incubate the reaction for an additional 2 h at 37 °C, and then heat inactivate the enzymes according to the manufacturer's guidelines.
4. Run the entire restriction digest on a 0.8 % w/v agarose gel and excise the DNA fragment corresponding to the digested vector. We strongly recommend the use of a blue-light transilluminator and a compatible stain (e.g., SYBR Safe) for this step rather than ethidium bromide and a UV transilluminator (*see* **Note 16**). Purify the excised DNA fragment using the gel extraction kit.
5. Quantify the gel-purified digested vector using a Nanodrop spectrophotometer or equivalent.
6. Aliquot vector and store at −20 °C. We typically aliquot 200–500 ng of the digested vector into multiple microcentrifuge tubes, so these can be thawed and used as required. This reduces the number of freeze/thaw cycles and maintains high-quality stocks of digested vector for short-term use (up to 2 weeks).
10. In parallel, digest and purify the overlap PCR product from Subheading 3.3, **step 8** (*see* **Note 25**). In our hands, the following steps provide sufficient digested gene insert to generate libraries of 100–2,000 individual clones if transforming into chemically competent cells or, if transforming into electrocompetent cells, $>10^3$clones. Digest 0.75 μg of the overlap product with 10 U of each of the restriction enzymes in 1× restriction enzyme buffer and a total volume of 50 μL. Incubate the reaction at 37 °C for 6–8 h or overnight. Heat and inactivate the restriction enzymes, as recommended by the manufacturer. Purify the digest using the DNA Clean & Concentrator kit, and elute in 20 μL pre-warmed MQ (*see* **Note 26**). Quantify using a Nanodrop spectrophotometer.
11. Prepare a ligation reaction containing 1× ligation buffer, 100 ng of digested vector DNA, 2 U T4 DNA ligase, a 3:1 molar ratio of digested gene insert DNA to vector DNA, and water. The exact quantity of water and the volume of the ligation will vary as final DNA concentration should be kept <20 ng/μL. Add the T4 DNA ligase last and mix gently (*see* **Note 26**). Prepare a control ligation reaction as above but without the insert DNA.
12. Incubate the ligation reactions at 16 °C for 12–18 h.

### 3.5 Transformation and Validation of the Library

The choice of *E. coli* strain to be used for transformation of the library is dependent upon the intended screening or selection method. The protocol chosen to generate and transform competent *E. coli* cells will depend on the number of variants that are desired to be screened. If more than 2,000 distinct clones are required, then the use of electrocompetent cells with high transformation efficiency is recommended. For a detailed protocol on how to generate and transform electrocompetent cells, the reader is referred to Chapter 1 of this volume by Copp et al. For most purposes, chemically competent *E. coli* cells should generate sufficient clones. Transformation of chemically competent cells allows us to reliably generate a minimum of 100 (more typically 400–500) distinct *E. coli* colonies, and the protocol can be scaled up if more are required. If making a large library or multiple small libraries in parallel, it is advisable to test both the quality of the digested vector and the efficiency of the competent cells before use.

1. Prepare [10] or purchase (e.g., DH5α (Invitrogen, Carlsbad, CA, USA).) chemically competent *E. coli* cells.
2. Transform both the ligation and the control ligation cells using a heat shock protocol. The exact volume of cells to transform will depend on the volume of the ligation as this should not exceed 10 % of the cell volume, e.g., if transforming a 15 μL ligation, we recommend the use of at least 150 μL of cells. The following steps are appropriate for the transformation of 200 μL of cells with a 20 μL ligation reaction: Thaw two frozen 200 μL aliquots of competent cells by incubating the aliquots on ice. Add 20 μL of the ligation reaction to one of the 200 μL aliquots of competent cells. Split the remaining 200 μL cell aliquot equally into two separate microcentrifuge tubes. Add 10 μL of the vector-only control ligation to one microcentrifuge tube and use the remaining 100 μL as a negative transformation control. Incubate on ice for 10 min. Transfer microcentrifuge tubes to a 42 °C water bath and incubate for 90 s. Recover microcentrifuge tubes on ice for 2 min. Add 800 μL of LB to the ligation reaction containing 200 μL cells. Add 400 μL of LB to each of the control reactions containing 100 μL cells. Incubate at 37 °C, 200 rpm for 45–60 min. Transfer four 250 μL aliquots of the ligation reaction cell suspension onto four LB agar plates containing appropriate antibiotics for selection of the expression plasmid. Transfer one 250 μL aliquot of each control reaction to separate agar plates containing appropriate antibiotics for selection of the expression plasmid. Incubate the plates at 37 °C overnight.
3. Examine the plates, and confirm that no colonies are present on the no-DNA transformation control plate. Compare colony numbers between the library ligation and vector-only control. While there may be a small number of colonies on the vector-only control, there should be hundreds of colonies on the ligation plates.
4. It is important to accurately calculate the percentage of colonies that do not carry a variant gene (i.e., vector-only background)

to ensure appropriate coverage of the variant library (*see* **Note 5**). In order to confirm background, vector-only numbers perform a PCR screen of at least 20 colonies from the library ligation plate. The PCR products should then be purified and sequenced to confirm library quality. For each colony, prepare a 20 μL PCR mix in thin-walled 0.2 ml tubes; also prepare one no-template negative control. To minimize consumable expenses, we typically use a standard PCR mix such as 2× BioMix Red (Bioline; Alexandria, NSW, Australia) or similar for this step. Our standard PCR set up using this reagent contains 8 μL sterile MQ, 10 μL 2× BioMix Red (Bioline; Alexandria, NSW, Australia), 1 μL forward primer (from a 10 μM stock solution; *see* **Note 27**), 1 μL reverse primer (from a 10 μM stock solution).

5. Use 2 μL pipette tips (or sterile toothpicks) to pick 20 colonies at random from the library ligation plates. Transfer each colony into one of the 20 μL PCR reactions prepared above.
6. Run a standard PCR reaction that will amplify the gene. (*See* the PCR reaction in Subheading 3.2, **step 2**. Adapt the protocol to the manufacturer's recommendations for the polymerase mix of choice, e.g., BioMix Red.)
7. Run a 2 μL aliquot of each PCR product on an agarose gel. The proportion of colony templates from which a gene can be amplified will provide an estimate of the proportion of the library that carries a plasmid with a variant gene insert. This should ideally be >90 % of the colonies.
8. To confirm mutagenesis, purify the remainder of the PCR products using the DNA Clean & Concentrator kit. Elute the purified DNA from each spin column in 20 μL sterile water. Quantify and sequence the PCR products, and confirm codon randomization at the correct site. We recommend sequencing all of the 20 PCR reactions if multiple sites have been targeted for mutagenesis. Researchers may sequence fewer reactions when carrying out single site mutagenesis.
9. Calculate the number of variants that should be screened; take into account the codon randomization, the proportion of the library mutants that carry a variant, and the desired level of library coverage (*see* **Note 5**). For example, if simultaneously randomizing two different codons using NDT codons (12 possible variants apiece), then 144 different gene sequences are possible and 430 colonies should be screened to give 95 % library coverage. If the PCR screen indicated that only 90 % of colonies carried a variant gene, then 478 (i.e., 430 × (100/90)) colonies should be screened.
10. For storage of the library, use pipette tips or sterile tooth picks to pick the determined number of individual colonies into

individual wells of a 96-well microplate containing 100 μL *E. coli* growth media and the appropriate antibiotic(s). Following overnight incubation at 200 rpm, 37 °C, add 100 μL of 80 % v/v sterile glycerol (i.e., to 40 % v/v glycerol final concentration) and mix thoroughly. Store the library at −80 °C.

## 4 Notes

1. It is essential to construct a library with a low proportion of "vector-only" background to minimize screening burden. We have found that it is helpful to make a large-scale preparation of digested vector for library construction and store it for short-term use at −20 °C, as this enables several libraries to be generated and tested, all using a vector preparation of known quality and quantity. However, the prepared vector should ideally be used within 1–2 weeks as we have noted a considerable drop in ligation efficiency after this point, even without repeated freeze/thaw cycles. Our protocol can easily be adapted to prepare smaller amounts of vector as required.
2. We find it particularly useful to use a plasmid with a "stuffer" fragment already inserted in the cloning cassette. This should be a DNA insert that can be removed using the same restriction sites that will be used to clone the amplified SSM library. The stuffer fragment should contain a unique internal restriction site not present in the native gene. Excision of this stuffer fragment from the vector enables easy monitoring of the restriction enzyme reaction, i.e., a small aliquot of the digestion reaction can be analyzed by gel electrophoresis to confirm complete digestion of the vector, the stuffer fragment, and cut vector being readily visualized as two distinct bands on the agarose gel. The use of a third restriction enzyme that cuts within the stuffer fragment minimizes the number of "vector-only" clones in the final library. Where possible, this third restriction enzyme should ideally utilize the same buffer as the other restriction enzymes used to prepare the vector for cloning. New England Biolabs provide an easy to use online tool for restriction analyses: http://tools.neb.com/NEBcutter2/.
3. We recommend the use of NEB High-Fidelity restriction enzymes. This range of enzymes has the same specificity as the native enzyme with reduced star activity and improved buffer compatibility (https://www.neb.com/products/restriction-endonucleases/hf-nicking-master-mix-time-saver-other/high-fidelity-restriction-enzymes/high-fidelity-restriction-endonucleases). The reduced star activity is particularly valuable in that it allows for extended (e.g., overnight) incubation times without risk of off-target digestion.

4. It may be desirable to target multiple codons in one round of evolution. However, the maximum number of target codons is primarily constrained by the throughput capacity of the end screen/selection as the number of possible variants will grow exponentially with each additional randomized codon. Practical considerations concerning library construction may also limit the number of target codons. If codons are too far apart to be simultaneously randomized on the same primer (*see* **Note 6**), then every additional targeted codon requires amplification of an additional gene fragment and increases the complexity of the overlap extension PCR. We have not encountered this as a limiting factor; our most complex scenario to date has been generation of variant libraries from five different gene fragments (ranging from 90 to 350 base pairs long) in which eight different codons were randomized.
5. We typically use the freely available CASTER tool, available on the Reetz laboratory homepage http://www.kofo.mpg.de/media/2/D1108347/0987095526/ISM_tools.zip [11].
6. When targeting multiple codons, we recommend keeping primers below 60-mers in length and including a region of approximately 20 nt identical to the gene at both the 3′ and 5′ end. If following these guidelines, codons within 21 nt of each other (i.e., seven amino acid positions) can be targeted using one Fwd primer that contains multiple degenerate codons. If sites are further than 21 nt apart, then they should be randomized on separate gene fragments. In some cases, this may require incorporation of (complementary) degenerate sequences into Rev primers, to avoid generation of gene fragments under 50 bp in length (*see* **Note 7**).
7. We do not recommend designing fragments to be <50 base pairs long, as when verifying successful amplification by agarose gel electrophoresis, it can be difficult to distinguish DNA fragments of this length from primer dimers.
8. When designing primer pairs in this manner, the resulting primers will often have melting temperatures (Tm) that differ from one another by substantially more than 5 °C. We have found that successful amplification of gene fragments is generally unhindered by the large difference in primer Tm and prefer this asymmetrical approach to primer design (in which only one primer contains degenerate sequence) as it allows the "Rev" primers to be reused in subsequent libraries that target alternative (adjacent or near-adjacent) codons. However, should the reader suspect that mismatched primer Tm values are the root cause of a failed PCR, the design of the "Rev" primers can readily be modified to make them the full reverse

| | |
|---|---|
| **Fwd primer** | 5'-CAATACAGCCCATCCAGCNNKAACTCCCAGCCGTGG-3' |
| **Rev primer** | 5'-CCACGGCTGGGAGTTMNNGCTGGATGGGCTGTATTG-3' |

**Fig. 4** Example of an internal primer pair designed to have similar Tm values for SSM of a target codon. The complementary degenerate sequences in the primer pair are *underlined*

complement of the "Fwd" primer. This simple expedient (as exemplified in Fig. 4) will ensure primer pairs with near-identical Tm values.

9. If targeted codon(s) is within 35 bp of the beginning or end of the gene, they can be randomized by including the degenerate sequences within the relevant flanking primer. The 35 bp limit is imposed under the assumptions that primer length should be kept under 60-mers, that a 6 nt restriction enzyme site is included, and that the 3′ primer end contains a 15 nt region complementary to the target gene.
10. The flanking primers should be noncomplementary and ideally have Tm values within 5 °C of each other. Primer Tm can be estimated accurately using the OligoAnalyzer tool from Integrated DNA Technologies (Coralville, IA, USA) (http://www.idtdna.com/analyzer/Applications/OligoAnalyzer/) or equivalent online programs, a comprehensive range of which can be found online: http://www.molbiol-tools.ca/PCR.htm. Prior to starting the overlap extension PCR protocol, it should be confirmed that the gene flanking primers can reliably and specifically amplify the parental gene from the plasmid template. If they cannot, then either attempt optimization of the PCR or design new flanking primers. The same PCR parameters that successfully amplify the parental gene should be used to amplify the overlap extension PCR product.
11. As an alternative to traditional restriction enzyme-based cloning, a library can be made by using the gene variants as mega-primers for the PCR amplification of a whole plasmid (an approach known as MEGAWHOP). The variant genes can act as mega-primers if the plasmid contains DNA regions homologous to the 3′ and 5′ ends of the variant gene [12] or contains the full length gene [9]. Using such approaches, we have generally not been able to generate large libraries without substantial PCR optimization and therefore have favored the use of traditional restriction enzyme-based cloning to construct libraries. However, when optimized, these techniques may lead to high-quality libraries (in which close to 100 % of clones contain a variant gene). As such, they may be of interest, particularly for research groups with experience using this cloning method, for those finding optimization of traditional

cloning problematic, or for those using very low-throughput screens in which only small but very high-quality libraries are required. If attempting whole plasmid PCR amplification, then the flanking primers must correspond to only the gene/plasmid sequence, i.e., they must not contain G/C clamps or additional restriction sites.

12. This quantity of polymerase is based on the 1 Unit/50 μL PCR reaction recommendation for Phusion high-fidelity polymerase (New England Biolabs; Ipswich, MA, USA); this may vary if using a different polymerase. Use of a high-fidelity polymerase to avoid undesired mutations is strongly recommended as multiple PCR cycles are required to generate a full length gene by overlap extension PCR. We recommend avoiding the use of any DNA polymerase that incorporates 3′ adenine overhangs to the ends of PCR products (e.g., *Taq* polymerase).
13. The 56 °C annealing temperature and 10 s extension time are our standard default parameters for PCR; the reader should use an extension time that will ensure the amplification of the largest gene fragment (this can be determined by consulting the manufacturer's instructions for the polymerase used) and an annealing temperature appropriate for the Tm of the specific primers. As a guideline, the optimal annealing temperature to use in a standard PCR is typically 5 °C cooler than the lowest primer Tm.
14. The presence of nonspecific PCR products or primer dimers can interfere with the subsequent overlap PCR. In our experience, if primer dimers or nonspecific PCR products cannot be visualized on a gel after examining 1 μL of the PCR product by agarose gel electrophoresis, then direct purification of the remaining PCR reaction using a DNA Clean & Concentrator column should suffice prior to overlap extension PCR. If, however, primer dimers or nonspecific PCR products can be seen on the gel, then the correct-sized gene fragment should be excised and purified using a commercial gel purification kit, e.g., Invitrogen's PureLink™ Quickgel extraction kit (*see* **Notes 15** and **16**). Polyacrylamide gel electrophoresis purification has also been reported to promote the excision and purification of an uncontaminated band [13], but we have not yet found this to be necessary. As an additional option, others have reported successful overlap extension PCR protocols using direct (non-purified) aliquots of the gene fragment PCRs [14].
15. DNA recovery following agarose gel electrophoresis purification is generally inefficient. If this purification step is required, then it may prove necessary to scale-up the PCR for fragment amplification (*see* Subheading 3.2, **step 1**) to 25–50 μL to ensure recovery of a sufficient quantity of gel-purified DNA.

16. It is well established that UV transillumination of ethidium bromide-stained DNA can induce damage [15]. Therefore, we recommend use of SYBR Safe stain and a blue-light transilluminator for preparation of the library vector. If a SYBR safe stain and blue-light transilluminator are not available, a small aliquot (2–5 μL of the PCR product or 200–300 ng digested plasmid vector) should be electrophoresed on an agarose gel (without ethidium bromide) next to a DNA ladder, with the remainder of the PCR electrophoresed on the opposite side of the ladder. Following electrophoresis, the lanes corresponding to the small aliquot of DNA and the ladder should be separated from the rest of the agarose gel and stained in ethidium bromide (0.5 μg/ml) for 10 min. The location of the DNA fragment corresponding to the correctly sized PCR product can then be visualized by UV transillumination and should be marked using a clean scalpel. The ethidium-stained agarose gel fragment can then be lined up next to the unstained portion of the agarose gel and the location of the DNA fragment can be determined from the scalpel lines and excised for purification.
17. We have routinely found that using sterile water preheated to approximately 50 °C enhances the DNA yield when using DNA Clean & Concentrator columns.
18. Overlap extension PCR is generally robust over a wide range of starting DNA concentrations. As a starting guide, we typically add 10 ng of DNA for every 100 bp of the gene in question, e.g., an overlap PCR to reconstruct a 800 bp gene from fragment A (100 bp), fragment B (400 bp), and fragment C (300 bp) would comprise 10 ng of fragment A, 40 ng of fragment B, and 30 ng of fragment C.
19. It is theoretically possible that there could be parental gene contamination if column purification (rather than gel purification) has been used to purify the gene fragments in the overlap PCR, due to carryover from the PCR template used for fragment amplification in Subheading 3.2. In practice we have not found this to occur due to the very small amount of parental gene template added originally (we typically use 100 pg) and the small quantity of the amplified gene fragments used in the overlap PCR reaction. However, it is possible to test for this potential source of contamination by setting up a negative control including only one gene fragment or multiple negative controls containing different (incomplete) sets of gene fragments. PCRs such as these cannot reconstruct a full length gene; therefore, if a product corresponding to the size of the full length gene is seen in these controls, there may be parental gene contamination. If identified, the parental gene can be removed by digesting the PCR fragments with DpnI restriction enzyme as described by Copp et al. in Chapter 1 of this volume.

20. The 50 °C annealing temperature is a standard default parameter for PCR. The Tm of the shared region between fragments can be calculated by entering the sequence of the fragment overlap into standard online Tm calculating tools (*see* **Note 10**), and an annealing temperature 5 °C lower than this Tm should be used. An alternative is touchdown PCR, e.g., set the annealing temperature of the first PCR cycle at 65 °C and then decrease this by 1 °C for each of the 14 subsequent PCR cycles [16].
21. This extension time is for a 750 bp gene using Phusion® DNA polymerase; the reader should adapt the extension time for their gene according to the manufacturer's instructions for their polymerase of choice.
22. If the desired fragment is not observed following agarose gel electrophoresis of 3–5 μL of PCR product, then the following troubleshooting steps (1–4) can be undertaken. Before proceeding with optimization, we recommend running the remaining 45 μL of the overlap PCR reaction on an ethidium bromide agarose gel to confirm that the overlap PCR has not yielded a small amount of appropriately sized DNA product that was not visible when only 3–5 μL of PCR product was examined.
    1. Confirm that the primers have been designed to give a 15–20 nt overlap between each fragment.
    2. If column-purified fragments were initially used, repeat the gene fragment amplification (*see* Subheading 3.2) and individually gel purify each fragment before repeating the overlap PCR steps.
    3. Adjust the annealing temperature of the overlap extension PCR protocol, or use a touchdown PCR protocol. If no-DNA bands are visible on the agarose gel, we suggest lowering the annealing temperature(s) of the overlap extension step (Subheading 3.3, **step 3**) in the first instance. Conversely, if only nonspecific DNA bands were observed, try increasing the annealing temperature in the first instance.
    4. Multi-fragment overlap extension PCRs can potentially be optimized by annealing and amplifying neighboring fragments in a separate overlap extension PCR reaction, prior to combining the products of these reactions for additional cycles of overlap extension PCR. For example, if combining a gene from three neighboring fragments A, B, and C, first carry out two separate overlap extension PCR reactions to generate fragments AB and BC; and then use a third overlap extension PCR reaction to combine fragments AB and BC [17].

23. Inappropriately sized DNA bands and smears can appear in the overlap PCR as a result of mis-priming. If the appropriately sized band can still be seen as a distinct entity, then the most straightforward option is to gel purify this band (*see* **Note 16**). If this does not yield sufficient DNA for downstream cloning applications, optimization of the overlap PCR should be attempted (*see* **Note 22**).
24. It may be necessary to generate more DNA for sequence analysis if the DNA yield from the overlap PCR is low. If additional DNA is required, use 50 ng of the overlap PCR product as a PCR template and amplify this using the gene flanking primers in a standard PCR (25–50 μL) using a cost-effective master mix such as 2× BioMix Red (Bioline; Alexandria, NSW, Australia).
25. Prior to digestion, it is prudent to remove an aliquot (100–200 ng) of the overlap PCR product and store it at –20 °C. If downstream cloning steps are unsuccessful and more variant gene product is required, this can be achieved by using the stored aliquot as a template for high-fidelity PCR amplification of the full length gene.
26. It is likely that more digested gene insert will be generated than is required for the suggested ligation. We recommend scaling up the downstream steps to utilize the entire amount of recovered insert DNA, provided the reader is confident of the quality of the digested vector and competent cells. Alternatively, the purified digested insert can be stored at –20 °C for 1–2 weeks without loss of cloning efficiency.
27. Ideally, one plasmid-based primer (corresponding to the plasmid sequence directly upstream or downstream of the gene) and one complementary gene-specific primer should be used. This is essential if conducting SSM of a native *E. coli* gene.

## References


1. Lutz S (2010) Beyond directed evolution—semi-rational protein engineering and design. Curr Opin Biotechnol 21(6):734–743
2. Chica RA, Doucet N, Pelletier JN (2005) Semi-rational approaches to engineering enzyme activity: combining the benefits of directed evolution and rational design. Curr Opin Biotechnol 16(4):378–384
3. Patrick WM, Firth AE (2005) Strategies and computational tools for improving randomized protein libraries. Biomol Eng 22(4):105–112
4. Reetz MT, Kahakeaw D, Lohmer R (2008) Addressing the numbers problem in directed evolution. Chembiochem 9(11):1797–1804
5. Nov Y (2013) Fitness loss and library size determination in saturation mutagenesis. PLoS One 8(7):e68069
6. Swe PM, Copp JN, Green LK, Guise CP, Mowday AM, Smaill JB, Patterson AV, Ackerley DF (2012) Targeted mutagenesis of the Vibrio fischeri flavin reductase FRase I to improve activation of the anticancer prodrug CB1954. Biochem Pharmacol 84(6):775–783
7. Hogrefe HH, Cline J, Youngblood GL, Allen RM (2002) Creating randomized amino acid libraries with the QuikChange multi site-directed mutagenesis kit. Biotechniques 33(5):1158–1165

8. Liu H, Naismith JH (2008) An efficient one-step site-directed deletion, insertion, single and multiple-site plasmid mutagenesis protocol. BMC Biotechnol 8(1):91
9. Miyazaki K, Takenouchi M (2002) Creating random mutagenesis libraries using megaprimer PCR of whole plasmid. Biotechniques 33(5):1033–1038
10. Sambrook J, Russell DW (2001) Molecular cloning: a laboratory manual, 3rd edn. Cold Spring Harbour Laboratory Press, Cold Spring Harbour, NY
11. Reetz MT, Carballeira JD (2007) Iterative saturation mutagenesis (ISM) for rapid directed evolution of functional enzymes. Nat Protoc 2(4):891–903
12. Bryksin AV, Matsumura I (2010) Overlap extension PCR cloning: a simple and reliable way to create recombinant plasmids. Biotechniques 48(6):463–465
13. Peng RH, Xiong AS, Yao QH (2006) A direct and efficient PAGE-mediated overlap extension PCR method for gene multiple-site mutagenesis. Appl Microbiol Biotechnol 73(1):234–240
14. Georgescu R, Bandara G, Sun L (2003) Saturation mutagenesis. In: Arnold FH, Georgiou G (eds) Directed enzyme evolution: screening and selection methods, vol 231, Methods in molecular biology. Humana Press, Totowa, NJ, pp 75–83
15. Hartman PS (1991) Transillumination can profoundly reduce transformation frequencies. Biotechniques 11(6):747–748
16. Don RH, Cox PT, Wainwright BJ, Baker K, Mattick JS (1991) "Touchdown" PCR to circumvent spurious priming during gene amplification. Nucleic Acids Res 19(14): 4008
17. An Y, Ji J, Wu W, Lv A, Huang R, Wei Y (2005) A rapid and efficient method for multiple-site mutagenesis with a modified overlap extension PCR. Appl Microbiol Biotechnol 68(6):774–778

# Chapter 7

# Iterative Saturation Mutagenesis: A Powerful Approach to Engineer Proteins by Systematically Simulating Darwinian Evolution

Carlos G. Acevedo-Rocha, Sabrina Hoebenreich, and Manfred T. Reetz

## Abstract

Iterative saturation mutagenesis (ISM) is a widely applicable and powerful strategy for the efficient directed evolution of enzymes. First, one or more amino acid positions from the chosen enzyme are assigned to multi-residue sites (i.e., groups of amino acids or "multisites"). Then, the residues in each multisite are mutated with a user-defined randomization scheme to all canonical amino acids or a reduced amino acid alphabet. Subsequently, the genes of chosen variants (usually the best but not necessarily) are used as templates for saturation mutagenesis at other multisites, and the process is repeated until the desired degree of biocatalyst improvement has been achieved. Addressing multisites iteratively results in a so-called ISM scheme or tree with various upward branches or pathways. The systematic character of ISM simulates in vitro the natural process of Darwinian evolution: variation (library creation), selection (library screening), and amplification (template chosen for the next round of randomization). However, the main feature of ISM that distinguishes it from other directed evolution methods is the systematic probing of a defined segment of the protein sequence space, as it has been shown that ISM is much more efficient in terms of biocatalyst optimization than random methods such as error-prone PCR. In addition, ISM trees have also shed light on the emergence of epistasis, thereby rationally improving the strategies for evolving better enzymes. ISM was developed to improve catalytic properties such as rate, substrate scope, stereo- and regioselectivity using the Combinatorial Active-site Saturation Test (CAST), as well as chemical and thermal stability employing the B-Factor Iterative Test (B-FIT). However, ISM can also be invoked to manipulate such protein properties as binding affinity among other possibilities, including protein-protein interactions. Herein, we provide general guidelines for ISM, using CAST as the case study in the quest to enhance the activity and regioselectivity of the monooxygenase $P450_{BM3}$ toward testosterone.

**Key words** Protein engineering, Directed evolution, Saturation mutagenesis, Library quality, Library statistics, Screening effort, CAST, B-FIT, P450-BM3, Steroids

## 1 Introduction

Iterative saturation mutagenesis (ISM) has emerged as a rational, efficient, and powerful strategy in protein engineering [1]. When employing ISM in the quest to improve one or multiple enzyme properties, the first step involves the identification and selection

Elizabeth M.J. Gillam et al. (eds.), *Directed Evolution Library Creation: Methods and Protocols*, Methods in Molecular Biology, vol. 1179, DOI 10.1007/978-1-4939-1053-3_7, 

of appropriate residues. These residues can be randomized individually, or they can be grouped into multi-residue sites (hereafter, simply designated as multisites), which are then subjected to randomization. When addressing activity, substrate scope, stereo-, and/or regioselectivity, the structure-based Combinatorial Active-site Saturation Test (CAST) method is called for, in which user-defined amino acid residues lining or near the binding pocket are selected and grouped prior to saturation mutagenesis [2]. If the goal is to enhance thermal and organic solvent stability, the B-Factor Iterative Test (B-FIT) facilitates the selection of amino acid residues for randomization [3, 4]. After reasonable multisites have been defined, the respective mutant libraries are generated and screened for the enzyme property of interest.

Several computer tools are available free of charge [5, 6], which can be used for designing mutant libraries and for assessing their degree of screening effort. The screening effort can be further reduced in several ways: (a) mixing specific degenerate primers to remove codon redundancy [7, 8], (b) using reduced amino acid alphabets to limit sequence space [9–11], and (c) pooling library variants to save time and resources [12, 13]. Depending on the defined goal of the ISM project, any one or all of these approaches can be employed. Irrespective of the method for library creation, it is crucial to routinely perform a "Quick Quality Control" (QQC) by sequencing a portion of the library to show whether the expected diversity has been created, avoiding futile screening for a mutant that may not exist, thus, saving time and resources (*see* below) [7].

Upon screening one or more multisites (designated as A, B, C, D, etc.), the gene of an improved variant from the first generation is then used as template for randomization at other previously defined multisites. Addressing these multisites iteratively results in a so-called ISM scheme or tree with various upward branches or pathways, the number of which depends on how many defined multisites are selected for randomization (*see* Fig. 1).

The systematic character of ISM simulates in vitro the natural process of Darwinian evolution: variation (library creation), selection (library screening), and amplification (template chosen for the next round of randomization). However, the main feature of ISM that distinguishes it from other directed evolution methods is the systematic probing of a defined segment of the protein sequence space. It has been shown that ISM is much more efficient in terms of biocatalyst optimization than random methods such as error-prone PCR [1]. Furthermore, the exploration of full ISM trees has contributed to our understanding of laboratory evolution [14, 15]. However, it is not necessary to explore completely a given ISM tree, nor to climb to the top of each branch because it has been shown in a large number of studies that both arbitrarily chosen pathways and "fit" mutants found in early generations generally lead to successful identification of mutants that are sufficiently

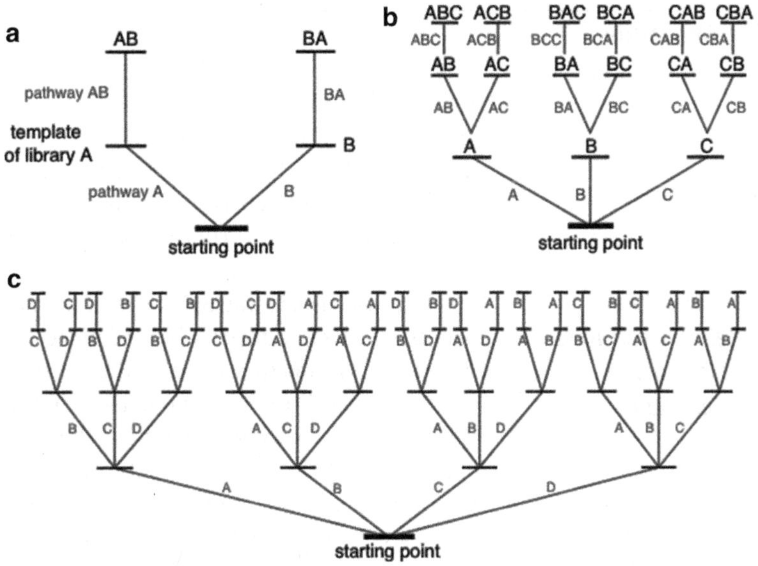


**Fig. 1** Illustration of ISM trees comprising (**a**) two (A/B), (**b**) three (A/B/C), or (**c**) four (A/B/C/D) multisites. The *black lines* represent the mutant chosen from the respective library used as template for the upward branch climbing (*gray lines*)

improved to be useful [1, 14]. Typical examples of ISM involve a variety of different enzyme types used for many practical biotransformations [16–24] or applications in metabolic engineering [17]. This method has been used for altering coenzyme-binding specificity [25] and for incorporating unnatural amino acids into proteins [26]. The ISM approach can also be used in other CAST-related structure-based approaches like the "Site-Saturation Mutagenesis of First-Sphere Residues" [27] or other methods combinatorially targeting the active site [28]. Importantly, it is not mandatory to have protein crystals or NMR structures when planning an ISM experiment because in a number of studies, 3D homology models have been used successfully to improve biocatalysts [1, 17, 18, 29] or other proteins [25]. Lastly, ISM has potential for the systematic exploration of large domain movements in allosteric enzymes [30] or for improving protein-protein interactions [31].

In this protocol, we provide general guidelines for ISM and exemplify the choice of important residues using CAST in the monooxygenase $P450_{BM3}$ from *Bacillus megaterium*, one of the most prominent candidates for practical biocatalytic transformations, in particular C–H bond activation and epoxidation of a wide range of substrates [32]. Wild-type (WT) $P450_{BM3}$ displays low activity and regioselectivity toward steroids, the hydroxylated products of which are useful intermediates for the synthesis of steroidal drugs [33]. For illustrative purposes, testosterone is chosen as the model substrate.

## 2 Materials

In general, good computer power with accessibility to the Internet and capability to install software is required. There are several indispensable tools for identification, selection, and randomization of amino acids as well as for library design and estimating screening effort (a combination thereof is advisable).

### 2.1 Structure Databases and Tools for 3D Protein Visualization

Software for visualizing 3D structures able to process PDB (Protein Data Base) files is indispensable (*see* **Note 1**). Structural 3D information can be found directly from the Research Collaboratory for Structural Bioinformatics Protein Data Bank (RCSB PDB) (http://www.rcsb.org/pdb/home/home.do) or from the Biological Magnetic Resonance Data Bank (BMR; http://www.bmrb.wisc.edu/). However, if a structure of the protein of interest is not available, a homology model can be built using various sources from the Protein Model Portal [34].

### 2.2 Tools for Comparing Protein Sequences (Optional)

In some cases, it is possible to gain knowledge about conserved or highly diverse amino acid residues using special "Multiple Sequence Alignment" (MSA) tools that are available at the EMBL (European Molecular Biology Laboratory) server (http://www.ebi.ac.uk/Tools/msa/) or from databases such as SALIGN (http://modbase.compbio.ucsf.edu/salign-cgi/index.cgi) [35]. The amino acid sequences of specific proteins can be also retrieved from the NCBI (National Center for Biotechnology Information) server (http://www.ncbi.nlm.nih.gov/).

### 2.3 Tools for Ligand Docking (Optional)

While some ligand docking tools are available free of charge (e.g., Autodock; http://autodock.scripps.edu/), other programs are license-based (e.g., Glide, Schrödinger, LLC). A comparison of different tools has been reported recently [36].

### 2.4 Randomization Scheme for Library Design

Computational tools for amino acid alphabet selection are available at:

- CASTER: http://www.kofo.mpg.de/media/2/D1108347/0987095526/ISM_tools.zip.
- AA-Calculator: http://guinevere.otago.ac.nz/cgi-bin/aef/AA-Calculator.pl.

### 2.5 Selection of Residues for Stability Improvement

Computational tools for selection of amino acids to target to improve stability are available at:

- B-FITTER: http://www.kofo.mpg.de/media/2/D1108347/0987095526/ISM_tools.zip.

### 2.6 Estimation of the Screening Effort of Designed Library

Various programs for calculating the size of libraries and the required screening effort can be used:

- CASTER: http://www.kofo.mpg.de/media/2/D1108347/0987095526/ISM_tools.zip.

- TopLib: http://stat.haifa.ac.il/~yuval/toplib/.
- GLUE-IT: http://guinevere.otago.ac.nz/cgi-bin/aef/glue-IT.pl.

## 3 Methods

The general ISM guidelines described below involve the following steps:

1. Identification and ranking of important residues.
2. Selection and grouping of target residues.
3. Choice of diversification method, screening effort, and amino acid alphabet.
4. The ISM process and its possible pitfalls.

Figure 2 summarizes these steps in a more detailed manner, and the application of the outlined ISM guidelines is demonstrated in the directed evolution of $P450_{BM3}$ in Subheading 3.5 below.

### *3.1 Identification of Important Residues*

Residue identification starts with the gathering of information; the more that is known about the function and structural relationships of the chosen enzyme, the more successful the outcome will be. The CAST approach facilitates the identification of residues involved in activity and stereoselectivity, whereas those involved in thermal stability and organic solvent tolerance can be found via the B-FIT method. In any case, the identification of important residues results in a large initial list of target residues; the number of target residues is thereafter rationally reduced until a feasible number is reached.

A reasonable strategy for information gathering is as follows. Start with review articles for a general overview of the enzyme if available. Deduce from the chemical reaction mechanism the residues that are directly involved in catalysis. Collect information about important residues, the so-called hot spots identified typically by previous mutagenesis studies as having a functional role, i.e., those playing a role in ligand anchoring or positioning, catalytic activity, stereoselectivity, substrate specificity, stability, etc. If, however, there is no such information on the residues, perform a Multiple Sequence Alignment (MSA) analysis to identify important ones. Of interest are those residues that are highly conserved but also diverse amino acid positions, especially if these are located in the binding site as in the case of CAST (*see* **Note 2**). Computational tools such as those mentioned in Chapter 20 of this volume can also aid to locate amino acids for targeting via ISM.

The next step is to gather or generate structural information. An available structure is the prime source of information for the selection of important residues. Structural data can be downloaded either from the RCSB PDB (crystal structures) or the BMR Data Bank

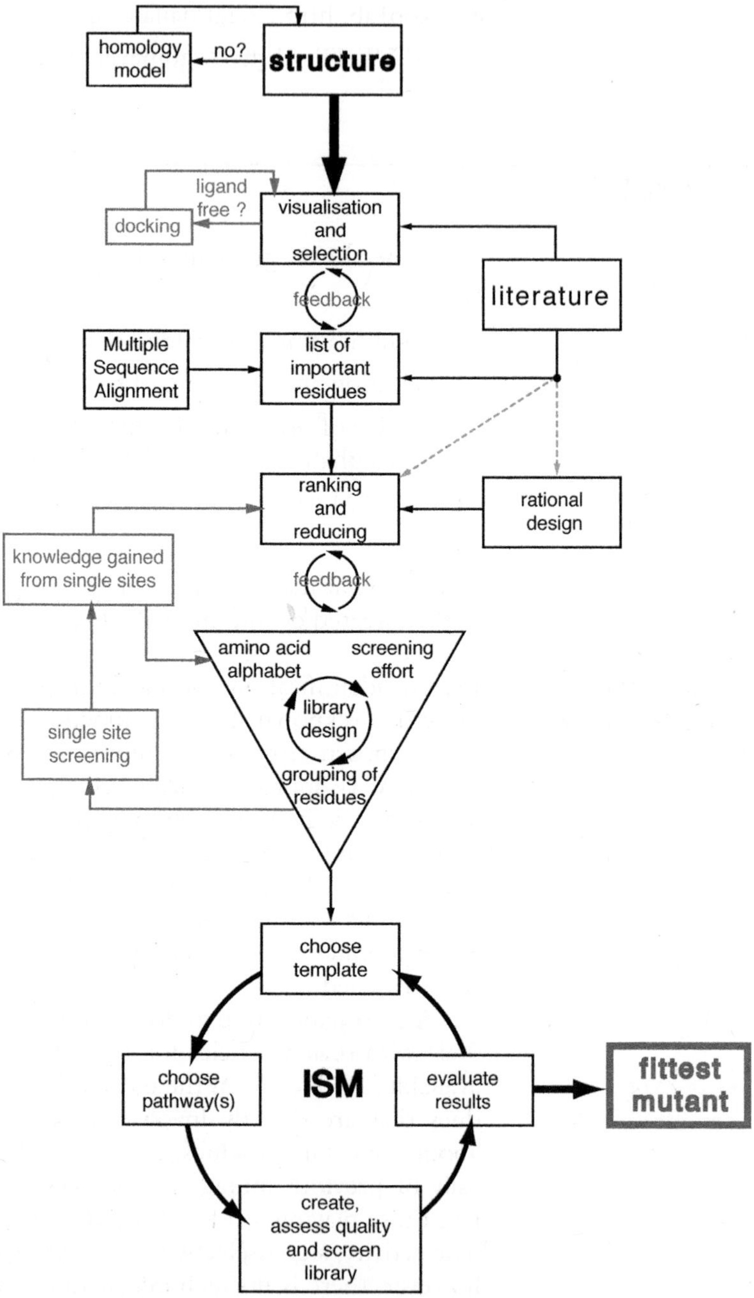


**Fig. 2** Flow diagram of structure-based directed evolution via ISM

(NMR structures). Alternatively, if there is no structure available, a 3D homology model can be generated. Finally, visualize and inspect the structural data using PyMOL or another similar program. Ligand-bound and/or ligand-free protein structures can be used since these are useful in different ways when selecting important residues via CAST or B-FIT.

*3.1.1 Combinatorial Active-Site Saturation Test (CAST)*

Ligand-bound structures are ideal for the identification of important residues. A ligand can be the natural substrate, an inhibitor, or any kind of small molecule that binds with specificity inside the binding pocket. Those residues in direct contact with ligands and closest to the catalytic center are of greatest interest (*see* **Note 3**). The residues in close proximity to the target substrate can also be identified via molecular docking (*see* **Note 4**); however, it is not mandatory to dock, since an empty binding pocket also allows the visual identification of important residues. Nevertheless, in two situations, molecular docking can provide additional information: if the target molecule is very different in shape and/or chemical properties from the bound ligand in a given structure or in cases when there are only ligand-free structures. In those cases in which more than one structure is available from the same enzyme or related family members, superimposing those structures is very useful in providing information about the side chains of those residues, which could show conformational changes, especially if these changes are induced by ligands (if present) of different chemical nature. In general, the residues forming the binding site should be considered, but depending on the shape, size, and electrostatic nature of the binding site, important residues can be also located relatively far away from the catalytic center (see **Note 5**). Whereas in small binding sites the number of target residues is likely to be limited to approximately 10 (especially those with contact to the ligand), in the case of deep cavities and large binding pockets, there may be 15–25 important residues or even more.

*3.1.2 B-Factor Iterative Test (B-FIT)*

In the case of B-FIT [3, 4], any type of structure is useful, whether ligand-bound or ligand-free, but it is important to have structures with good resolution for the identification of B-Factors available from X-ray data (*see* **Note 6**). The B-factors give a measure of the relative vibrational motion of atoms and thus, electron density. Because it is known that thermostable enzymes are usually rigid, targeting flexible residues displaying the highest B-factors can be a useful strategy to enhance the thermal stability or organic solvent tolerance of enzymes. The number of flexible residues can be very large depending on the enzyme, but those displaying the highest B-factors are generally preferred. As with CAST, the total number of important residues could be 10–25 or more.

***3.2 Ranking of Important Residues***

The maximal number of final target residues is limited by some practical constraints. The number of multisites and target residues per multisite is directly proportional to the library screening effort (*see* Subheading 3.3). In turn, the screening effort is often determined by the speed (low-, medium-, or high-throughput) of the assay method. Thus, the screening capacity and method will limit the maximal number of target residues. If many residues have been labeled as "important" ones, the assignment of scores for their ranking is very useful; subsequently, those residues with the lowest scores can be eliminated. The assignment of empirical scores to each residue is based on evidence from the literature and/or hints obtained from the structure.

#### 3.2.1 Evidence from the Literature

The following criteria can be used to weight each residue:

1. The residue was reported to enhance activity.
2. The residue was reported to influence substrate scope.
3. The residue was reported to act in determining stereoselectivity.
4. The residue was reported to enhance stability (thermo- and/or solvent stability).

By contrast, the residues directly involved in the reaction mechanism can be excluded (*see* **Note 7**).

#### 3.2.2 Hints from the Structure

The following criteria indicate residues should be prioritized for mutagenesis:

1. Residues in direct proximity to the catalytic center (approx. 4 Å or less)
2. Residues that are bulky or small (these are good targets to reshape the binding site significantly)
3. The residues proline and to lesser extent glycine that are located in loops (these residues have a high potential for reshaping the binding site)
4. Residues that have a high probability of functioning as a substrate anchor or positioner (especially those residues forming salt bridges, hydrogen bonds, and the different types of $\pi$-interactions with the substrate)

By contrast, residues for which the side chain is pointing away from the substrate may be less useful targets for mutagenesis, but their inclusion or exclusion may depend on the total number of target residues.

Each time a residue fulfills one or more of the aforementioned criteria, a point is assigned to it. Similarly, the more a given residue is reported in the literature (up to four times), the more points the residue should receive. After listing and allocating points to all of the important residues, these are ranked, and those with the lowest scores are eliminated. If two or more residues get the same score, the ISM project goal should be considered and those residues eliminated that are least likely to contribute to the enzyme parameter in question. For example, suppose that the goal is to enhance stereoselectivity and two residues were equally scored for stereoselectivity. If one has been additionally reported to influence substrate scope (evidence 2) despite being located far from the catalytic center, whereas the second residue is located very close to the catalytic center (hint 1), the former residue will be preferred for mutagenesis because experimental evidence should carry more weight than a hint. Upon reaching a reasonable number of target residues, these are grouped into multisites. These guidelines are the result from two decades of experience and represent the current state of the art. Nevertheless, successful ISM studies have been reported which did not fully comply with these guidelines [1].

### 3.3 Grouping of Target Residues

It is not mandatory to group target residues into multisites, as the targeting of individual residues has worked in many cases [16–21]. However, multisites including two to three amino acid residues are usually more successful in biocatalyst optimization [1, 22–24]. The grouping of residues strongly depends on their spatial distribution and the location of their corresponding codons along the gene sequence. Most important is the three-dimensional distribution because residues close in space have a high potential to act cooperatively rather than simply additively [1]. A limiting factor though is the location of the target codons relative to each other in the gene sequence. Therefore, the grouping of residues into multisites also depends on the library design. Library design involves the method to create diversity and a statistical model to assess the effort that will be required to screen (or select) a particular library. The screening effort in turn depends on the randomization scheme employed for saturation mutagenesis of the target residues.

#### 3.3.1 Creation of Sequence Diversity

Although manifold library creation techniques are available [37], the localization and number of target codons can impose time and/or economical constraints for the successful creation of one or more libraries with the aimed diversity. Whereas PCR-based methods are easy to use, emergent strategies based on DNA assembly may be more useful, especially when traditional methods such as QuikChange [38] fail to provide the desired diversity [39]. Furthermore, if many target codons that are considerably separated from each other along the gene sequence have to be randomized simultaneously, instead of doing stepwise time-consuming PCR-based strategies, it is advisable to look for methods that can introduce the desired mutations in a single step; for more details, *see* Chapter 13 on "Assembly of Designed Oligonucleotides" in this book. Selection systems based on conferring a growth advantage on the host organisms can entail very large numbers of variants [40], but in these cases, the generation of smart libraries should also be considered due to the physical limitations imposed by the amount of DNA and transformation efficiency [41].

#### 3.3.2 Assessment of the Screening Effort Required Using Statistical Models

Screening effort refers to the number of samples required to analyze partially or entirely the targeted protein sequence space. Due to the stochastic nature of the process, the screening effort of combinatorial libraries rises dramatically as the number of target residues, and of codons per residue, increases. One of the bottlenecks of directed evolution is therefore the "numbers problem" [10]. The statistical model evolved by Patrick and Firth for 95 % library coverage in a multisite experiment consisting, for example, of two residues targeted simultaneously with NNK/S (N = A/T/G/C; K = T/G and S = G/C) degeneracy predicts a screening effort of 3,068 transformants [5]. This number increases to 98,164 when the same probability is considered for a multisite of three residues [7]. This number may be too high for practical screening, as in the case of directed evolution of many stereoselective enzymes [1]. Nevertheless, numerous published ISM studies have yielded hits without reaching 95 % library coverage. In a different statistical approach described recently by Nov, the probability (also 95 %) of finding at least one out of a certain number of the most fit mutants is calculated [6]; this entails a dramatically lower screening effort for the same scenario as above (*see* Chapter 18 of this volume). If the aim of the study is to find a mutant that fulfills the given fitness threshold, for example, high enantioselectivity, and not to explore every mutant on the targeted sequence landscape, it is sufficient to apply this statistical approach [6]. In our experience, Nov's statistical approach is successful in achieving the set ISM goal independent of the enzyme or the randomization schemes (unpublished results). Ideally, it is desirable to obtain the fittest mutant, but any fit mutant is generally sufficient for a directed evolution study aimed at finding an improved mutant with minimal screening.

#### 3.3.3 Choosing the Amino Acid Alphabet

The most important strategy to reduce the screening effort is to limit the combinatorial space by using reduced amino acid alphabets [9, 10, 23]. This leads to the creation of focused, "smart" mutant libraries with a high hit rate. Most directed evolution studies including those based on ISM [16–18, 20–23] have used NNK/S degeneracy for library creation because the 20 canonical amino acids are encoded by a single codon, thus allowing full chemical diversity to be explored at the targeted position (*see* **Note 8**). However, since NNK/S encodes for 32 possible codons, several amino acids are overrepresented (e.g., Leu, Arg) while others are underrepresented (e.g., Met, Trp). This codon redundancy means that more efforts must be invested in screening, since the total number of codons and not amino acids determines the combinatorial space. Two approaches have been used to solve this problem, which arises from the natural degeneracy of the genetic code: (1) mix appropriate degenerate and/or nondegenerate codons to achieve a user-defined amino acid alphabet and/or (2) use predefined amino acid alphabets encoded in a single degenerate codon. Computational tools, such as CASTER and AA-Calculator, are available that facilitate the choice of degenerate codons for the latter approach (see Subheading 2 above). In both cases, it is important to be sure that there is no redundancy, i.e., only one codon specifies each amino acid. For example, the codon NDT (N = A/T/G/C; D = A/G/T) encodes without redundancy a set of 12 amino acids (Phe, Leu, Ile, Val, Tyr, His, Asn, Asp, Cys, Arg, Ser, and Gly) displaying a balanced mixture of chemical properties (large, small, hydrophobic, hydrophilic, polar, positively and negatively charged). The NDT degeneracy has been used with success in many examples [10, 42, 43]. However, if the desired amino acid(s) (based on previous information from the literature) is (are) not encoded by NDT, it is possible to mix specific primers [44] or use other reduced amino acid alphabets (*see* **Note 9**) [9–11]. Every ISM project is unique, and the amount of information available will vary from enzyme to enzyme. Consequently, the grouping and degeneracy of residues is very case dependent and usually differs between specific ISM projects. Therefore, different multisite grouping and amino acid alphabet randomization schemes are appropriate depending on the screening effort that can be invested (*see* Table 1).

**Table 1**
**Amino acid randomization schemes with varying numbers of residues per site or multisite, which are recommended for two different types of screening effort**

| Number of residues per site or multisite | Amino acid alphabet randomization (codons: amino acids) | Screening effort[a] | |
|---|---|---|---|
| | | 95 % library coverage[b] | 95 % probability of finding 1 of the 3 best[c] |
| 1[d, e] | 22c-Trick: NDT/VHG/TGG (22:20) | 63 | 20 |
| 2[d] | 22c-Trick: NDT/VHG/TGG (22:20) | 1,450 | 424 |
| 3[f] | NDT (12:12) | 5,176 | 1,726 |
| 4[g] | XXX (6:6) | 3,882 | 1,294 |
| 5[g] | YYY (4:4) | 3,067 | 1,023 |

[a]Assuming a perfect library creation with only the statistical expected distribution, which originates from the probabilistic nature of the process (*see* **Note 10**)
[b]These numbers represent a high probability to encounter 95 % of the generated mutants according to the by W. Patrick and A. Firth [5] and exemplified in [7]. These numbers can be also calculated using TopLib settings for discovering only the top variant [6]
[c]These numbers provide a 95 % probability to find at least one out of the three fittest mutants according to the statistical approach recently published by Y. Nov and used in TopLib [6]
[d]Instead of using NNN or NNK/S degeneracy, it is advisable to decrease the screening effort by reducing the genetic code redundancy using a combination of three [7] or four [8] primers
[e]If the researcher is interested in gaining catalytic information of a given residue for mechanistic or practical purposes, instead of just finding a fit mutant within a combinatorial library, site-directed mutagenesis to separately generate all 19 mutants is recommended
[f]NDT does not contain Ala and Pro, which are included in the equivalent NHC (H = A/C/T) alphabet. The latter might be more suitable for loop randomization
[g]There are various reduced, redundancy-free alphabets of four and six amino acids (*see* CASTER). If there is no suitable degenerate codon available encoding only the chosen reduced amino acid alphabet, the user can order them separately and mix in proper equimolar ratios

### 3.4 Iterative Saturation Mutagenesis (ISM)

#### 3.4.1 Template and Multisites: Where to Start?

Once the multisites have been defined (*see* **Note 11**), the user has to choose first the starting template. One of the best starting points is of course the WT enzyme, but enzymes with only a few (one or two) mutations can be used as well (*see* **Note 12**). Once the starting enzyme has been selected, the user has next to decide which pathway(s) to follow, remembering that the overall number of pathways increases exponentially with the number of multisites (Fig. 1). In our experience, it is advisable first to create the library of each multisite to gain information about its fitness (*see* **Note 13**). Importantly the Quick Quality Control (QQC) analysis should be applied to each library prior to its screening. The QQC consists of performing a single sequencing reaction on the pool of plasmids, followed by an assessment of the degree of degeneracy introduced into the target codon at the DNA level by comparing the observed and expected peak intensities of the DNA bases in the sequencing electropherograms (*see* **Note 14**) [7]. If the QQC corroborates the presence of the desired diversity, colony picking and library screening can be started. The libraries are then screened using a suitable and robust yet sensitive assay (*see* **Note 15**).

#### 3.4.2 First ISM Generation: How to Continue?

Usually, the multisite libraries of the first ISM generation will cover different degrees of fitness, most likely from minimal to large improvement. If an excellent hit fulfilling the target goal is identified in the first generation, further investment of effort may not be necessary. If this is not the case, however, each library may still include a group of somewhat improved mutants. This means that several mutants could be candidates for use as the template for the next ISM step. The sequencing of these mutants is important because it provides information, which will be necessary for making decisions when addressing other multisites in subsequent iterations of the evolution project; however, it is advisable to test the best mutants (10 or less) under controlled conditions prior to selecting one as a template (*see* **Note 16**). From the group of the best mutants, any one can be randomly chosen as the starting template for the next round as has been shown in a large number of ISM studies [1, 44, 13]. If fitness is defined by two catalytic parameters such as rate and stereoselectivity, it is advisable to make a compromise, i.e., not to choose the best mutant corresponding to only one of the properties (e.g., rate) but to select one that shows improvement in both parameters [13].

#### 3.4.3 Subsequent ISM Generation(s): Whither to Proceed?

Next comes the selection of the upward pathway(s). Arbitrarily chosen pathways have led to success [1, 14], but it is advisable to visit first the multisites with the highest individual fitness values found in the previous ISM generation(s), especially as found in the first generation (*see trajectory template* → C → CB → CBD → CBDA in Fig. 3). Although there is no guarantee that this assumption will always work, the rational use of the gathered knowledge

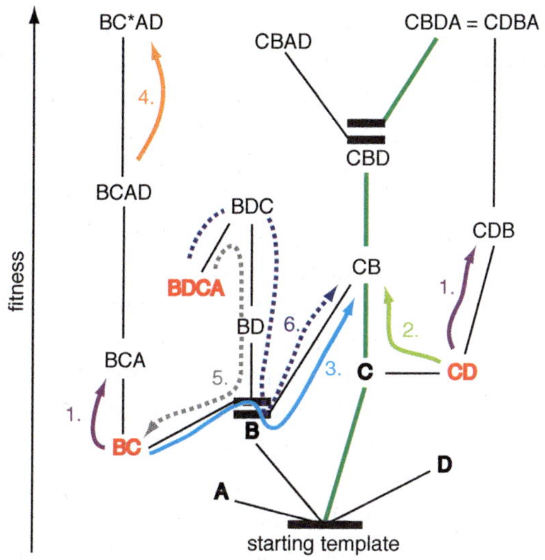


**Fig. 3** Hypothetical scenario of an ISM study with various degrees of fitness exemplifying the optional rescuing strategies (RS). *RS1*: if a local minimum is encountered in a given library (e.g., BC), choose any template from that library and address another multisite (e.g., A → pathway BCA). In another case, if exploring a given library (e.g., CD) results in no gain or loss of improvement, any template from that library can be used to continue upward (e.g., pathway CDB). *RS2*: if no more fit mutants are found in a given library (e.g., CD), return to the previous parental template and visit another multisite (e.g., B → pathway CB). *RS3*: if a local minimum is encountered in a given library (e.g., BC), return to the parental library and choose the second best or any less fit mutant and visit another multisite (e.g., C → pathway BC). *RS4*: if a complete branch has been explored without reaching the defined fitness value (e.g., BCAD), it is possible to revisit a previous multisite (e.g., C) to find better matching mutations (e.g., BC*AD). *RS5*: if a local minimum is encountered at a late stage in a given library (e.g., BDCA), step back two or more generations (e.g., until B) and visit another multisite (e.g., C → pathway BC). *RS6*: if a local minimum is encountered at a late stage in a given library (e.g., BDCA), step back two or more generations (e.g., until B), change template (e.g., second best), and visit another multisite (e.g., C → pathway BC). The *green branch* indicates the usual trajectory followed in many ISM studies, but it is not the only possibility for success

from the fitness landscape increases the probability of success. After deciding what pathway to follow for visiting a given multisite, another library must be created using the chosen mutant from the previous ISM generation as template. Again, it is important not to forget to apply the QQC to ensure that the desired diversity has been introduced at the DNA level. The reliable library can now be screened to find improved variants. If, however, in the chosen pathway a local minimum is encountered, i.e., when the screened library fails to contain an improved mutant, there is no reason to terminate the project at an unacceptable stage of catalyst improvement because ISM provides several rescuing strategies (*see below* Subheading 3.5).

#### 3.4.4 Complete ISM Branches and Trees: What About All Pathways?

It is not compulsory to explore either a complete branch or all pathways because success is often achieved in the first or second ISM generations [1, 44]. Nevertheless, the exploration of complete ISM trees has been extremely useful for the study of fitness landscapes, contributing greatly to our understanding of laboratory evolution and epistasis, the major driver of natural evolution [45]. Epistasis is a phenomenon that describes the interactions between sequence variations within the same gene (intramolecular) or between different genes (intermolecular) [46]. In intramolecular epistasis, i.e., proteins, the mutational effects may be additive or nonadditive. In additivity, the sum of the individual fitness values of the mutations equals the total fitness value of those together. In nonadditivity, the sum of the individual fitness values is less (antagonistic) or greater (cooperative) than the total. The exploration of ISM trees composed of multisites with two or three residues has shown that both additive and nonadditive pathways may be present along fitness landscapes [14, 15]. The complete exploration of ISM trees sheds light on the origin of epistasis, thereby rationally improving the strategies for evolving better enzymes.

### 3.5 Possible Pitfalls in ISM

#### 3.5.1 Lack of Diversity

The most serious pitfall in ISM and any other protein engineering or directed evolution study is the generation of a DNA library that lacks the desired combinatorial diversity. If diversity is not created, no oversampling will ever result in the identification of a fit mutant. For this purpose, the QQC was introduced [7] as a cheap, precautionary test that saves resources and avoids incorrect conclusions being drawn concerning the usefulness of targeting certain residues (*see* **Note 17**).

#### 3.5.2 Lack of Hits

If the QQC ensures that the library diversity is as expected, but no hits are found in the first generation even after investing considerable screening effort (*see* Table 1), it is likely that the multisite grouping and/or the amino acid alphabet is not optimal. If this happens, change the amino acid alphabet for randomization. If this does not improve the results, regroup and re-rank the target residues in appropriate multisites following the evidence and hints described above.

#### 3.5.3 Local Minima

Another potential pitfall is the encountering of a local minimum, which may occur at any step of the ISM process. If this happens, ISM provides several rescuing strategies (*see* Fig. 3):

1. Choose a non-improved or even inferior mutant as new template from the relevant library and continue upward to the next multisite (*see* **Note 18**).
2. Step back *one* generation, and follow a different upward pathway using the same template.
3. Step back *one* generation, and follow the same pathway but start from another template, for example, the second best mutant or even an inferior one.

4. Continue and revisit a previously addressed multisite.
5. Step back *two or more* generations without changing parental template(s) and follow another upward pathway.
6. Step back *two or more* generations, and choose another template and upward pathway.

It is important to remember that the likelihood of encountering local minima is decreased as more information becomes available, for example, after targeting all multisites during the first ISM generation.

#### 3.5.4 Other Potential Pitfalls

Other problems may appear during the screening process, for example, when attempting to evolve rate enhancement (*see* **Note 16**). Practical problems possibly caused by the physical properties of the plastic microtiter plate may have to be considered (*see* **Note 19**).

#### 3.5.5 Example of ISM Using CAST to Evolve $P450_{BM3}$

We recently reported the successful evolution of $P450_{BM3}$ mutants for the efficient stereoselective hydroxylation of steroids [44], a reaction that still represents a major challenge in organic synthesis [47]. In the following example, we refer to the previously mentioned study using testosterone as the target substrate. Keep in mind that the identification and selection of target residues in $P450_{BM3}$ are not trivial due to its large binding pocket and considerable large number of important residues.

The first step requires the identification of important residues. An excellent review about $P450_{BM3}$ is now available, including information about its mechanism, substrate acceptance, and mutations of amino acids that serve as a complementary guide for the identification of target residues [32]. For example, the residue most frequently mutated for many diverse purposes corresponds to Phe87, followed by Ala82, Ala328, Arg47, Leu75, Val78, Leu74, Leu188, Tyr51, Leu437, and many more, including Ala264 and Thr268. However, the two latter are catalytic residues involved in water and proton shuttling. In addition, Cys400 plays an important catalytic role, serving as the axial ligand of the heme cofactor. Therefore, Ala264, Thr268, and Cys400 were excluded as target residues. Although an MSA could be performed to identify "hot spots" using P450 gene sequences [48], structural alignments (*see* **Note 20**) are preferred because there is very little gene sequence conservation (15 %) among P450 family members [49].

The next step involves the collection of 3D structures. Using the UniProt accession ID of $P450_{BM3}$ (P14779), any database search will retrieve all its entries. As of November 2012, there were 52 crystal ligand-free or ligand-bound structures available (8 pending) in the PDB, from which 11 correspond to the WT and the rest to mutants. Superimposing all these 3D structures (*see* **Note 20**) provides a good overview about conformational changes of key residues occurring upon interaction with a ligand. However, since testosterone (steroid) differs significantly from the encountered

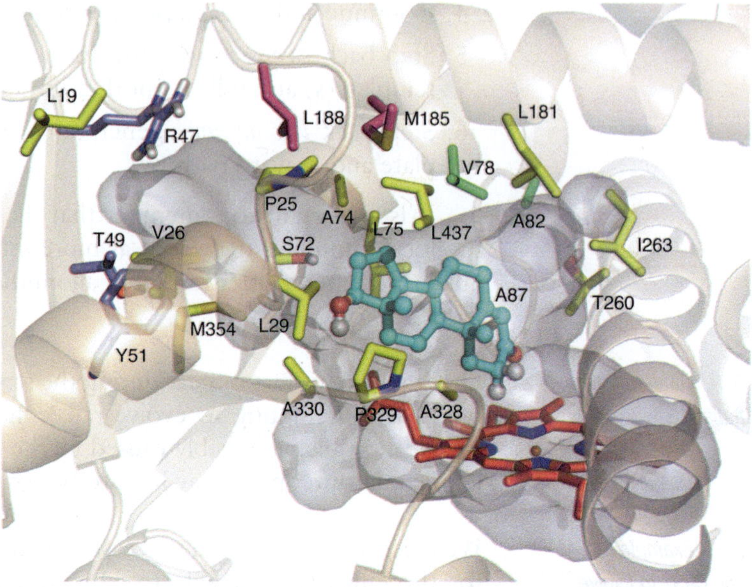


**Fig. 4** CASTing was performed on $P450_{BM3}$ mutant F87A as the starting point for the stereoselective hydroxylation of testosterone. The binding site of BM3 is relatively large, extending up to several angstroms in distance. The Cα atom of Arg47 and the iron atom are separated by a distance of 26 Å. The C-terminal β4 loop is not shown for clarity. Following the stated guidelines, residues of interest were identified, and the 23 residues ultimately selected are shown as sticks. The multi residue sites A (Arg47/Thr49/Tyr51), B (Val78/Ala82), and C (M185/L188) that were targeted are displayed in *blue*, *green*, and *pink*, respectively. The remaining residues are shown in *yellow* and F87A in *black* behind testosterone (cyan). A pose, potentially resulting in 2β-hydroxylation, is shown as an exemplar. The catalytic center of BM3 corresponds to the iron atom (*orange sphere*) bound within a porphyrin ring (*red*). The image was created with PyMOL Molecular Graphics System, version 1.5.0.4 (Schrödinger, LLC)

ligands (C-16 fatty acids and related derivatives), molecular docking with the target substrate to identify important residues was additionally performed (*see* **Note 21**). At the end, 28 residues of interest were identified from three different ligand-bound structures (*see* **Note 22** *and* Fig. 4).

Next comes the ranking of all residues. Table 2 lists the identified 28 residues of BM3, labeled as "important" in this example, together with their overall scores based on evidence from the literature and/or structural hints. The list was reduced to 23 residues by eliminating those involved in the mechanism and others with lower scores. Further reduction was difficult because many residues have been reported to be of functional importance (*see* Table 2).

Due to the outstanding importance of residue 87 in substrate scope and stereoselectivity, a single-site saturation library was created, from which F87A was identified as the most active

**Table 2**
**Listing of "important residues" and their ranking in descent order for the hydroxylation of testosterone**

| | | | Evidence from literature | | | | Structural hints | | | | | |
|---|---|---|---|---|---|---|---|---|---|---|---|---|
| Entry | Sequence order | Important residues | Activity | Stereo-governor | Substrate scope | Involved in catalysis | Within 4 Å of testosterone | Bulky or small | Proline and glycine | Anchors | Backbone interactions | Score |
| 1 | 13 | Phe87 | 3 | 5 | 5 | | 1 | 1 | 0 | 0 | | 15 |
| 2 | 12 | Ala82 | 2 | 2 | 3 | | 0 | 2 | 0 | 1 | | 10 |
| 3 | 5 | Arg47 | 1 | 0 | 3 | | 0 | 1 | 0 | 1 | | 6 |
| 4 | 6 | Tyr51 | 1 | 0 | 3 | | 0 | 1 | 0 | 1 | | 6 |
| 5 | 21 | Ala328 | 4 | | | | 1 | 1 | 0 | 0 | | 6 |
| 6 | 16 | Leu188 | 4 | | | | 0 | 1 | 0 | 0 | | 5 |
| 7 | 14 | Leu181 | 3 | | | | 1 | 0 | 0 | 0 | | 4 |
| 8 | 27 | Leu437 | 3 | | | | 1 | 0 | 0 | 0 | | 4 |
| 9 | 7 | Ser72 | 1 | | | | 1 | 1 | 0 | 1 | | 4 |
| 10 | 9 | Ala74 | 1 | | | | 1 | 1 | 0 | 0 | | 3 |
| 11 | 10 | Leu75 | 1 | | | | 1 | 1 | 0 | 0 | | 3 |
| 12 | 11 | Val78 | 1 | | | | 1 | 1 | 0 | 0 | | 3 |
| 13 | 18 | Ile263 | –1 | 2 | 1 | | 1 | 0 | 0 | 0 | | 3 |
| 14 | 23 | Ala330 | 0 | 0 | 1 | | 1 | 1 | 0 | 0 | | 3 |
| 15 | 2 | Pro25 | 1 | | | | 0 | 0 | 1 | 0 | | 2 |
| 16 | 17 | Thr260 | 1 | | | | 0 | 1 | 0 | 0 | | 2 |
| 17 | 22 | Pro329 | 0 | 0 | 0 | | 1 | 0 | 1 | 0 | | 2 |

(continued)

**Table 2 (continued)**

| Entry | Sequence order | Important residues | Evidence from literature | | | | Structural hints | | | | | Score |
|---|---|---|---|---|---|---|---|---|---|---|---|---|
| | | | Activity | Stereo-governor | Substrate scope | Involved in catalysis | Within 4 Å of testosterone | Bulky or small | Proline and glycine | Anchors | Backbone interactions | |
| 18 | 1 | Leu19 | 1 | | | | 0 | 0 | 0 | 0 | | 1 |
| 19 | 3 | Val26 | 1 | | | | 0 | 0 | 0 | 0 | | 1 |
| 20 | 4 | Leu29 | 1 | | | | 0 | 0 | 0 | 0 | | 1 |
| 21 | 15 | Met185 | 0 | 0 | 0 | | 0 | 1 | 0 | 0 | | 1 |
| 22 | 24 | Met354 | 0 | 0 | 0 | | 1 | 0 | 0 | 0 | | 1 |
| 23 | 8 | Gln73 | – | – | – | | – | – | – | – | –1 | –1 |
| 24 | 19 | Ala264 | – | – | – | –1 | – | – | – | – | | –1 |
| 25 | 20 | Thr268 | – | – | – | –1 | – | – | – | – | | –1 |
| 26 | 25 | Asp435 | – | – | – | | – | – | – | – | –1 | –1 |
| 27 | 26 | Thr436 | – | – | – | | – | – | – | – | –1 | –1 |
| 28 | 28 | THr438 | – | – | – | | – | – | – | – | –1 | –1 |

A total number of 28 "important residues" were listed (entry number). The order along the gene sequence is important for the creation of the mutant library. Based on literature evidences, one or more points were assigned to each residue. In some cases, the point is involved in more than one parameter or could not be clearly assigned. The most highly ranked residue is F87, but in the example described herein, F87A was used as the template. Those important residues listed as 1, 2, 3, 4, 6, 12, 21 and 23 to 28 were selected and excluded, respectively mutant for testosterone hydroxylation. The starting enzyme was therefore $P450_{BM3}$ F87A instead of the WT. From Table 2, it was decided to target Arg47 and Tyr51, due to their importance as governors of substrate scope [32], while residue A82 was chosen because it has been reported to greatly influence the reshaping of the active site to increase catalysis and stereoselectivity. Residue Leu188 was included due to its appearance in several useful, evolved mutants at that time [32]. Residues Thr49 and Met185 were added because these were considered to be important for enabling cooperative effects between residues Arg47 and Tyr51 (Thr49) and Leu188 (Met185), respectively. These decisions led to a total number of seven target residues, which were grouped into three multisites taking into account their relative physical proximity (Fig. 4): (A) Arg47/Thr49/Tyr51, (B) Val78/Ala82, and (C) Met185/Leu188. Coincidentally, the respective codons of the target residues are located close to each other along the gene sequence. Using a modified Mega-Primer method [50], a single degenerate primer was sufficient to create each respective library [44]. Whereas multisites A and C were randomized with the degenerate codon NDC (*see* Table 1), multisite B was fully randomized using NNK degeneracy (*see* **Note 23**). The successfully created libraries were screened with a medium-throughput HPLC screening system. Restricting the number of variants via a reduced amino acid alphabet (multisites A and C) led to lower screening effort of 8,800 colonies for the first generation (95 % library coverage).

The evolution of regio- and stereoselective mutants for the 2β- or 15β-hydroxylation of testosterone was achieved with high selectivity (94 or 92 %, respectively) at the first ISM stage [44]. Moreover, mutants with medium stereoselectivity for 16β-hydroxytestosterone (74 %) were also identified [51]. Although the mutant initially identified for 15β-hydroxytestosterone was already very selective, the exploration of pathway AB with a rationally reduced amino acid alphabet sufficed to identify a mutant with higher regio- and stereoselectivity for 15β hydroxylation (96 %) after screening only 600 additional transformants (*see* **Note 24**).

### 3.6 Conclusion

When using ISM, there is no general rule where to start and where to go, but the more information that is initially available concerning the system and that accumulates during the ISM process, in terms of key residues mutated to specific properties, the more successful the strategy becomes. ISM is one of the most rational ways to perform directed evolution, whether in the embodiments of CAST or B-FIT or when the focus is directed toward other protein properties.

## 4 Notes

1. Although PyMOL (Schrödinger, LLC) is usually the visualization program of choice, recent versions require a license. Nevertheless, older versions are available free of charge from the Internet. Similarly, the "Swiss-PdbViewer" is free of charge [52].
2. MSAs are a powerful approach by which to study gene families for which sequences are highly conserved. In fold-related families that are not related by gene sequence (i.e., those which arise by convergent evolution), the information content derived from an MSA may be limited.
3. Using PyMOL, these residues are found by selecting the ligand with the mouse, using the right click to open the menu and navigating to actions → around → residues around 4 Å. Repeat the procedure, but select the catalytic center to identify residues within about 4 Å distance of the catalytic center. A distance longer than 4 Å may be advisable in some cases, depending upon the enzyme.
4. Induced-fit docking into a substrate-bound structure is preferred. If conformational flexibility upon ligand binding is observed and if there are many structures available, multiple docking experiments should be performed with the different 3D structures. Perform a control by re-docking the observed ligand into the respective crystal structure to test if your calculations are correctly set up.
5. Normally, zones within 4–10 Å around the ligand are used to identify important residues.
6. If X-ray data is not available, error-prone PCR can be performed leading to the identification of putative hot spots of mutants that are more stable than WT. The mutated residues of less stable variants may also be hot spots. In both cases, the respective residues can then be grouped for subsequent ISM.
7. If the goal is to change or implement a reaction mechanism as in de novo protein design, the catalytic residue(s) can be replaced by other(s), for instance, amino acids serving as nucleophiles or involved in acid/base catalysis.
8. Several possibilities to encode all 20 amino acids are possible. When using NNN or NNK/S degenerate codons, only one single primer needs to be bought, whereas other techniques involve the purchase of many, three [7] or four [8], primers.
9. A particular reduced amino acid alphabet can be selected using two main approaches. One involves the generation of an MSA to identify "consensus-like mutations" at specific locations. Another one requires knowledge about the target molecule and binding site physicochemical properties, e.g., hydrophobic, polar, large, small, etc.

10. The library generation process follows rules from statistics and probability theory because primer annealing and colony picking are random sampling processes, among others. For a more detailed explanation, *see* [7] and references therein.

11. The formation of multisites is an important process, which may prove to be crucial. Sometimes, information regarding "hot spot" residues is available and should be exploited, but usually this is not the case. In such cases, it is advisable to first learn more about the system by performing, for example, single-site saturation mutagenesis with all 20 canonical amino acids at those target residues with highest interest. The screening of a few dozen to hundreds of mutants may give important information about the role of those particular residues. We have followed successfully this strategy in a number of studies (unpublished). Sometimes the availability of this additional information can be useful in regrouping the target residues in a different manner. It can also serve as a basis for the design/selection of an appropriate reduced amino acid alphabet.

12. Starting with highly evolved mutants carries the risk of remaining in a local fitness minimum. Crossing the high-energy barrier may be only possible, if at all, via rescuing strategy five or six (Fig. 3).

13. The exploration of all multisites is not mandatory, for example, if there are only two multisites and within one multisite there are one or more residues contributing to fitness (e.g., enhanced rate and substrate scope), it is advisable to start with this multisite.

14. The plasmids can be extracted from cells grown either in liquid or on solid cultures. In the former case, all the mutant plasmids transformed into the host and amplified after cell growth can be extracted in a straightforward manner. However, in the latter case, a more limited number of mutant plasmids corresponding to a partial aliquot of the total cell culture after transformation will be amplified upon host growth (unless the whole culture is plated onto an agar dish). In this case, usually 2 mL of water are added to the resulting microbial "carpet," which is then removed using a Drigalski spatula. The cells are collected by centrifugation and the plasmids extracted. In either case, only one sequencing reaction is needed. If the changes at the DNA level have been introduced correctly, the resulting electropherograms will display peak intensities with the same relative percentages of each base as expected. For example, if NNK was introduced at a given position, a perfect degeneracy will correspond to 25 % of each nucleotide (A/T/G/C) at both the first and second positions (N) and to 50 % of T/G at the third position (K). The observed percentages can be calculated from the intensities (height/area) of the respective peaks.

15. It is very important to generate a clear signal-to-noise ratio by controlling carefully the screening or selection conditions. Owing to the high standard deviations in multi-titer plate assays (10–25 %), a high signal-to-noise ratio will lead to easy loss of improved mutants in the background. Activity usually has a higher standard deviation than selectivity values. It is advisable to screen a plate containing 96 copies of the positive control to assess the case-specific error range for the user-designed screening assay.
16. The mutant actually performing best under screening conditions is not always the best one identified due to three possible reasons: first, screening assays have per se a high error range (*see* **Note 15**). Second, protein levels may vary significantly among expressed mutants and among clones in a plate, which can lead to incorrect conclusions especially when evolving enzyme activity. And third, large-scale process conditions often differ from screening conditions [53]. Therefore, it is recommended to test the mutants in a larger scale under controllable conditions [39]. This means working in a flask where cell growth, target enzyme induction, and concentration can be controlled carefully.
17. If the desired diversity is not achieved in a given library, the experimental conditions should be optimized. If the method relies on PCR, a critical parameter is the annealing temperature between the template and primers (*see* **Note 10**). For a detailed discussion *see* [7].
18. This strategy involves no additionally screening effort compared to the other rescuing strategies.
19. One example is high substrate hydrophobicity. Undesired adsorption to the plastic wall causes loss of reproducibility and a high frequency of false-negative results. In these cases, glass multi-titer plates or plates with glass inserts are recommended.
20. In PyMOL, the command "align_all" performs a sequence alignment followed by a structural alignment, while "super_all" will perform sequence-independent structure-based alignment of multiple objects to a reference object. Both require the installation of a PyMOL script, written by R.L. Campbell (http://pldserver1.biochem.queensu.ca/~rlc/work/pymol/).
21. The mutant structure was manually created from PDB entry 1JPZ, chain B, using Maestro and refined with the Protein Preparation Wizard (Schrödinger, LLC). Testosterone was docked using Glide with a modified force field for the heme cofactor [54]. Docking resulted in several poses, from which one is shown in Fig. 4.

22. Structures 3BEN, 1JPZ, and 1FAG (from each chain A) were superimposed with super_all. About 28 residues were located within a distance of 4 Å around the three ligands (omega-imidazolyl fatty acid, N-palmitoylglycine, and palmitoleic acid, respectively). From these residues, the following were found in 4 Å distance to the docked testosterone: Ser72, Ala74, Leu75, Val78, Phe87, Leu181, Ile263, Ala328, Pro329, Ala330, Met354, and Leu437. Residues Ala264 and Thr268 were excluded on basis of their involvement in the reaction mechanism, whereas residues Gln73, Glu435, Thr436, and Thr438 were eliminated from the list due to the fact that they only made contacts with the ligands via atoms in the peptide backbone, which might be expected to be similar regardless of the residues inserted at those positions. The remaining residues are Leu19, Pro25, Val26, Leu29, Arg47, Tyr51, Ala82, Met185, Leu188, and Thr260. Thr49 was added to the list due to its close proximity to Arg47 and Tyr51 to enable cooperative effects among these residues. This procedure resulted in a list of 23 important residues.
23. Remember that using NNN or NNK/S is no longer state of the art: it is possible to reduce or totally eliminate the redundancy of the genetic code by mixing a series of degenerate and nondegenerate codons to achieve a ratio of 22 to 20 [7] or 20 to 20 [8] codons to amino acids, respectively.
24. A highly active, but unselective mutant from library A was chosen as template for addressing multisite B, which was found to yield the most 15β-selective variants, especially when bulky aliphatic and aromatic residues were present in residue 82. Therefore, a reduced amino acid alphabet for position 82 consisting of Gly, Val, Leu, Met, Phe, and Trp as well as NNK degeneracy for position 78 was used. Starting from a template of library A and decreasing rationally, the diversity reduced the screening effort from 5,100 to 600 samples, using 95 % library covering statistics.

## Acknowledgments

We thank Rubén Agudo-Torres, Gheorghe-Doru Roiban, Zhi-Gang Zhang, and Pamela Torres-Salas for constructive discussions and the critical review of this manuscript. We are also grateful to the editors for their excellent editorial work. This work was supported by the Max Planck Society and the Arthur C. Cope Foundation.

## References


1. Reetz MT (2011) Laboratory evolution of stereoselective enzymes: a prolific source of catalysts for asymmetric reactions. Angew Chem Int Ed Engl 50(1):138–174
2. Reetz MT, Bocola M, Carballeira JD, Zha D, Vogel A (2005) Expanding the range of substrate acceptance of enzymes: combinatorial active-site saturation test. Angew Chem Int Ed Engl 44(27):4192–4196
3. Reetz MT, Carballeira JD (2007) Iterative saturation mutagenesis (ISM) for rapid directed evolution of functional enzymes. Nat Protoc 2(4):891–903
4. Reetz MT, Soni P, Fernandez L, Gumulya Y, Carballeira JD (2010) Increasing the stability of an enzyme toward hostile organic solvents by directed evolution based on iterative saturation mutagenesis using the B-FIT method. Chem Commun (Camb) 46(45):8657–8658
5. Patrick WM, Firth AE (2005) Strategies and computational tools for improving randomized protein libraries. Biomol Eng 22(4):105–112
6. Nov Y (2012) When second best is good enough: another probabilistic look at saturation mutagenesis. Appl Environ Microbiol 78(1):258–262
7. Kille S, Acevedo-Rocha CG, Parra LP, Zhang ZG, Opperman DJ, Reetz MT, Acevedo JP (2013) Reducing codon redundancy and screening effort of combinatorial protein libraries created by saturation mutagenesis. ACS Synth Biol. http://pubs.acs.org/doi/abs/10.1021/sb300037w
8. Tang L, Gao H, Zhu X, Wang X, Zhou M, Jiang R (2012) Construction of "small-intelligent" focused mutagenesis libraries using well-designed combinatorial degenerate primers. Biotechniques 52(3):149–158
9. Reetz MT, Wu S (2008) Greatly reduced amino acid alphabets in directed evolution: making the right choice for saturation mutagenesis at homologous enzyme positions. Chem Commun (Camb) 43:5499–5501
10. Reetz MT, Kahakeaw D, Lohmer R (2008) Addressing the numbers problem in directed evolution. Chembiochem 9(11):1797–1804
11. Chaparro-Riggers JF, Polizzi KM, Bommarius AS (2007) Better library design: data-driven protein engineering. Biotechnol J 2(2):180–191
12. Polizzi KM, Parikh M, Spencer CU, Matsumura I, Lee JH, Realff MJ, Bommarius AS (2006) Pooling for improved screening of combinatorial libraries for directed evolution. Biotechnol Prog 22(4):961–967
13. Bougioukou DJ, Kille S, Taglieber A, Reetz MT (2009) Directed evolution of an enantioselective enoate-reductase: testing the utility of iterative saturation mutagenesis. Adv Synth Catal 351(18):3287–3305
14. Gumulya Y, Sanchis J, Reetz MT (2012) Many pathways in laboratory evolution can lead to improved enzymes: how to escape from local minima. Chembiochem 13(7):1060–1066
15. Reetz MT, Sanchis J (2008) Constructing and analyzing the fitness landscape of an experimental evolutionary process. Chembiochem 9(14):2260–2267
16. De Groeve MR, De Baere M, Hoflack L, Desmet T, Vandamme EJ, Soetaert W (2009) Creating lactose phosphorylase enzymes by directed evolution of cellobiose phosphorylase. Protein Eng Des Sel 22(7):393–399
17. Bhuiya MW, Liu CJ (2010) Engineering monolignol 4-O-methyltransferases to modulate lignin biosynthesis. J Biol Chem 285(1):277–285
18. Tang WL, Li Z, Zhao H (2010) Inverting the enantioselectivity of P450pyr monooxygenase by directed evolution. Chem Commun (Camb) 46(30):5461–5463
19. Norrgard MA, Mannervik B (2011) Engineering GST M2-2 for high activity with indene 1,2-oxide and indication of an H-site residue sustaining catalytic promiscuity. J Mol Biol 412(1):111–120
20. Blum JK, Ricketts MD, Bommarius AS (2012) Improved thermostability of AEH by combining B-FIT analysis and structure-guided consensus method. J Biotechnol 160(3–4):214–221
21. Wen S, Tan T, Zhao H (2012) Improving the thermostability of lipase Lip2 from Yarrowia lipolytica. J Biotechnol 164(2):248–253
22. Okrasa K, Levy C, Wilding M, Goodall M, Baudendistel N, Hauer B, Leys D, Micklefield J (2009) Structure-guided directed evolution of alkenyl and arylmalonate decarboxylases. Angew Chem Int Ed Engl 48(41):7691–7694
23. Jochens H, Aerts D, Bornscheuer UT (2010) Thermostabilization of an esterase by alignment-guided focussed directed evolution. Protein Eng Des Sel 23(12):903–909
24. Engstrom K, Nyhlen J, Sandstrom AG, Backvall JE (2010) Directed evolution of an enantioselective lipase with broad substrate scope for hydrolysis of alpha-substituted esters. J Am Chem Soc 132(20):7038–7042
25. Liang L, Zhang J, Lin Z (2007) Altering coenzyme specificity of Pichia stipitis xylose

reductase by the semi-rational approach CASTing. Microb Cell Fact 6:36

26. Kaya E, Vrabel M, Deiml C, Prill S, Fluxa VS, Carell T (2012) A genetically encoded norbornene amino acid for the mild and selective modification of proteins in a copper-free click reaction. Angew Chem Int Ed Engl 51(18): 4466–4469
27. Zhang K, Shafer BM, Demars MD, Stern HA, Fasan R (2012) Controlled oxidation of remote sp(3) C-H bonds in artemisinin via P450 catalysts with fine-tuned regio- and stereoselectivity. J Am Chem Soc 134(45):18695–18704
28. Chen MM, Snow CD, Vizcarra CL, Mayo SL, Arnold FH (2012) Comparison of random mutagenesis and semi-rational designed libraries for improved cytochrome P450 BM3-catalyzed hydroxylation of small alkanes. Protein Eng Des Sel 25(4):171–178
29. Nair NU, Zhao H (2008) Evolution in reverse: engineering a D-xylose-specific xylose reductase. Chembiochem 9(8):1213–1215
30. Wu S, Acevedo JP, Reetz MT (2010) Induced allostery in the directed evolution of an enantioselective Baeyer-Villiger monooxygenase. Proc Natl Acad Sci U S A 107(7): 2775–2780
31. Bosshart A, Panke S, Bechtold M (2013) Systematic optimization of interface interactions increases the thermostability of a multimeric enzyme. Angew Chem Int Ed Engl 52(37):9673–9676
32. Whitehouse CJ, Bell SG, Wong LL (2012) P450(BM3) (CYP102A1): connecting the dots. Chem Soc Rev 41(3):1218–1260
33. Fernandes P, Cruz A, Angelova B, Pinheiro HM, Cabral JMS (2003) Microbial conversion of steroid compounds: recent developments. Enzyme Microb Technol 32(6):688–705
34. Arnold K, Kiefer F, Kopp J, Battey JN, Podvinec M, Westbrook JD, Berman HM, Bordoli L, Schwede T (2009) The protein model portal. J Struct Funct Genomics 10(1):1–8
35. Madhusudhan MS, Webb BM, Marti-Renom MA, Eswar N, Sali A (2009) Alignment of multiple protein structures based on sequence and structure features. Protein Eng Des Sel 22(9):569–574
36. Plewczynski D, Lazniewski M, Augustyniak R, Ginalski K (2011) Can we trust docking results? Evaluation of seven commonly used programs on PDBbind database. J Comput Chem 32(4):742–755
37. Siloto RMP, Weselake RJ (2012) Site saturation mutagenesis: methods and applications in protein engineering. Biocatal Agric Biotechnol 1(3):181–189
38. Papworth C, Bauer JC, Braman J, Wright DA (1996) Site-directed mutagenesis in one day with >80 % efficiency. Strategies 9(3):3–4
39. Agudo R, Roiban GD, Reetz MT (2012) Achieving regio- and enantioselectivity of P450-catalyzed oxidative CH activation of small functionalized molecules by structure-guided directed evolution. Chembiochem 13(10):1465–1473
40. Taylor SV, Kast P, Hilvert D (2001) Investigating and engineering enzymes by genetic selection. Angew Chem Int Ed Engl 40(18):3310–3335
41. Denault M, Pelletier JN (2007) Protein library design and screening: working out the probabilities. Methods Mol Biol 352:127–154
42. Liu J, Cropp TA (2012) A method for multi-codon scanning mutagenesis of proteins based on asymmetric transposons. Protein Eng Des Sel 25(2):67–72
43. Bonnet J, Subsoontorn P, Endy D (2012) Rewritable digital data storage in live cells via engineered control of recombination directionality. Proc Natl Acad Sci U S A 109(23): 8884–8889
44. Kille S, Zilly FE, Acevedo JP, Reetz MT (2011) Regio- and stereoselectivity of P450-catalysed hydroxylation of steroids controlled by laboratory evolution. Nat Chem 3(9):738–743
45. Breen MS, Kemena C, Vlasov PK, Notredame C, Kondrashov FA (2012) Epistasis as the primary factor in molecular evolution. Nature 490(7421):535–538
46. Lehner B (2011) Molecular mechanisms of epistasis within and between genes. Trends Genet 27(8):323–331
47. Wong LL (2011) P450(BM3) on steroids: the Swiss Army knife P450 enzyme just gets better. Chembiochem 12(17):2537–2539
48. Nelson DR (2009) The cytochrome p450 homepage. Hum Genomics 4(1):59–65
49. Munro AW, Leys DG, McLean KJ, Marshall KR, Ost TW, Daff S, Miles CS, Chapman SK, Lysek DA, Moser CC, Page CC, Dutton PL (2002) P450 BM3: the very model of a modern flavocytochrome. Trends Biochem Sci 27(5):250–257
50. Sanchis J, Fernandez L, Carballeira JD, Drone J, Gumulya Y, Hobenreich H, Kahakeaw D, Kille S, Lohmer R, Peyralans JJ, Podtetenieff J, Prasad S, Soni P, Taglieber A, Wu S, Zilly FE, Reetz MT (2008) Improved PCR method for the creation of saturation mutagenesis

libraries in directed evolution: application to difficult-to-amplify templates. Appl Microbiol Biotechnol 81(2):387–397
51. Kille S (2010) Flavoproteins in directed evolution: iterative CASTing to evolve YqjM and P450-BM3. Ruhr-Universität Bochum, Bochum
52. Guex N, Peitsch MC (1997) SWISS-MODEL and the Swiss-PdbViewer: an environment for comparative protein modeling. Electrophoresis 18(15):2714–2723
53. Michener JK, Smolke CD (2012) High-throughput enzyme evolution in Saccharomyces cerevisiae using a synthetic RNA switch. Metab Eng 14(4):306–316
54. Seifert A, Tatzel S, Schmid RD, Pleiss J (2006) Multiple molecular dynamics simulations of human p450 monooxygenase CYP2C9: the molecular basis of substrate binding and regioselectivity toward warfarin. Proteins 64(1):147–155

# Chapter 8

# Generating Targeted Libraries by the Combinatorial Incorporation of Synthetic Oligonucleotides During Gene Shuffling (ISOR)

**Liat Rockah-Shmuel, Dan S. Tawfik, and Moshe Goldsmith**

## Abstract

Protein engineering by directed evolution relies on the use of libraries enriched with beneficial variants. Such libraries should explore large mutational diversities while avoiding high loads of deleterious mutations. Here we describe a simple protocol for incorporating synthetic oligonucleotides that encode designed, site-specific mutations by assembly PCR. This protocol enables a researcher to "hedge the bets," namely, to explore a large number of potentially beneficial mutations in a combinatorial manner such that individual library variants carry a limited number of mutations.

**Key words** Directed evolution, Protein engineering, Targeted mutations, Combinatorial library, Designed library, Saturation library, Gene shuffling

## 1 Introduction

Selections and screens of collections of randomly mutated variants are still the methods of choice for many laboratory evolution experiments aimed at protein engineering [1–3]. However, several factors account for the fact that the frequency of "hits" in random libraries tends to be very low (for details, *see* ref. 4), the frequency of beneficial mutations in randomly mutated variants is very low (in the order of $10^{-3}$), combinations of multiple beneficial mutations are often needed to improve a property, and >40 % of mutations have deleterious effects. Random mutagenesis libraries, therefore, demand exploration by high throughput selections or screens that are not applicable in many cases.

Rational and computational library designs can help focus mutagenesis to smaller regions of the protein sequence in order to improve the likelihood of finding a beneficial variant. Although such methods greatly minimize the number of sequence permutations that need to be explored, they are hampered by the complexity of

Elizabeth M.J. Gillam et al. (eds.), *Directed Evolution Library Creation: Methods and Protocols*, Methods in Molecular Biology, vol. 1179, DOI 10.1007/978-1-4939-1053-3_8, © Springer Science+Business Media New York 2014

proteins and our limited knowledge regarding sequence–function relationships. The nonadditive or epistatic effects of combined mutations further complicate the search [5]. In particular, many mutations may be neutral or even deleterious on their own, but can become advantageous when combined with others.

To enable the simultaneous exploration of a large number of positions, thus increasing the chances of obtaining variants with beneficial combinations of mutations while at the same time restricting the number of mutants to those that can be screened with low- to medium-throughput assays, targeted mutagenesis libraries are created in a combinatorial manner [4]. A convenient way of making such libraries is to "spike" genes with oligonucleotides encoding randomized substitutions during gene shuffling [6]. In this way, a large number of positions can be explored (as many as 50 in a 300 amino acid protein), with individual library variants containing only few mutations (e.g., *see* refs. 2, 7–10). Each library variant carries a random and different subset of mutated residues, while the entire set of mutations and targeted positions is represented in the complete library. The aim of this method, dubbed "Incorporating Synthetic Oligonucleotides via Gene Reassembly" (ISOR), is to generate variants with random combinations of substitutions at a certain frequency, in a facile and general manner. In fact, it is a simple variation of the gene shuffling protocol [11], modified from a previous version [6] and thus allows diversification by substitutions, insertions, or deletions of large sets of predetermined residues.

A description of library design methods is beyond the scope of this chapter. It is important, however, to indicate that the design of small yet smart libraries is based on structural and sequence information and sometimes on computation of active-site designs. These libraries are characterized by a large fraction of active variants despite the presence of a surprisingly large number of active-site mutations (e.g., *see* refs. 2, 7, 12–16). Such libraries can yield results that are comparable to several rounds of evolution with conventional random mutagenesis libraries [14]. A crucial factor in library design is the mutational load, i.e., the number of mutated residues per gene. This and other design factors are discussed in detail in ref. [4].

## 2 Materials

Prepare all solutions using ultrapure water (deionized water with a resistivity of 18 MΩ cm at 25 °C) and analytical grade reagents. Prepare and store all reagents at room temperature unless indicated otherwise.

#### 2.1 Reagents for DNA Preparation

1. Gene of interest in a vector or other DNA template.
2. DpnI restriction endonuclease (e.g., New England Biolabs, Beverley, MA, USA).
3. PCR purification kit (e.g., Qiagen, QIAquick PCR Purification Kit, Hilden, GMBH).

#### 2.2 Solutions and Equipment Required for Digestion

1. Tris–HCl: 1 M pH 7.5.
2. $MnCl_2$: 100 mM (*see* **Note 1**).
3. DNaseI (e.g., Takara Bio Inc., Shiga, Japan) diluted to 0.05 U/μl (*see* **Note 2**).
4. EDTA: 500 mM pH 8.2 (*see* **Note 1**).
5. Stopwatch.
6. A 37 °C Thermoblock (*see* **Note 3**).
7. A 80 °C Thermoblock (*see* **Note 3**).

#### 2.3 Solutions and Equipment Required for Purification of Fragments

1. 5× DNA loading buffer: 30 % (v/v) glycerol, 0.1 M EDTA, pH 8, 0.25 % (w/v) xylene cyanol FF (*see* **Note 4**).
2. DNA ladder (e.g., GeneRuler 50 bp DNA Ladder, ready-to-use, 50–1,000 bp, Thermo Scientific, Pittsburgh PA, USA).
3. TAE buffer: 40 mM Tris–HCl, 20 mM acetic acid, and 1 mM EDTA, final pH 8.0.
4. A 2 % (w/v) agarose gel prepared in TAE buffer, stained with ethidium bromide (0.5 μg/ml). Use medium-size wells (enough to hold 100 μl of solution).
5. DNA electrophoresis device.
6. UV transilluminator.
7. Gel extraction kit (e.g., 3.5 kDa MWCO D-tube dialyzers from Novagen™ for electroelution).

#### 2.4 Solutions and Equipment Required for Assembly PCR

1. Purified DNA fragments from Subheading 2.3.
2. All spiking synthetic oligonucleotides introducing the diversified residues at appropriate concentrations (*see* **Note 5**).
3. High-fidelity polymerase (e.g., PFU Ultra II-Agilent™ or KOD—EMD Millipore™) and polymerase reaction buffers as recommended and/or supplied by the manufacturer.
4. dNTPs.
5. PCR thermocycler.

## 3 Methods

Carry out all procedures at room temperature unless otherwise specified.

### 3.1 DNA Preparation

1. Amplify your gene of interest by standard PCR using external primers to achieve about 25 μg of starting DNA (*see* **Notes 6** and 7).
2. Treat with 20 units of DpnI (NEB™) per 50 μl of PCR for 1 h at 37 °C to remove the residual template plasmid DNA.
3. Purify and concentrate the DNA to a final concentration of ≥200 ng/μl, using the PCR purification kit, according to the manufacturer's instructions.

### 3.2 Gene Fragmentation

1. Prepare the reaction solution: In a 500 μl tube, mix 20–25 μg DNA with Tris–HCl and $MnCl_2$ to generate a 200 μl volume solution containing 100 mM Tris–HCl pH 7.5 and 10 mM $MnCl_2$. Preheat to 37 °C for 5 min using a heating block or PCR machine (*see* **Note 3**).
2. Prepare stop solutions: Four microtubes (200 μl) containing 25 μl of 500 mM EDTA pH 8.2 equilibrated to 85 °C in a heating block (*see* **Note 3**).
3. Add the dilute DNaseI to the reaction solution tube (kept at 37 °C) to a final concentration of 0.015 U/μg of DNA (e.g., Table 1), start the stopwatch, mix, and incubate at 37 °C (*see* **Note 8**).
4. Remove 50 μl aliquots every 0.5 min (i.e., after 0.5, 1, 1.5, and 2 min) and mix each immediately with an aliquot of the EDTA stop solution kept at 85 °C.
5. Incubate the tubes at 85 °C for 15 min.
6. Let the samples cool down and then add DNA loading buffer (*see* **Note 4**).
7. Electrophorese the samples on a 2 % (w/v) agarose-TAE gel next to a DNA ladder comprised of 50 or 100 bp bands.

**Table 1**
**Example of a DNaseI digestion reaction mixture**

| Component | Volume (μl) | Final |
|---|---|---|
| PCR DNA mix 219 ng/μl | 91.4 | 20 μg |
| Tris/Cl 1 M pH 7.5 | 20 | 0.1 M |
| $MnCl_2$ 155 mM | 13 | 10 mM |
| Dilute DNaseI 0.05 U/μl | 6 | 0.015 U/μg DNA |
| Sterile, ultrapure water | 69.6 | – |
| Total | 200 | |

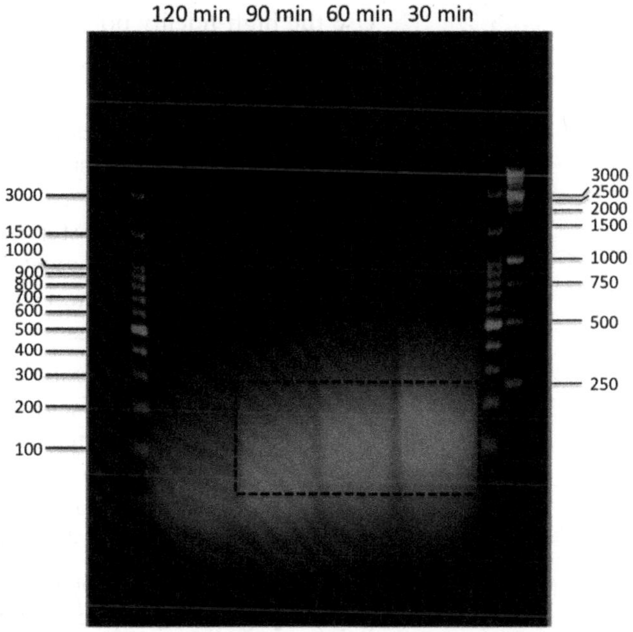


**Fig. 1** Agarose gel electrophoresis of PON1 DNA following DNaseI digestion. The genes of nine rePON1 variants (*see* ref. 2) were PCR amplified and mixed in an equimolar ratio (20 μg total DNA). The mixture was digested with DNaseI (0.015 U/μg) according to the protocol detailed in this chapter, and then samples were taken at four time points (30, 60, 90, and 120 s—as indicated above the lanes), inactivated with EDTA at 85 °C, and separated by electrophoresis on a 2 % (w/v) TAE agarose gel stained with ethidium bromide. Molecular weight markers of 100 bp were run in the two lanes flanking the fragments. A 1 kb molecular weight marker was run in the *right lane*. The *dashed square* indicates the region containing 50–250 bp DNA fragments that was removed from the gel for extraction of the fragments by electroelution

Each time point sample should be run in a different well. This is done in order to evaluate the best digestion conditions (*see* **Note 9**). You should be able to see a smear of DNA fragments that changes in size range as a function of the DNaseI digestion time (Fig. 1).

8. Excise the gel containing the digested DNA on a UV transilluminator while aiming for DNA fragments that are 50–250 bp long.
9. Extract the DNA fragments from the gel using a gel extraction kit or by electroelution according to the manufacturer's instructions (*see* **Note 10**).

### 3.3 Incorporation of Oligonucleotides by Assembly PCR

1. Reassemble your gene by PCR. Mix 100 ng of the purified fragments, in the presence of 1–10 nM of library oligonucleotides (*see* **Note 5**). Do not add the external primers at this point!

Use the high-fidelity polymerase with its recommended buffer and dNTP concentration, in a final volume of 50 μl.

2. Use the following PCR program: One denaturation step at 95 °C for 2 min; 40 cycles composed of a denaturation step at 94 °C (30 s); 13 successive hybridization steps separated by 2 °C each, from 65 °C to 41 °C for 1.5 min each (total 19.5 min); an elongation step of 1 min/kb gene at 72 °C (the optimal temperature for most PCR DNA polymerases); and then a final 7 min elongation step at 72 °C to allow full elongation of all assembled genes.
3. Amplify the full-length assembly product by a "nested" PCR using external primers. Use different dilutions of the reassembly reaction as a template in a standard 50 μl PCR (1:25–1:1,000; *see* **Note 11**). In the case of failure to assemble, or no incorporation of the oligonucleotide-encoded library mutations, an optional protocol is described in **Note 12**.
4. Purify the nested-PCR product using a PCR purification kit (e.g., Qiagen, QIAquick PCR Purification Kit, Hilden, GMBH).
5. Restrict the purified DNA using the appropriate restriction enzymes and ligate to a vector to create the plasmid library (e.g., as described by Copp et al., in Chapter 1 of this volume). Alternatively, use restriction-free DNA cloning methods for the same purpose [17, 18].

## 4 Notes

1. DNaseI requires a divalent metal ion for its activity. It randomly produces nicks in double-stranded DNA in the presence of $Mg^{2+}$, but in the presence of $Mn^{2+}$, both strands are cleaved into fragments. The enzyme loses its activity reversibly with EDTA and irreversibly by heat treatment at 80 °C for 10 min.
2. DNaseI is supplied at high concentrations (2,000–50,000 U/ml) since it is mainly used to rapidly degrade large amounts of genomic DNA. One unit of DNaseI is defined as the amount of enzyme required to completely degrade 1 μg of plasmid DNA in 10 min at 37 °C in 10 mM Tris–HCl, pH 7.5, 50 mM $MgCl_2$, 13 mM $CaCl_2$. In order to control the amount of degraded DNA, the DNaseI stock solution is diluted to 0.05 U/μl in its specified storage buffer (although different concentrations may apply for different DNaseI preparations or different templates).
3. Because digestion times are relatively short, it is advisable to place the stop solution tubes in one heating block or PCR machine, and the digestion reaction in another heating block

or PCR machine located next to it, in order to enable rapid and immediate inactivation of the reaction.

4. Make sure the DNA loading buffer does not contain bromophenol blue as dye, since it will run with your fragments on the gel. Use xylene cyanol as dye instead.
5. The diversification oligonucleotides should be ~33 nt long (~15 bases from each side of the mutated codon). It is generally of no importance whether the synthesized oligonucleotides correspond to the sense or antisense strand. However, in cases where neighboring residues need to be targeted independently, the usage of oligonucleotides complementing opposing strands is recommended. A mixture of all diversifying oligonucleotides is made. The mixture is usually equimolar, i.e., the concentration of all mutations is equal (meaning that the concentration of an oligonucleotide encoding two substitutions is held twofold higher than that of an oligonucleotide encoding for only one substitution). However, non-equimolar mixtures, in which certain library mutations are overrepresented in the oligonucleotide mixture, can also be used (e.g., stabilizing, compensatory mutations). The oligonucleotide mixture is added to the PCR such that the total concentration of all spiking oligonucleotides in the final PCR ranges from 1 to 10 nM. This concentration range usually results in the incorporation of one to four substitutions per gene and should enable the generation of libraries with different mutational loads (for detailed discussion of the effect of mutational loads, *see* Section 5.4 in ref. 4).
6. Start with enough DNA (typically above 20 μg) as the digestion conditions may be hard to reproduce. Additionally, the more DNA that is used for the DNaseI digestion, the better is the final yield of purified fragments. Larger DNA amounts can be obtained by increasing the volume of the PCR to 200 μl. Use external primers that extend ≥100 bp from the gene's ends to enable the application of another set of external primers ("nested primers"; *see* **Note 12**) after the assembly PCR.
7. If the library is based on several starting genes, it is possible to mix their templates and co-amplify, or mix their PCR products, in equal amounts, or in any desired ratio, to a final amount of 20–25 μg DNA. The outcome would be shuffling of these starting genes in parallel with the incorporation of the diversification oligonucleotides.
8. A final DNaseI concentration of 0.015 U/μg DNA was found to be optimal for digestion of 20–25 μg of 1–1.5 kb DNA templates. This concentration may be increased or decreased for different DNA templates. Alternatively, the incubation times with the DNaseI may be altered. In general, the digestion conditions need to be initially calibrated using 20–25 μg

of template and a range of reaction times and/or DNaseI concentrations. Scaling up the reaction may prove problematic as the conditions found for small amounts of template DNA may not apply for greater amounts.

9. Do not overrun the gel as the fragments of the desired size range (50–250 bp) will be distributed throughout a large block of agarose (reducing the efficiency of gel extraction). Adjust the running time so that the distance between the 100 bp and the 300 bp marker bands does not exceed 0.5 cm.
10. Gel extraction kits tend to produce low purification yields when applied to short DNA fragments (typically 2–10 %; including kits that are optimized for small fragments). We recommend the use of electroelution devices that greatly increase the yields of purified fragments (typically 20–50 %). The most frequent cause of failure of ISOR is low quantity and/or purity of the fragments. It is therefore recommended that the purified fragments be analyzed by gel electrophoresis allowing their size and concentration to be assessed.
11. For the "nested PCR", use external primers that nest in from the 3′ and 5′ ends of the DNA fragment used for digestion. Usually, 1–2 μl of the assembly PCR provide enough template for the "nested PCR"; however, diluting the assembly PCR mix may improve the yields of the "nested PCR".
12. An alternative protocol is based on the target gene being amplified in separate PCRs with each of the library mutations. For each PCR, use the mutation coding oligonucleotide (*see* **Note 5**) together with an external primer on the complementary strand. The mutagenesis oligonucleotides can be designed as either sense or antisense oligonucleotides, and the complementary oligonucleotide should be an external primer, extending ≥100 bps from the gene's end. The PCRs therefore result in a series of different length fragments each carrying a different library mutation, while covering part of the gene length. The PCR fragments are then mixed such that each fragment concentration is proportional to the number of mutations it encodes (as with the oligonucleotides, *see* **Note 5**), digested with DNaseI and reassembled (for a detailed protocol, *see* Supplementary Methods in ref. 19).

## Acknowledgments

Financial support by the Defense Threat Reduction Agency (DTRA) of the US Department of Defense (contract HDTRA1-11-C-0026) and by the CounterACT Program, National Institutes of Health Office of the Director, and the National Institute of Neurological Diseases and Stroke (Grant Number U54-NS058183-06) are gratefully acknowledged.

## References


1. Dalby PA (2011) Strategy and success for the directed evolution of enzymes. Curr Opin Struct Biol 21(4):473–480
2. Goldsmith M, Tawfik DS (2012) Directed enzyme evolution: beyond the low-hanging fruit. Curr Opin Struct Biol 22(4):406–412
3. Cobb RE, Si T, Zhao H (2012) Directed evolution: an evolving and enabling synthetic biology tool. Curr Opin Chem Biol 16(3–4):285–291
4. Goldsmith M, Tawfik DS (2013) Enzyme engineering by targeted libraries. Methods Enzymol 523:257–83
5. Wang XJ, Minasov G, Shoichet BK (2002) Evolution of an antibiotic resistance enzyme constrained by stability and activity trade-offs. J Mol Biol 320(1):85–95
6. Herman A, Tawfik DS (2007) Incorporating Synthetic Oligonucleotides via Gene Reassembly (ISOR): a versatile tool for generating targeted libraries. Protein Eng Des Sel 20(5):219–226
7. Alcolombri U, Elias M, Tawfik DS (2011) Directed evolution of sulfotransferases and paraoxonases by ancestral libraries. J Mol Biol 411(4):837–853
8. Berger I, Guttman C, Amar D, Zarivach R, Aharoni A (2011) The molecular basis for the broad substrate specificity of human sulfotransferase 1A1. PLoS One 6(11)
9. Khersonsky O, Rothlisberger D, Wollacott AM, Murphy P, Dym O, Albeck S, Kiss G, Houk KN, Baker D, Tawfik DS (2011) Optimization of the in-silico-designed kemp eliminase KE70 by computational design and directed evolution. J Mol Biol 407(3):391–412
10. Scanlon TC, Teneback CC, Gill A, Bement JL, Weiner JA, Lamppa JW, Leclair LW, Griswold KE (2010) Enhanced antimicrobial activity of engineered human lysozyme. ACS Chem Biol 5(9):809–818
11. Stemmer WP (1994) Rapid evolution of a protein in vitro by DNA shuffling. Nature 370(6488): 389–391
12. Rothlisberger D, Khersonsky O, Wollacott AM, Jiang L, DeChancie J, Betker J, Gallaher JL, Althoff EA, Zanghellini A, Dym O, Albeck S, Houk KN, Tawfik DS, Baker D (2008) Kemp elimination catalysts by computational enzyme design. Nature 453(7192):190–195
13. Chen F, Gaucher EA, Leal NA, Hutter D, Havemann SA, Govindarajan S, Ortlund EA, Benner SA (2010) Reconstructed evolutionary adaptive paths give polymerases accepting reversible terminators for sequencing and SNP detection. Proc Natl Acad Sci U S A 107(5): 1948–1953
14. Chen MMY, Snow CD, Vizcarra CL, Mayo SL, Arnold FH (2012) Comparison of random mutagenesis and semi-rational designed libraries for improved cytochrome P450 BM3-catalyzed hydroxylation of small alkanes. Protein Eng Des Sel 25(4):171–178
15. Jochens H, Bornscheuer UT (2010) Natural diversity to guide focused directed evolution. Chembiochem 11(13):1861–1866
16. Khare SD, Kipnis Y, Greisen PJ, Takeuchi R, Ashani Y, Goldsmith M, Song YF, Gallaher JL, Silman I, Leader H, Sussman JL, Stoddard BL, Tawfik DS, Baker D (2012) Computational redesign of a mononuclear zinc metalloenzyme for organophosphate hydrolysis. Nat Chem Biol 8(3):294–300
17. Ulrich A, Andersen KR, Schwartz TU (2012) Exponential megapriming PCR (EMP) cloning-seamless DNA insertion into any target plasmid without sequence constraints. PLoS One 7(12)
18. Miyazaki K (2003) Creating random mutagenesis libraries by megaprimer PCR of whole plasmid (MEGAWHOP). In: Arnold FH, Georgiou G (eds) Directed evolution library creation: methods and protocols, vol 231, Methods in molecular biology. Humana Press, Totowa, NJ, pp 23–28
19. Gupta RD, Tawfik DS (2008) Directed enzyme evolution via small and effective neutral drift libraries. Nat Methods 5(11):939–942

# Chapter 9

# OmniChange: Simultaneous Site Saturation of Up to Five Codons

**Alexander Dennig, Jan Marienhagen, Anna Joëlle Ruff, and Ulrich Schwaneberg**

## Abstract

Multi-site-saturation mutagenesis allows altering of "localizable" properties such as activity and selectivity and enables the discovery of cooperative amino acid substitutions which are unlikely to be discovered by saturating single codons individually or iteratively. The herein presented method "OmniChange" does not require any DNA modifying enzyme (e.g., endonucleases or ligases), and diverse mutant libraries with up to five simultaneously saturated positions are generated in a robust and technically simple manner in four steps. The key feature of the OmniChange method is a highly efficient chemical cleavage of phosphorothiolated nucleotides by ethanol-iodine to generate 12-nucleotide-long 5′ overhangs in double-stranded DNA. The generated vector and inserts can be hybridized in a one-pot assembly leading to fully functional mutagenic plasmids, and the employed *E. coli* host can easily ligate up to 10 DNA nicks without any further enzymatic treatment. OmniChange is furthermore a reliable and general tool for multi-DNA fragment assembly which is DNA sequence independent.

**Key words** Focused mutagenesis, Directed evolution, Multi-site-saturation mutagenesis, Ligase-independent assembly, Sequence-independent hybridization

## 1 Introduction

Multi-site-saturation mutagenesis (MSSM) enables generation of mutant libraries containing all possible mutations or a subset of mutations at multiple codons within a gene sequence [1, 2]. MSSM has the potential to discover cooperative effects of amino acid substitutions and combinations of substitutions which cannot be discovered by saturating single codons individually or in an iterative manner [1]. In addition, MSSM is most beneficial to improve "localizable" enzyme properties (activity, selectivity, substrate profiling) where amino acid site chains are in close contact to each other, for example, within the active site of a biocatalyst.

Current multi-site-saturation methods are often limited by the distance and number of simultaneously targeted codons [3].

Elizabeth M.J. Gillam et al. (eds.), *Directed Evolution Library Creation: Methods and Protocols*, Methods in Molecular Biology, vol. 1179, DOI 10.1007/978-1-4939-1053-3_9, © Springer Science+Business Media New York 2014

CASTing has proven to be a very successful tool for saturation of multiple neighboring codons located on a short stretch of DNA [4]. However, targeted codons have to fit on one primer, which limits freedom to select codons independently. State-of-the-art methods, such as the QuikChange Multi Site-Directed Mutagenesis Kit, enable saturation mutagenesis of up to three independent codons simultaneously, by using Multiplex-PCR; however this becomes more difficult with increasing amount of targeted codons [5]. Several MSSM methods have been developed in the past, but most of them require multiple PCR steps followed by restriction/ligation cloning, which decreases simplicity and ease of handling [3].

OmniChange offers simultaneous and efficient saturation of up to five independent codons in a simple and robust protocol without a purification step prior to transformation into *Escherichia coli* [3]. The OmniChange multi-site-saturation method comprises four steps: *Step 1*: Vector and insert generation by standard PCR with phosphorothiolated oligonucleotides (PTO) (12 on the 5′ end of each primer) containing the degenerate codons; *Step 2*: Chemical DNA-cleavage reaction by addition of iodine, to generate complementary 5′-overhangs via breaking of the phosphorothiodiester bonds; *Step 3*: Assembly of mutated plasmids via complementary DNA hybridization; and *Step 4*: Transformation and nick repair in *E. coli*.

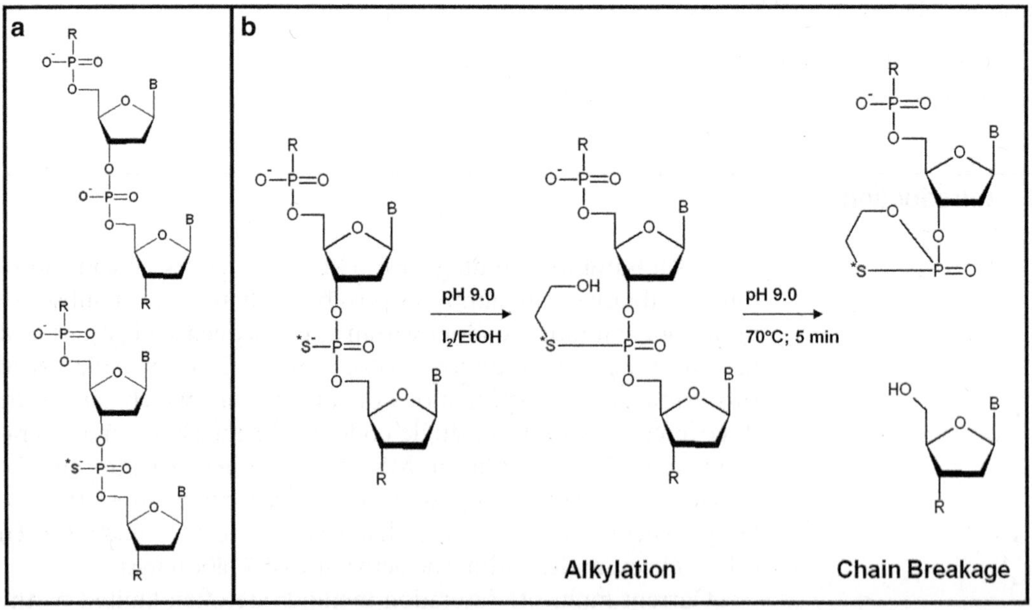


**Fig. 1** (**a**) Representation of natural nucleotide backbone connected via phosphodiester (*upper*) and a phosphorothiolated DNA backbone (*lower*, sulfur atom indicated with an *asterisk*). (**b**) Chemical cleavage of phosphorothiolated nucleotides in the presence of iodine and ethanol ($I_2$/EtOH). In Step 1, the sulfur atom is alkylated leading to an unstable intermediate which releases the phosphorothiolated nucleotide from the DNA. The detailed mechanism was introduced by Eckstein and Gish [6]

During the chemical cleavage in Step 2, all 12 phosphorothioate nucleotides are cleaved in a multiple manner so that the generated vector and insert can hybridize with the 12-nt-long overhangs [6] (Fig. 1). The long specific DNA overhangs ensure selective hybridization of DNA molecules.

The OmniChange method is sequence independent, does not require a minimal distance between mutated codons, and can be accomplished within a day including transformation. Furthermore, OmniChange can be used in a modular manner, expanding/decreasing the amount of targeted codons with the same subset of oligonucleotides.

## 2 Materials

### 2.1 DNA Amplification by PCR

1. Program for visualization of DNA sequences, primer design, and evaluation of sequencing data (e.g., Clone Manager 9 Professional Edition, Scientific & Educational Software, Cary, NC, USA).
2. Standard PCR oligonucleotides each containing 12 phosphorothioates at the 5′-end at HPLC-purity and in a salt-free form (e.g., from Eurofins MWG Operon, Ebersberg, Germany). Received primers have to be diluted in $ddH_2O$ to a final concentration of 20 μM (store at −20 °C).
3. Standard PCR oligonucleotides flanking the gene of interest to perform colony PCR.
4. 20 ng/μl target expression plasmid containing gene of interest (store at −20 °C).
5. 50× dNTP mixture (10 mM of each dATP, dCTP, dGTP, and dTTP; e.g., from Fermentas (St. Leon-Rot, Germany) (store at −20 °C)).
6. High-fidelity DNA polymerase (e.g., Phusion DNA Polymerase, New England Biolabs, Frankfurt, Germany) and corresponding polymerase buffer (store at −20 °C).
7. Restriction enzyme DpnI (New England Biolabs, Frankfurt Germany) for removal of methylated template DNA.
8. *Taq* DNA polymerase and corresponding polymerase buffer (store at −20 °C).
9. Agarose gel: 1 % (w/v) agarose solubilized at 65 °C in TAE buffer (40 mM Tris–acetate, 1 mM EDTA; pH 8.3).
10. Suitable DNA staining solution (e.g., 0.5 μg/ml or 0.001 % ethidium bromide).
11. PCR cleanup kit (e.g., NucleoSpin Extract II DNA purification kit (Macherey-Nagel, Dueren, Germany)).

12. Plasmid extraction kit (e.g., QIAGEN QIAprep Spin Miniprep Kit, Hilden, Germany).
13. Sterile $ddH_2O$.

### 2.2 Transformation Components

1. Chemical competent cells (e.g., *E. coli* DH5α or any BL21 strain can be used for transformation).
2. Sterile SOC regeneration media: 2 % tryptone, 0.5 % yeast extract, 8.6 mM NaCl, 2.5 mM KCl,10 mM $MgCl_2$, 10 mM $MgSO_4$, 20 mM glucose; adjust to pH 7.0 using NaOH.
3. Appropriate antibiotics.
4. Sterile LB agar plates: 1 % tryptone, 0.5 % yeast extract, 1 % NaCl, 1.5 % agar with appropriate antibiotics; adjust to pH 7.0 using NaOH.
5. LB medium: 1 % tryptone, 0.5 % yeast extract, 1 % NaCl with appropriate antibiotics; adjust to pH 7.0 using NaOH.

### 2.3 Iodine Cleavage Components

1. Tris–HCl buffer: 500 mM; pH 9.0.
2. Elemental iodine, $I_2$: 100 mM, solubilized in EtOH.
3. $ddH_2O$.

### 2.4 Equipment

1. PCR thermocycler (e.g., Eppendorf Mastercycler proS, Hamburg, Germany) for DNA polymerization and iodine cleavage.
2. Agarose gel electrophoresis system (e.g., Bio-Rad, Munich, Germany).
3. UV-scanning system for DNA visualization after agarose gel electrophoresis (e.g., UGenius, Syngene, Cambridge, UK).
4. Micro centrifuge (e.g., Centrifuge 5424, Eppendorf, Hamburg, Germany).
5. NanoDrop 1000 UV spectrophotometer (NanoDrop Technologies, DE, Wilmington, USA).
6. Ice box (polystyrene or similar material) filled with ice.
7. Thermoblock (Thermoblock plus, Eppendorf, Hamburg).
8. 37 °C incubator with shaking function (e.g., Multitron Standard, Infors HT, Switzerland).

## 3 Methods

The design of oligonucleotides will be illustrated using the P450 BM3 gene from *Bacillus megaterium* [7] cloned into the widely used pET28a plasmid (Novagen, La Jolla, CA, USA) (Fig. 2).

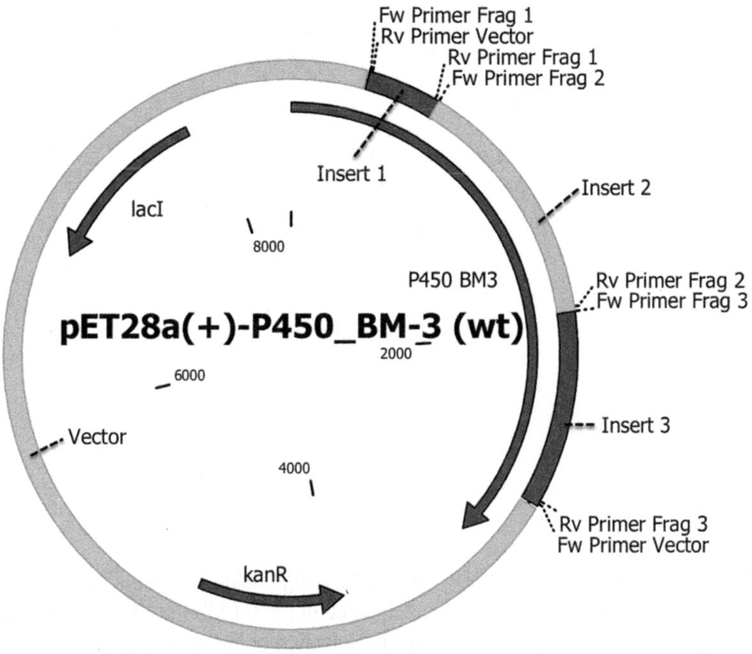


**Fig. 2** pET28a-vector containing the P450 BM3 gene [7]. In this example, the plasmid was separated into four sections (Inserts 1–3 and the vector backbone). Vector and inserts are amplified with twelve 5′-PTO-modified nucleotides on each primer. NNK-codons can be integrated into reverse (Rv) or forward (Fw) primers

### 3.1 Selection of Codons and Hybridization Areas (Fig. 3a)

1. Select two or more codons in your gene of interest and mark positions in your plasmid sequence.
2. Select 12 nucleotides upstream or downstream of targeted position for hybridization. Label hybridization areas on your plasmid sequence. These nucleotides will later represent the 12 PTOs on every 5′ end of the designed primers.

### 3.2 Design of Oligonucleotides (Fig. 3b, c)

1. Use a calculation tool for design of oligonucleotides (e.g., http://eu.idtdna.com/analyzer/Applications/Oligo-Analyzer/) (*see* **Note 1**).
2. Ensure that your hybridization areas (corresponding to the PTOs) are unique for each vector and insert (in particular, ensure that there are no complementary regions between hybridization areas).

### 3.3 PCR (Fig. 4; Step 1)

All solutions should be prepared on ice unless stated otherwise.

1. Place high-fidelity DNA polymerase, corresponding buffer, plasmid template, 20 μM forward and reverse primers, and 10 mM dNTP mix on ice.

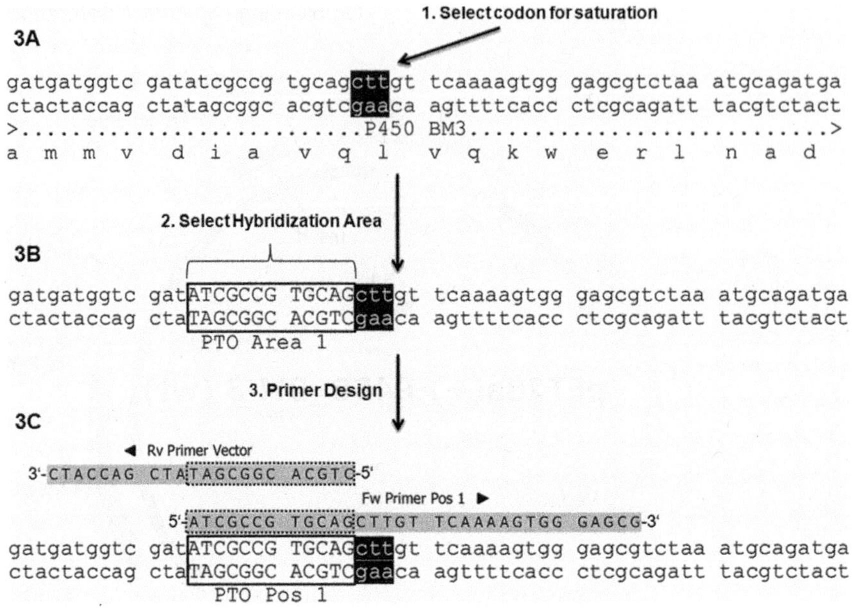


**Fig. 3** Illustration for efficient oligonucleotide design using the OmniChange method. (**a**) Selection of the codon for saturation (exemplified using the sequence of P450 BM3 from *Bacillus megaterium* [7]). The target codon is highlighted in *black with white letters*. (**b**) Selection of the area that serves later for hybridization of vector and insert. The selected hybridization area is indicated as *capital letters within the black frame*. (**c**) Design of oligonucleotides (full sequence highlighted in *light gray shading*) with 12 consecutive PTO nucleotides (sequence highlighted in *light gray shading framed in black dotted box*) introduced on the 5′ ends of each primer

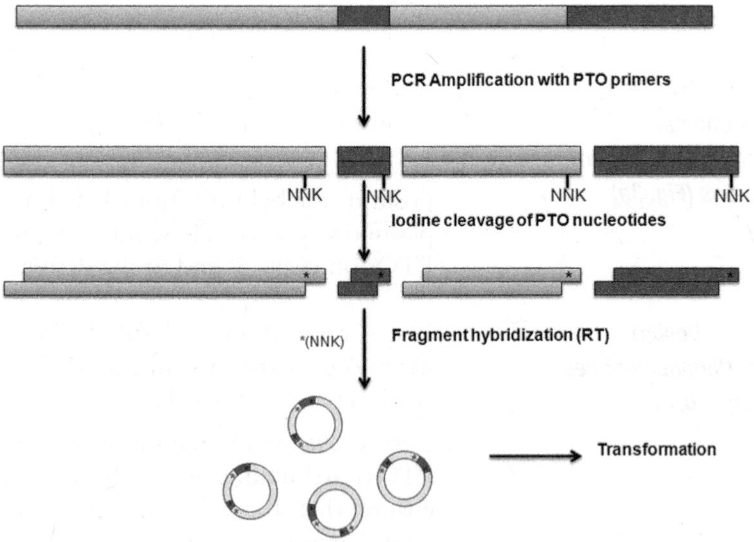


**Fig. 4** Four steps of the OmniChange protocol for saturation of four independent codons simultaneously. Step 1: Amplification of vector and inserts with 5′-PTO-modified primers. Step 2: Iodine cleavage for removal of the 12 PTO nucleotides and generation of specific 12-nt-long hybridization areas. Step 3: Hybridization of the vector and inserts to reassemble full length plasmids containing four NNK-codons at selected positions (with nicks in the DNA backbone at each hybridization site). Step 4: Transformation of the assembled plasmids into *E. coli* to enable nick repair and plasmid replication. NNK-saturated codons are indicated with *stars*

2. For each target codon, prepare a PCR sample containing the following components:
   (a) 5 μl 5× high-fidelity polymerase buffer
   (b) 1 μl 50× dNTP mixture
   (c) 1 μl of vector- or insert-specific forward primer
   (d) 1 μl of vector- or insert-specific reverse primer
   (e) 0.5 μl high-fidelity DNA polymerase (5 U)
   (f) 1 μl plasmid DNA template
   (g) Add $ddH_2O$ to a final volume of 50 μl.
3. Prepare an additional tube containing the same reaction mixture but without primers, serving as a DpnI control for each vector/insert being amplified.
4. Place tubes in PCR in thermocycler and proceed with following program:

   98 °C for 1 min; 98 °C for 15 s, appropriate annealing temperature for 15 s, 72 °C for 1 min/kb, 25 cycles; 72 °C for 3 min final elongation (for a 1–2 kb amplimer) (*see* **Note 2**).
5. Check the quality of PCR products by running a small aliquot from each reaction on an agarose gel (*see* **Note 3**).
6. Stain agarose gel with ethidium bromide solution (15 min) and visualize bands on UV transilluminator (*see* **Note 4**).
7. Add 20 U DpnI to the remaining PCR samples and incubate according to manufacturer's recommended conditions to remove residual template DNA (commonly 3 h at 37 °C).
8. Purify PCR products with PCR cleanup kit (*see* **Note 5**).
9. Quantify purified DNA with NanoDrop spectrophotometer.
10. Calculate the molar quantity of the vector and each insert in pmol/μl (*see* **Note 6**).

### 3.4 Iodine Cleavage (Fig. 4; Step 2)

Store all components on ice unless stated otherwise.

1. Dilute vector to 0.03–0.04 pmol/μl concentration. Dilute each insert to 0.11 pmol/μl concentration (*see* **Note 7**).
2. Prepare two thin wall PCR tubes, each containing 4 μl of the PCR product corresponding to the amplified vector. One of these tubes will provide vector for the experimental reaction, while the other will serve as a vector religation control. For each insert fragment also prepare two PCR tubes, each containing 4 μl of amplified PCR product corresponding to that insert. One of these tubes will be used in the experimental reaction, and one will serve as a DpnI digestion control. Prepare additional tubes (equal to the number of amplified inserts) each filled with 4 μl $ddH_2O$, for $ddH_2O$ cleavage reactions that will be added to the vector religation control (*see* **Note 8**).

3. Prepare iodine cleavage mixture as follows: combine 50 μl 500 mM Tris–HCl buffer pH 9.0, 30 μl 100 mM iodine EtOH solution and 20 μl dd$H_2O$ in a single test tube, mix thoroughly, and store the mixture on ice.
4. Add 2 μl of the iodine cleavage mixture to each of the PCR tubes from **step 2** (make sure that the solution remains at least faintly yellow in color) (*see* **Note 9**).
5. Place tubes into PCR thermocycler with a preheated lid (80 °C).
6. Heat samples to 70 °C for 5 min.
7. After incubation solutions should be transparent (*see* **Note 10**).
8. Store cleaved vector and inserts on ice prior to hybridization.

#### 3.5 Hybridization of Vector and Inserts to Generate Full Length Plasmids (Fig. 4, Step 3)

1. All steps are performed at room temperature.
2. Place the two PCR tubes containing cleaved vector DNA at room temperature.
3. Add cleaved Insert 1 to the first tube (OmniChange tube). Add one of the cleaved dd$H_2O$ reactions to the second tube (religation control tube). Mix each solution thoroughly and incubate for 5 min.
4. Add cleaved Insert 2 to the OmniChange tube. Add a second cleaved dd$H_2O$ reaction to the religation control tube. Repeat until all inserts have been added to the OmniChange tube and all dd$H_2O$ reactions have been added to the religation control tube.
5. Place the OmniChange tube, the religation control tube, and all the DpnI digestion control tubes for 5 min on ice prior to transformation.

#### 3.6 Transformation of the Assembled OmniChange Library (Fig. 4; Step 4)

1. Preheat the thermoblock to 42 °C.
2. The following samples have to be transformed: the OmniChange reaction, the religation control, and each of the DpnI digestion controls (*see* **Note 11**).
3. For each transformation, prewarm 1 ml SOC media to 42 °C in the thermoblock.
4. For each transformation, thaw one tube containing 100 μl chemically competent *E. coli* cells on ice.
5. Add 5–10 μl of hybridization mix from the OmniChange reaction, the religation control, or each of the DpnI digestion controls to separate tubes containing 100 μl thawed competent cells. Incubate each tube for 15 min on ice.
6. Place the tubes containing competent cells and DNA for 45 s into a preheated thermoblock (42 °C).
7. Incubate the tubes for 10 min on ice.
8. Add 900 μl prewarmed SOC media to each tube.

9. Place tubes at 37 °C with shaking at 250 rpm for 45 min to recover transformed cells.
10. Plate cells on selective agar plates.
11. Store plates at 37 °C overnight to grow recombinant colonies.

### 3.7 Statistical and Quality Evaluation of OmniChange Library

1. Count colonies on each plate. Only cells that were transformed with the OmniChange reaction containing the vector and all inserts should form colonies. If colonies are obtained on the religation control, these should not be more than 10 % of the number of colonies on the OmniChange plate (*see* **Note 12**).
2. Pick 10–20 colonies with a sterile toothpick into sterile tubes, each filled with 3 ml LB medium containing appropriate antibiotics. Grow cell cultures overnight at 37 °C with 250 rpm shaking.
3. Perform plasmid extraction using plasmid purification kit. Quantify plasmid DNA with NanoDrop spectrometer.
4. Prepare a PCR containing the following components:
   (a) 10 μl 10× *Taq* DNA polymerase buffer
   (b) 1 μl 50× dNTP mixture
   (c) 1 μl of colony PCR forward primer (0.2 μM)
   (d) 1 μl of colony PCR reverse primer (0.2 μM)
   (e) 0.5 μl *Taq* DNA polymerase (5 U)
   (f) 1 μl plasmid DNA template (20 ng)
   (g) Add $ddH_2O$ to a final volume of 50 μl
5. Place tubes in PCR thermocycler and proceed with following program:
   (a) 94 °C for 0.5 min; 94 °C for 15 s, appropriate annealing temperature for 15 s, 72 °C for 1 min/kb for 20 cycles; 72 °C for 3 min final elongation (*see* **Note 13**).
6. Check the quality and length of the PCR products on a 1 % agarose gel by electrophoresis and subsequent staining with ethidium bromide.
7. Send correctly assembled plasmids containing the full length gene of interest for sequencing.
8. Align the sequence of wild-type plasmid and sequenced OmniChange variants (e.g., using Clone Manager 9 software from Scientific & Educational Software, Cary, NC, USA).

## 4 Notes

1. Primers are designed as normal oligonucleotides for standard PCR. Generally, primers should have a GC content of around 40–60 % similar melting temperatures and should not form unwanted secondary structures (heterodimers or loops).

If your vector backbone is very large (>3 kb), it is possible to split the vector into an additional insert by introducing another hybridization area.

2. Incubation for 1 min at 98 °C is recommended for hot-start DNA polymerases such as Phusion (New England Biolabs, Frankfurt, Germany). Temperatures for denaturation as well as amplification velocity can vary strongly depending on the polymerase used and manufacturer. Please follow the instructions of the manufacturer to ensure efficient amplification.
3. It is recommended to analyze DNA molecules smaller than 500 bp on a 2 % (w/v) agarose gel.
4. Only one distinct band should be visible on the agarose gel. If there are multiple bands, gel extraction and cleanup can be performed to obtain desired vector or insert. It is recommended to also perform PCR again with a modified program (i.e., modify annealing temperature or elongation time and recheck primer properties).
5. It is recommended to elute purified DNA molecules with dd$H_2O$ from purification kit columns. Commercial elution buffers with high salt content might interfere with the chemical cleavage conditions in Step 2 of the OmniChange protocol.
6. The online program http://molbiol.ru/eng/scripts/01_07.html is helpful for conversion of DNA concentrations to molar quantities.
7. If required amount of DNA is not obtained after purification, repeat PCRs to produce more DNA. Purified samples can be concentrated using a vacuum centrifuge (SpeedVac) to remove dd$H_2O$ from samples. Further optimization of PCR protocols could also improve DNA yields significantly.
8. This control is essential to perform statistical analysis of the OmniChange library and to make sure that plates obtained after transformation contain inserts.
9. If solution becomes transparent, add additional 1 or 2 μl of cleavage mixture.
10. Solutions in all tubes should be transparent. If not, repeat heat incubation step for additional 5 min.
11. All these controls are essential to evaluate the quality of the OmniChange library.
12. Plates with DpnI controls must be empty (<10 colonies). When >10 colonies are obtained on agar plates, repeat the PCRs with a decreased amount of plasmid template. Colonies from the OmniChange transformation with vector and inserts should have around 5–10 times more colonies than the religation

control. If no colonies are obtained, check the quality of your competent cells and increase the amount of insert (initially recommended is doubling the concentration to 0.22 pmol/μl for all inserts at Step 2 of Subheading 3.4).

13. It is recommended to read and follow the manufacturer's instructions for efficient *Taq* DNA polymerase amplification.

## Acknowledgments

The authors acknowledge BASF SE (Ludwigshafen, Germany) for financial support during development of the OmniChange method.

### References

1. Reetz MT, Carballeira JD, Vogel A (2006) Iterative saturation mutagenesis on the basis of B factors as a strategy for increasing protein thermostability. Angew Chem Int Ed Engl 45: 7745–7751
2. Shivange AV, Marienhagen J, Mundhada H et al (2009) Advances in generating functional diversity for directed protein evolution. Curr Opin Chem Biol 13:19–25
3. Dennig A, Shivange AV, Marienhagen J, Schwaneberg U (2011) OmniChange: the sequence independent method for simultaneous site-saturation of five codons. PLoS One 6:e26222
4. Reetz MT, Bocola M, Carballeira JD et al (2005) Expanding the range of substrate acceptance of enzymes: combinatorial active-site saturation test. Angew Chem Int Ed Engl 44:4192–4196
5. Hogrefe HH, Cline J, Youngblood GL, Allen RM (2002) Creating randomized amino acid libraries with the QuikChange Multi Site-Directed Mutagenesis Kit. Biotechniques 33: 1158–1160
6. Eckstein F, Gish G (1989) Phosphorothioates in molecular biology. Trends Biochem Sci 14:97–100
7. Whitehouse CJ, Bell SG, Wong LL (2012) P450(BM3) (CYP102A1): connecting the dots. Chem Soc Rev 41:1218–1260

# Chapter 10

# Random Insertional–Deletional Strand Exchange Mutagenesis (RAISE): A Simple Method for Generating Random Insertion and Deletion Mutations

Ryota Fujii, Motomitsu Kitaoka, and Kiyoshi Hayashi

## Abstract

Although proteins can be artificially improved by random insertion and deletion mutagenesis methods, these procedures are technically difficult. Here we describe a simple method called random insertional–deletional strand exchange mutagenesis (RAISE). This method is based on gene shuffling and can be used to introduce a wide variety of insertions, deletions, and substitutions. RAISE involves three steps: DNA fragmentation, attachment of a random short sequence, and reconstruction. This yields unique mutants and can be a powerful technique for protein engineering.

**Key words** Random mutagenesis, Insertion, Deletion, Substitution, Recombination, Directed evolution

## 1 Introduction

After several billion years of evolution, living organisms have acquired an enormous amount of diversity. The driving force for this diversity is random mutation. In recent decades, protein evolution arising from random mutations has been reproduced in vitro using random mutagenesis and screening and selection techniques [1–9]. These methods provide mutants with evolved properties that are similar to those created through natural evolution by random mutation and selection. This so-called "directed" evolution is one of the most powerful methods for artificially improving the properties of proteins.

Creating mutant libraries is important for directed molecular evolution [10]. The most popular methods used for library construction involve random point mutagenesis, e.g., error-prone PCR [11]. Such methods are easy to perform because they only involve a few steps using standard techniques, which makes them useful for improving protein properties.

Elizabeth M.J. Gillam et al. (eds.), *Directed Evolution Library Creation: Methods and Protocols*, Methods in Molecular Biology, vol. 1179, DOI 10.1007/978-1-4939-1053-3_10, © Springer Science+Business Media New York 2014

In 2004, random mutagenesis methods were developed for introducing insertions or deletions of fixed lengths into DNA sequences [10]. Introducing insertions and deletions (indels) can change the structures and properties of proteins more dramatically than point mutations [12–14]. However, none of these methods have surpassed the efficacy of traditional point mutations. This may be due to the limited variety of mutations due to the limitations of inserting fixed sequences or deletions of fixed lengths [13, 15–18]. Therefore, a novel method that can generate random indels with variable sequences and lengths is desired.

To enhance the capability to introduce random mutations that contain indels into proteins, we developed a simple method: random insertional–deletional strand exchange mutagenesis (RAISE) [19]. The protocol used for RAISE is similar to that for DNA shuffling [20] and involves only three steps that do not require any specialized skills. We used terminal deoxynucleotidyl transferase (TdT), a DNA polymerase [21, 22] that is essential for introducing random insertions into the antibody hypervariable regions in mammals [23], to introduce random nucleotides at the 3′ termini of digested DNA fragments prior to a self-priming PCR step. By adding this step, we successfully introduced random insertions, deletions, and substitutions of varying lengths into an entire target gene.

This method involves only three steps: DNA fragmentation, attachment of a random short sequence, and reconstruction (Fig. 1). First, target DNA is randomly fragmented using DNaseI. Second, several random nucleotides are attached to the 3′ termini of DNA fragments using TdT. Finally, each fragment with a tail of random nucleotides is reconstructed into a full-length sequence by self-priming PCR [20]. We define RAISE as a series of manipulations using DNaseI digestion, TdT modification, and self-priming PCR.

We assume that the additional sequences cause random mutations by the following mechanisms. A single-strand fragment (primer) with a flanking tail introduced by TdT (random sequence) is annealed with one of its complementary fragments (template) at the proper position on the primer. The 3′ terminus of the random sequence is randomly annealed close to the position to be extended by DNA polymerase during the self-priming PCR step (Fig. 2a). Then, the region of the template between the primer and the random sequence (strand X) is substituted with the random sequence (strand Y) by extending the primer with DNA polymerase. If the random sequence tail is longer than the replaced region, an insertion is included (Fig. 2b). In contrast, a deletion will occur if the sequence tail is shorter. If the fragment lengths are the same, this results in a simple substitution. We call these mutations "region-exchanged mutations." The occurrence of region-exchanged mutations is the most notable feature of the RAISE method.

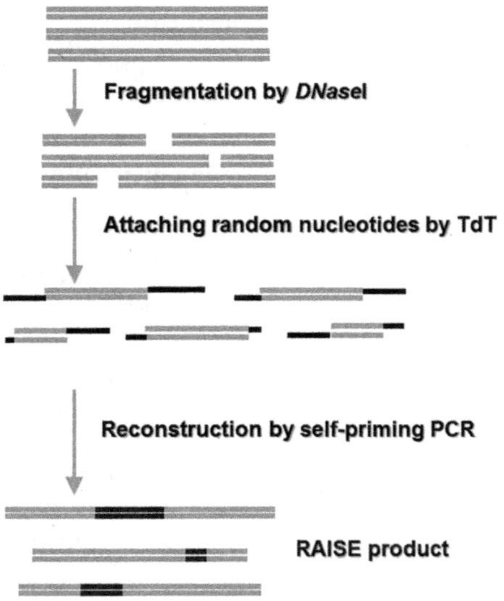


**Fig. 1** Schematic of the RAISE method. Reproduced from [19] with permission of Oxford University Press

RAISE has an advantage over conventional mutation methods in that it can produce a wide variety of mutations, including insertions, deletions, and substitutions. The sequences of these mutations are random and the lengths of these mutations can vary from one to several amino acids. Because amino acid indels are often found during natural protein evolution [12, 24], the various indels generated by RAISE appear to be appropriate for protein evolution. In addition, indels and substitutions are often combined. For example, replacing the amino acid sequence "IPNDERD" with "KKMRA" involves two amino acid deletions and five amino acid substitutions. These long substitutions can greatly expand the mutation variations.

In our previous work, to verify that RAISE could be used for in vitro evolutionary experiments, we tested its capability to enhance TEM β-lactamase activity for the hydrolysis of the antibiotic ceftazidime [19]. We found that the best region-exchanged mutations exhibited higher activities than those generated by point mutations, which indicated the superiority of the RAISE method.

RAISE can introduce not only indels but also point mutations and long substitutions. Furthermore, it can be used for DNA recombination because it includes DNA fragmentation and reconstruction steps as well as DNA shuffling [20]. This indicates that all types of mutations (substitution, insertion, deletion, and recombination) can be generated using the RAISE method.

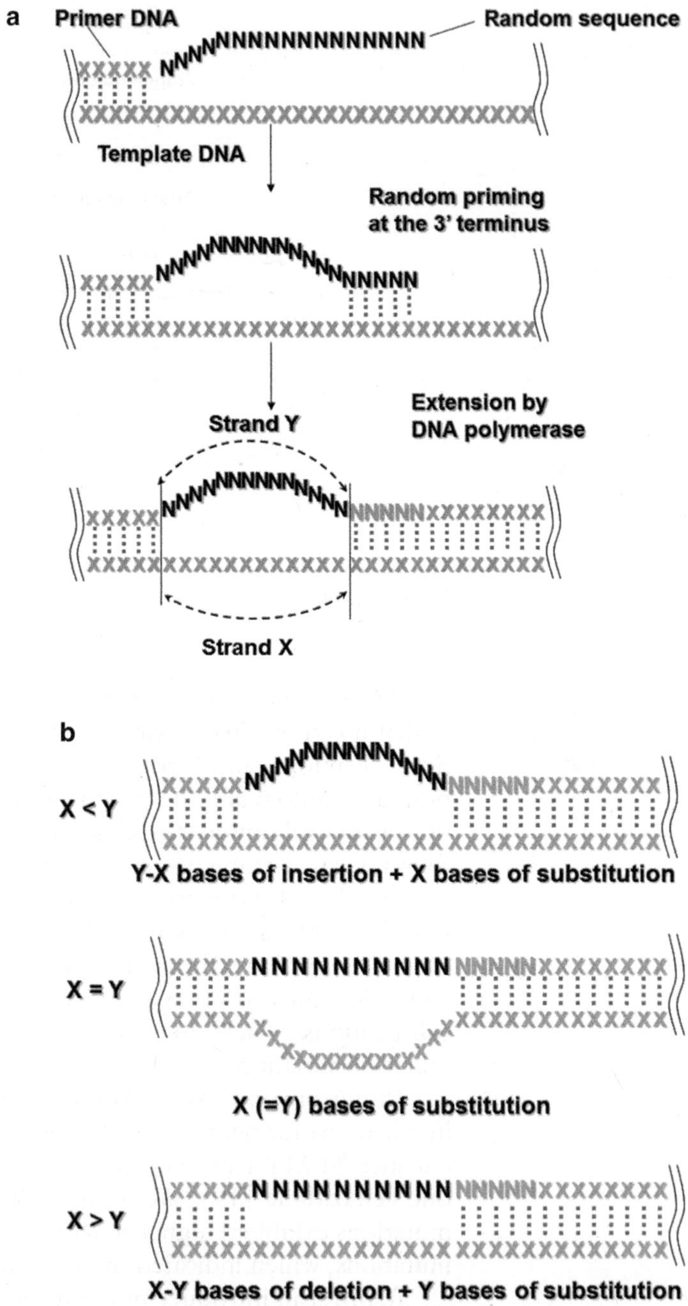


**Fig. 2** Region-exchanged mutation. (**a**) Mechanism to generate a region-exchanged mutation. (**b**) Types of region-exchanged mutations. Reproduced from [19] with permission of Oxford University Press

## 2 Materials

1. RNase-free DNaseI from Takara (Otsu, Japan).
2. Terminal deoxynucleotidyl transferase (TdT) from Toyobo (Osaka, Japan). TdT is also available from New England Biolabs (Beverly, MA, USA).
3. KOD Dash and KOD plus DNA polymerases from Toyobo (Osaka, Japan).
4. Deep Vent (exo$^-$) DNA polymerase from New England Biolabs (Beverly, MA, USA).
5. MinElute Reaction Cleanup and Gel Extraction Kits from QIAGEN (Hilden, Germany).
6. Gel electrophoresis apparatus, such as Mupid-2plus from Advance (Tokyo, Japan).
7. PCR machine, such as Gene Amp PCR System 9700 from Applied Biosystems (Foster, CA, USA).
8. Forward and reverse primers to amplify the target gene.

## 3 Methods

1. Amplify the target gene by PCR (a volume of >500 μL is recommended).
2. Purify the product with a MinElute Reaction Cleanup Kit and elute in 100 μL to obtain a final product concentration of 0.5–1 μg/μL.
3. Digest 20 μg of the product with 0.5 U of DNaseI (RNase-free) in a buffer of 50 mM Tris–HCl, pH 7.0, and 10 mM $MnCl_2$ at 16 °C for 10 min (desired average length is 100–300 bp by agarose gel electrophoresis). Stop reaction by adding 40 μL of 0.5 M EDTA (*see* **Note 1**).
4. Purify DNA fragments with a MinElute Reaction Cleanup Kit. Determine fragment concentrations by measuring their absorbance at 260 nm and calculate molar concentrations.
5. Tail DNA fragments (2 pmol) at the 3′ end by adding 2.5 U of TdT in a 20-μL solution containing buffer and dNTPs at 37 °C for 1 h. Adjust the dNTP concentration to 10 times the molar concentrations of the fragments (20 pmol), a ratio which yields tails that average 5 bases at the 3′ terminus of each fragment (*see* **Note 2**).
6. Purify the 3′-tailed fragments with MinElute Reaction Cleanup Kits. Determine fragment concentrations by measuring their absorbance at 260 nm.
7. Reassemble the purified products by self-priming PCR using a DNA polymerase with weak or no proofreading activity

(*see* **Note 3**). Incubate purified fragments (50–100 ng) with 2 U of Deep Vent (exo$^-$) DNA polymerase or 2.5 U of KOD Dash in 10 μL containing 200 μM of each dNTP and buffer for the corresponding polymerase. Use an amplification protocol with an initial denaturation at 96 °C for 2 min, followed by 40 cycles of denaturation at 96 °C for 30 s, annealing at 60 °C for 30 s, and extension at 75 °C for 30 s.

8. Amplify the full-length gene by PCR. A 1-μL aliquot of the self-priming PCR mixture is mixed with 1 U of KOD plus in a 50-μL solution containing 200 μM of each dNTP, 1 mM $MgSO_4$, 5 μL of 10× Buffer for KOD plus, and 0.3 pmol/μL of primers. Amplify using a protocol that involves an initial denaturation at 96 °C for 2 min, followed by 10 cycles of denaturation at 96 °C for 15 s, annealing at 56 °C for 30 s, and extension at 68 °C for 1 min.
9. Separate PCR products by 2 % (w/v) agarose gel electrophoresis and extract DNA fragments with a MinElute Gel Extraction Kit to yield the RAISE products.
10. Confirm mutation frequency by gene cloning and DNA sequencing (*see* **Notes 4** and **5**).

## 4 Notes

1. The desired average length of the DNA fragments is 100–300 bp. $MnCl_2$ can be used as a DNaseI cofactor to control the sizes of the digestion products. If the fragment lengths are too short (e.g., <50 bp), too many point mutations may be introduced during the subsequent self-priming PCR step [25].
2. Fragment size can be changed by 3′-tailing. This can be confirmed by agarose gel electrophoresis.
3. If a DNA polymerase with high proofreading activity is used (e.g., KOD plus), no region-exchanged mutation will occur.
4. In our previous work [19], two types of mutations were detected: region-exchanged mutations and point mutations (Table 1). Their mutation frequencies in a DNA sequence were 0.7 ± 0.1/kb and 2.1 ± 0.2/kb, respectively. The frequencies of significant amino acid region-exchanged mutations and amino acid point mutations were 0.2 ± 0.1 and 0.6 ± 0.1 amino acids/kb DNA, respectively, calculated by consideration of silent mutations (nearly two-thirds of point mutations are silent) and frameshift mutations (two-thirds of region-exchanged mutations may include frameshifts).
5. It is recommended that the RAISE method is combined with high-throughput screening/selection of >10,000 mutants [26–29]. This increases the likelihood that improved mutants

**Table 1**
**Examples of mutations introduced by RAISE**

| Length of strand X[a] (bp) | Length of strand Y[a] (bp) | Sequence |
|---|---|---|
| 20 | 10 | 5′-ATGAACGAAATAGACAGATC-3′ → 5′-CTGGTTTATT-3′ |
| 10 | 16 | 5′-GTGGGTCTCG-3′ → 5′-CGCTGCCGTCCCGCTC-3′ |
| 8 | 2 | 5′-ATCTCAAC-3′ → 5′-CT-3′ |
| 7 | 3 | 5′-TATTGAC-3′ → 5′-GGA-3′ |
| 7 | 3 | 5′-AACAATT-3′ → 5′-CTC-3′ |
| 6 | 6 | 5′-TACACG-3′ → 5′-GCTCCT-3′ |
| 6 | 6 | 5′-ATGAAC-3′ → 5′-GTGGAA-3′ |
| 5 | 6 | 5′-ACGTT-3′ → 5′-TGGAAG-3′ |
| 2 | 2 | 5′-GC-3′ → 5′-AA-3′ |
| 2 | 2 | 5′-GA-3′ → 5′-AT-3′ |
| 2 | 2 | 5′-TA-3′ → 5′-CC-3′ |
| 2 | 2 | 5′-AC-3′ → 5′-GT-3′ |
| 2 | 1 | 5′-AG-3′ → C |
| 2 | 1 | 5′-GT-3′ → C |
| 1 | 2 | T → 5′-AA-3′ |
| 1 | 0 | G deletion (×2 mutations) |
| 1 | 0 | C deletion (×1 mutations) |
| 1 | 0 | A deletion (×4 mutations) |
| 1 | 0 | T deletion (×3 mutations) |
| 0 | 1 | G insertion (×2 mutations) |
| 0 | 1 | T insertion (×2 mutations) |

[a] *See* Fig. 2a. Reproduced from [19] with permission of Oxford University Press

will be found, as this method can introduce a nearly infinite variety of mutations. If the number of assay mutants is insufficient, this method will not be effective due to the low frequency of region-exchanged mutations (0.3 amino acids/kb) and the high frequency of frameshifts (0.7 amino acids/kb).

## Acknowledgments

This study was supported in part by a grant from the Program for Promotion of Basic Research Activities for Innovative Biosciences (PROBRAIN).

## References


1. Antikainen NM, Martin SF (2005) Altering protein specificity: techniques and applications. Bioorg Med Chem 13:2701–2716
2. Otten LG, Quax WJ (2005) Directed evolution: selecting today's biocatalysts. Biomol Eng 22:1–9
3. Robertson DE, Steer BA (2004) Recent progress in biocatalyst discovery and optimization. Curr Opin Chem Biol 8:141–149
4. Powell KA, Ramer SW, del Cardayre SB et al (2001) Directed evolution and biocatalysis. Angew Chem Int Ed Engl 40:3948–3959
5. Brakmann S (2001) Discovery of superior enzymes by directed molecular evolution. Chembiochem 2:865–871
6. Farinas ET, Bulter T, Arnold FH (2001) Directed enzyme evolution. Curr Opin Biotechnol 12:545–551
7. Goldsmith M, Tawfik DS (2013) Enzyme engineering by targeted libraries. Methods Enzymol 523:257–283
8. Goldsmith M, Tawfik DS (2012) Directed enzyme evolution: beyond the low-hanging fruit. Curr Opin Struct Biol 22:406–412
9. Wang M, Si T, Zhao H (2012) Biocatalyst development by directed evolution. Bioresour Technol 115:117–125
10. Neylon C (2004) Chemical and biochemical strategies for the randomization of protein encoding DNA sequences: library construction methods for directed evolution. Nucleic Acids Res 32:1448–1459
11. Leung DW, Chen E, Goeddel DV (1989) A method for random mutagenesis of a defined DNA segment using a modified polymerase chain reaction. Technique 1:11–15
12. Shortle D, Sondek J (1995) The emerging role of insertions and deletions in protein engineering. Curr Opin Biotechnol 6:387–393
13. Jones DD (2005) Triplet nucleotide removal at random positions in a target gene: the tolerance of TEM-1 b-lactamase to an amino acid deletion. Nucleic Acids Res 33:e80
14. Baldwin AJ, Busse K, Simm AM et al (2008) Expanded molecular diversity generation during directed evolution by trinucleotide exchange (TriNEx). Nucleic Acids Res 36:e77
15. Murakami H, Hohsaka T, Sisido M (2002) Random insertion and deletion of arbitrary number of bases for codon-based random mutation of DNAs. Nat Biotechnol 20:76–81
16. Pikkemaat MG, Janssen DB (2002) Generating segmental mutations in haloalkane dehalogenase: a novel part in the directed evolution toolbox. Nucleic Acids Res 30:e35
17. Hayes F, Hallet B (2000) Pentapeptide scanning mutagenesis: encouraging old proteins to execute unusual tricks. Trends Microbiol 8: 571–577
18. Kim D, Rhee Y, Rhodes D et al (2005) Directed evolution and identification of control regions of ColE1 plasmid replication origins using only nucleotide deletions. J Mol Biol 351:763–775
19. Fujii R, Kitaoka M, Hayashi K (2006) RAISE: a simple and novel method of generating random insertion and deletion mutations. Nucleic Acids Res 34:e30
20. Stemmer WPC (1994) Rapid evolution of a protein *in vitro* by DNA shuffling. Nature 370: 389–391
21. Lewin B (1994) Gene. Oxford University Press, Oxford
22. Lewis SM (1994) The mechanism of V(D)J joining: lessons from molecular, immunological, and comparative analyses. In: Dixon FJ (ed) Advances in immunology, vol 56. Academic, San Diego, pp 27–150
23. Komori T, Okada A, Stewart V et al (1993) Lack of N regions in antigen receptor variable region genes of TdT-deficient lymphocytes. Science 261:1171–1175
24. Pascarella S, Argos P (1992) Analysis of insertions/deletions in protein structures. J Mol Biol 224:461–471
25. Lorimer IAJ, Pastan I (1995) Random recombination of antibody single chain Fv sequences after fragmentation with DNaseI in the presence of $Mn^{2+}$. Nucleic Acids Res 23:3067–3068
26. Aharoni A, Griffiths AD, Tawfik DS (2005) High-throughput screens and selections of enzyme-encoding genes. Curr Opin Chem Biol 9:210–216
27. Goddard JP, Reymond JL (2004) Recent advances in enzyme assays. Trends Biotechnol 22:363–370
28. Goddard JP, Reymond JL (2004) Enzyme assays for high-throughput screening. Curr Opin Biotechnol 15:314–322
29. Schmidt M, Bornscheuer UT (2005) High-throughput assays for lipases and esterases. Biomol Eng 22:51–56

# Chapter 11

# Transposon-Based Approaches for Generating Novel Molecular Diversity During Directed Evolution

D. Dafydd Jones, James A.J. Arpino, Amy J. Baldwin, and Matthew C. Edmundson

## Abstract

This chapter introduces a set of transposon-based methods that were developed to sample trinucleotide deletion, trinucleotide replacement, and domain insertion. Each approach has a common initial step that utilizes an engineered version of the Mu transposon called MuDel. The inherent low sequence specificity of MuDel results in its random insertion into target DNA during in vitro transposition. Removal of the transposon using a type IIS restriction endonuclease generates blunt-end random breaks at a frequency of one per target gene and the concomitant loss of 3 bp. Self-ligation or insertion of another DNA cassette results in the sampling of trinucleotide deletion or trinucleotide substitution/domain insertion, respectively.

**Key words** Directed evolution, Molecular diversity, Mu transposon, Trinucleotide deletion, Trinucleotide exchange, Domain insertion, Expanded genetic code

## 1 Introduction

Generating molecular diversity lies at the heart of directed evolution [1–4]. With an ever-expanding directed evolution toolbox, different mutational approaches are required. These will access sequence space beyond that sampled by simple gene point mutations that ultimately restrict sequence, structural, and functional diversity [1, 5, 6]. Sampling randomly introduced point mutations severely limits the amino acid range that can be tested at a particular residue position as statistically only one base per codon will ever be mutated [4]. Furthermore, many traditional directed evolution approaches ignore mutations that alter the protein backbone. Such mutations, termed indels (insertion/deletion), are commonly observed among natural protein homologues [7, 8] and range from single amino acid deletions to the insertion of extended protein segments or domains. Mutation of the protein backbone changes the structure of a protein in a manner distinct to that of

Elizabeth M.J. Gillam et al. (eds.), *Directed Evolution Library Creation: Methods and Protocols*, Methods in Molecular Biology, vol. 1179, DOI 10.1007/978-1-4939-1053-3_11, © Springer Science+Business Media New York 2014

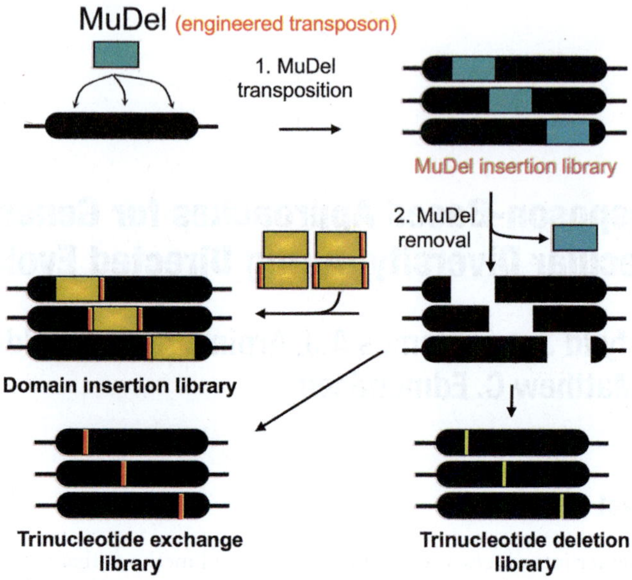


**Fig. 1** Simple outline of the transposon-based random mutagenesis approach. The common first step involves the generation of a MuDel whole-gene insertion library followed by its removal to generate random breaks in the target DNA. From this point, the methods diverge so that different mutational events can be sampled, namely, trinucleotide deletion, trinucleotide exchange, and domain insertion

side chain substitutions, sampling new conformational and hence functional space [9]. Until recently, generating indel mutations for directed evolution libraries was a major challenge for the field. Rational site-directed approaches can be used but generally require detailed information regarding structure and function to guide mutagenesis. Even armed with such information, the local and global structural consequences of rational indel mutagenesis can be difficult to predict.

To address these challenges, a set of transposon-based approaches was developed to sample trinucleotide deletion [10, 11], trinucleotide replacement [12, 13], and domain insertion [14–16] (Fig. 1). Each of these approaches relies on a common first step that ultimately defines genetic diversity sampled across a target gene and thus across the protein backbone: transposition with an engineered version of the Mu transposon, termed MuDel [10].

The Mu transposon, or more accurately mini-Mu [17], is a DNA construct that can be inserted randomly into a target DNA sequence with high efficiency and accuracy but with low target site preference. Transposition is performed in vitro using the commercially available MuA transposase. To implement the various mutagenesis strategies, mini-Mu is modified symmetrically close to both termini to introduce the recognition sequence for MlyI, a type IIS restriction endonuclease that cuts outside its recognition sequence.

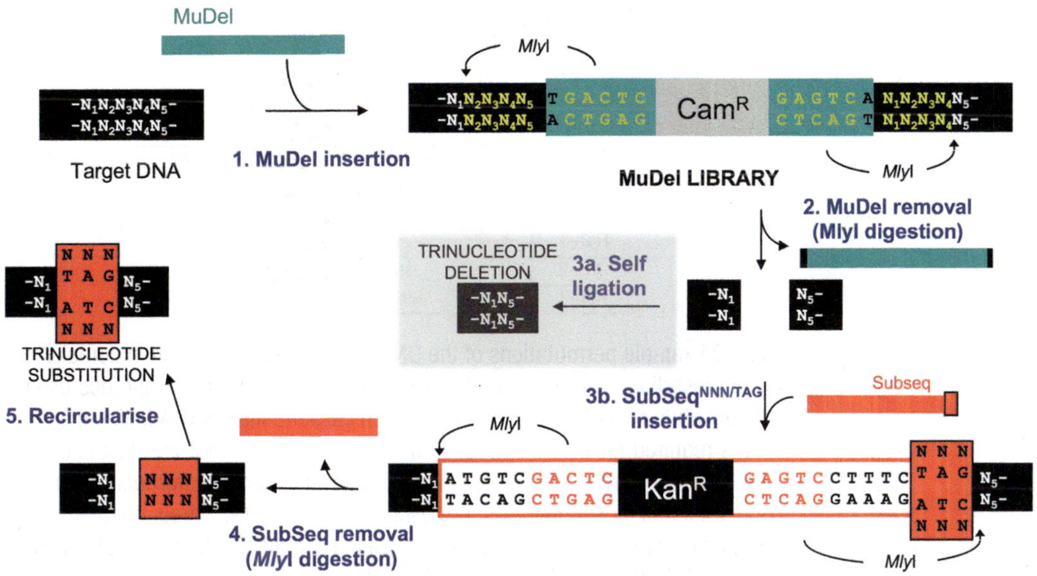


**Fig. 2** Mechanism of MuDel-based mutagenesis. *Step 1.* MuDel insertion. Two MlyI recognition sites (5′GAGTC(N)$_5$↓) are placed 1 bp away from the site of transposon insertion. Insertion of MuDel results in the duplication of 5 bp ($N_1N_2N_3N_4N_5$) of the target gene at the insertion point. *Step 2.* Digestion with MlyI removes MuDel together with 8 bp of the target DNA (4 bp at each end), which equates to removal of a contiguous 3 bp sequence from the starting target gene ($N_2N_3N_4$). *Step 3a.* TND. Self-ligation at this stage results in the reconstitution of the full-length gene but with 3 bp ($N_2N_3N_4$) removed. *Step 3b.* TriNEx. SubSeq is ligated into the gap vacated by MuDel. SubSeq contains two MlyI recognition sites strategically placed toward the ends of the cassette. One site is located so that MlyI will cut at the exact point where the target DNA joins SubSeq. The second site will cut 3 bp into SubSeq so donating 3 bp ($N_XN_YN_Z$) to the target DNA. *Step 4.* Digestion with MlyI removes SubSeq but with 3 bp of its sequence now replacing the 3 bp deleted from the target gene. *Step 5.* Intramolecular ligation reforms the target gene but with one contiguous trinucleotide sequence replaced with another

Placement of the MlyI sites within MuDel, in combination with the transposon insertion mechanism, results in the removal of 3 bp at random positions within a target gene upon MlyI digestion (Fig. 2). Self-ligation of the resulting DNA generates a library of proteins with single amino acids deleted at random positions.

Deletion of trinucleotide sequences opens up the possibility of their replacement with a new trinucleotide sequence and is the basis for trinucleotide exchange (TriNEx) ([13]; Fig. 2). This approach uses the expanded amino acid range of site-directed saturation mutagenesis in combination with the broader sampling capacity of whole-gene mutagenesis. To implement TriNEx an additional DNA cassette termed SubSeq is used, which donates a new trinucleotide sequence back to the target gene. MlyI is used to achieve trinucleotide exchange by means of the mechanism outlined in Fig. 2. The donating sequence in SubSeq is determined by the user and can be randomized [13] or defined [12]. The latter has been used to perform nonnatural amino acid whole-gene directed evolution.

| Cassette Library | ORF | Cassettes |
|---|---|---|
| None-1 | I | GCA DOMAIN AGG |
| None-2 | II | CGCA DOMAIN AG |
| None-3 | III | CTGCA DOMAIN A |
| GGS-1 | I | GGCGGTAGCGCA DOMAIN AGGGGTGGCAGC |
| GGS-2 | II | CGGCGGTAGCGCA DOMAIN AGGGGTGGCAG |
| GGS-3 | III | CTGGCGGTAGCGCA DOMAIN AGGGGTGGCA |
| Ran-1 | I | NNSGAT DOMAIN CGTNNS |
| Ran-2 | II | SNNSGAT DOMAIN CGTNNSNN |
| Ran-3 | III | NSNNSGAT DOMAIN CGTNNSN |

**Fig. 3** Example permutations of the DNA insert cassette for protein domain/segment insertion. ORF refers to the different open reading frame libraries used to reconstitute an in-frame chimeric gene on cassette insertion. Additional nucleotides required for frame correction are *underlined*. The first and last codons of the protein insertion coding region are shown on a *black background*. The rest of the intervening sequence is represented by DOMAIN. Three domain linking sequence variations are shown: (1) no linking sequences (None-X), (2) GlyGlySer linker (GGS-X), (3) random short linker (Ran-X)

Domain insertion [18–20] is a powerful approach to generate new protein scaffolds in which the functions of normally disparate proteins can be coupled, generating new components for use in, for example, biosensing, synthetic biology, and bionanotechnology. Domain insertion is achieved using the MuDel approach by replacing SubSeq with a DNA cassette encoding a protein segment of choice [14–16] (Fig. 1). As well as sampling a variety of insertion positions across a target gene, the nature of the DNA cassette itself can form the basis of a mini-library (Fig. 3). The ends of the DNA cassette can be modified to (1) overcome reading frame artifacts due to cross-codon deletion events during MuDel excision so increasing the diversity of insertion positions sampled and (2) vary the nature of the linking sequences between the host protein and the introduced protein segment/domain.

## 2 Materials

### 2.1 Transposition and Mutagenesis Components

1. pNOM [10] (*see* **Note 1**).
2. pNOM-XP3 [15] (*see* **Notes 1–3**).
3. Transposons. The key sequence details of the MuDel transposon are shown in Fig. 2 (*see* **Notes 4** and **5**).
4. MuA transposase from Finnzymes (now Thermo Scientific, Waltham, MA, USA).
5. MuA transposase reaction buffer: 25 mM Tris–HCl, pH 8.0, 10 mM $MgCl_2$, 110 mM NaCl, 0.05 % Triton, and 10 % (v/v) glycerol. This is supplied as 5× reaction buffer.

6. SubSeq DNA cassette. The SubSeq DNA cassette is required for TriNEx; the essential sequence features are shown in Fig. 2 (*see* **Note 6**).

### 2.2 General Materials

1. Deionized or Milli-Q™ (MQ) water was used throughout. Water was sterilized by autoclave. Ampicillin and kanamycin were made as 100 and 25 mg/ml stock solutions, respectively, and filter sterilized with a 0.22 μm filter unit. Chloramphenicol was made as a 34 mg/ml stock solution in ethanol (without filter sterilization). The stock solutions were used to supplement bacterial growth media by dilution to an appropriate working concentration.
2. *E. coli* cell strains (electrocompetent or chemically competent) with high transformation efficiencies (>$10^9$ cfu per μg pUC18 DNA) are required. DH5α and NovaBlue (Merck KGaA, Darmstadt, Germany) strains are used for library construction, and various versions of BL21 (DE3) including BL21-Gold (DE3) (Stratagene, La Jolla, CA, USA) and Tuner™ (DE3) (Novagen, Merck Biosciences, Darmstadt, Germany) are used for protein expression.
3. Luria Bertani broth (LB) medium: per liter, mix 5 g yeast extract, 10 g tryptone and 10 g NaCl. Autoclave at 121 °C for 25 min and let cool to ambient temperature.
4. LB agar medium: per liter, mix 5 g yeast extract, 10 g tryptone, 10 g NaCl, and 15 g agar. Autoclave at 121 °C for 25 min and then cool to 50 °C in a warm water bath. Add appropriate antibiotic selection marker and gently shake to homogenize the solution, then pour plates and let agar solidify at room temperature. Plates can be stored at 4 °C for up to 4 weeks.
5. Molecular biology grade agarose: prepare gels by boiling 0.6–2.0 % agarose (w/v) in 1× TAE buffer (40 mM Tris–acetate, 1 mM EDTA), cool and then add 0.5 μg/ml ethidium bromide.
6. Restriction endonucleases (with reaction buffers), e.g., MlyI, NdeI, and XhoI.
7. 100× BSA stock solution: 10 mg/ml.
8. Quick T4 DNA ligase system (New England Biolabs, Ipswich, MA, USA).
9. T4 polynucleotide kinase.
10. 10× T4 polynucleotide kinase reaction buffer: 70 mM Tris–HCl, 10 mM $MgCl_2$, 5 mM DTT pH 7.6.
11. 10× ATP solution: 10 mM ATP in water.
12. APex™ heat-labile alkaline phosphatase (1 unit/μg DNA) (Epicentre® Biotechnologies, Madison, WI, USA).
13. Oligonucleotide primers: 100 μM stock solutions.

14. GoTaq® DNA polymerase (Promega, Madison, WI, USA) system.
15. Phusion® high-fidelity DNA polymerase (New England Biolabs, Ipswich, MA, USA) system.
16. PCR, gel extraction, MinElute, plasmid miniprep, and plasmid midiprep purification kits (Qiagen, Hilden, Germany), used according to the manufacturer's guidelines. DNA concentrations and 260/280 nm ratios after purification were estimated using a NanoDrop® ND-1000 UV–Vis spectrophotometer (Thermo Scientific, Pittsburgh, PA, USA).

## 3 Methods

### 3.1 MuDel Insertion Library Construction

Prior to transposition, clone the target gene within the multiple cloning site of pNOM or a pNOM-based vector (*see* **Notes 7** and **8**).

1. Transposition with MuA transposase (Thermo Scientific). Add target plasmid DNA (100 ng equivalent to pNOM) and MuDel (20–100 ng) to MuA transposase reaction buffer containing 0.22 μg of MuA transposase in a final volume of 20 μl. Leave reactions at 30 °C for 3 h and then heat inactivate at 75 °C for 10 min.
2. Transform chemically competent *E. coli* cells with the equivalent of 5–20 ng of plasmid DNA from the deactivated transposition reaction (*see* **Note 9**).
3. Plate a small proportion of transformed cells (~1 %) on LB agar plates supplemented with 20 μg/ml chloramphenicol to assess transposition efficiency and calculate the total number of target plasmid molecules containing MuDel. After overnight incubation at 37 °C, extrapolate the number of colonies back to the number present in the total transformation mixture (*see* **Notes 10** and **11**).
4. Use the remaining portion of transformed cells to inoculate 1 L LB broth supplemented with 20 μg/ml chloramphenicol and incubate in a shaker at 37 °C for 12–16 h. Purify the pooled MuDel insertion plasmid DNA library from liquid culture using Qiagen Plasmid Midi Kit and store at −20 °C (*see* **Note 12**).
5. Isolate MuDel inserted within the target gene, and remove any insertions that have occurred within the plasmid, by digesting the pooled plasmid DNA with the original restriction enzymes used to clone the target gene into pNOM or its derivatives (e.g., NdeI and XhoI). Separate the resulting fragments by agarose gel electrophoresis. Restriction digestion will produce four DNA fragments: (A) plasmid backbone with MuDel, (B)

plasmid backbone without MuDel, (C) target gene with MuDel, and (D) target gene without MuDel.

6. Isolate the two bands corresponding to fragments B and C and purify the DNA from the gel matrix using a QIAquick gel extraction kit (Qiagen).
7. Ligate the two purified fragments using the Quick Ligation™ Kit (NE Biolabs) and transform into *E. coli*. As the two DNA fragments originated from the same restriction digestion, they may be presumed to be equimolar, so use appropriate volumes of each fragment in the ligation reaction to sample from 1:1 to 3:1 insert to vector ratios.
8. To calculate the total number of clones containing MuDel that comprise the library, plate a small proportion of transformed cells (~1 %) on LB agar plates supplemented with 20 μg/ml chloramphenicol. After overnight incubation at 37 °C, extrapolate the number of colonies back to the number present in the total transformation mixture (*see* **Note 13**).
9. Use the remaining portion of transformed cells to inoculate 1 L LB broth supplemented with 20 μg/ml chloramphenicol and incubate in a shaker at 37 °C for 12–16 h. Purify the pooled MuDel target gene insertion plasmid DNA library from liquid culture and store at −20 °C (*see* **Notes 14** and **15**).

### 3.2 Construction of Trinucleotide Deletion Library (TND)

The initial MuDel library constructed as outlined in Subheading 3.1 is the basis for TND library generation. At this stage, all the clones present in the MuDel library are assumed to have MuDel inserted within the target gene.

1. Remove MuDel from transposon-insert library by digestion with MlyI (New England Biolabs). Perform digestion with 1 U of enzyme per μg DNA in NEBuffer 4 supplemented with 0.1 mg/ml BSA in a total reaction volume of 50 μl. Incubate the reaction at 37 °C for 1 h/μg DNA (*see* **Note 16**).
2. Separate the resulting DNA fragments by agarose gel electrophoresis. Extract the DNA fragment equivalent to the size of the original starting plasmid (pNOM or derivative) plus the target gene using a QIAquick gel extraction kit (Qiagen). Discard the MuDel band.
3. Recircularize the purified linear library DNA (50 ng) by intramolecular ligation with the Quick Ligation™ Kit (NE Biolabs). Incubate ligation reactions at 25 °C for 20 min (*see* **Note 17**).
4. Use 1 μl of the ligation reaction mixture to transform electrocompetent or chemically competent *E. coli* cells. Grow transformed cells on LB agar plates supplemented with 100 μg/ml ampicillin (*see* **Note 18**).
5. The library can now be screened to identify protein variants with desired properties.

### 3.3 Construction of a Trinucleotide Exchange (TriNEx) Library

The initial MuDel library constructed as outlined in Subheading 3.1 will be the basis for generating the TriNEx library. At this stage, all the variants present will be deemed to have MuDel inserted within the target gene.

1. Generate the SubSeq DNA cassette (Fig. 2) by PCR as outlined in Subheading 2.1, **item 4**. Phosphorylate the SubSeq PCR product using 10 U T4 polynucleotide kinase (New England Biolabs) in 1× T4 polynucleotide kinase reaction buffer (New England Biolabs) and 1 mM ATP. Incubate the reaction at 37 °C for 1 h then heat inactivate at 65 °C for 20 min. Purify the resulting DNA using the QIAquick PCR purification kit.
2. Remove MuDel from the pooled plasmid DNA by digestion with MlyI as outlined in Subheading 3.2, **step 1**. Dephosphorylate digested DNA by adding APex heat-labile alkaline phosphatase (1 unit/μg DNA) (Epicentre Biotechnologies) to the restriction endonuclease reaction and incubate at 37 °C for the duration of the restriction digest. Heat inactivate the phosphatase and MlyI by incubating the reaction mixture at 70 °C for 5–10 min.
3. Separate the resulting DNA digest by agarose gel electrophoresis and extract the fragment equivalent to the size of the original starting plasmid (pNOM or derivative) plus the target gene from the gel.
4. Ligate the SubSeq DNA cassette into linear plasmid DNA with T4 DNA ligase for 30 min at room temperature. In a total volume of 20 μl, use a molar ratio of 1:3 (pNOM–SubSeq) with a total of 200–250 ng DNA in each reaction (*see* **Note 19**).
5. Electroporate 2 μl of the ligation reaction (1/10th) into *E. coli* DH5α cells. Calculate the number of transformed cells containing SubSeq incorporated within the host plasmid, by plating the equivalent of 2–3 % of the electroporated cells on LB agar plates supplemented with 25 μg/ml kanamycin. Incubate the plates overnight at 37 °C and then count the number of colonies observed to calculate ligation and transformation efficiency. The observed transformation efficiency is expected to be 3–4 × $10^4$ cfu/μg DNA (*see* **Note 20**).
6. Add the remaining portion of transformed cells to 50 ml of LB broth containing 200 μg/ml kanamycin and incubate in a shaking incubator at 37 °C for 16 h.
7. Isolate plasmid DNA from the culture using a QIAprep spin miniprep kit (Qiagen) and store at −20 °C (*see* **Note 21**).
8. Digest the purified plasmid DNA (2 μg) with MlyI for 1 h at 37 °C (Subheading 3.2, **step 1**) to remove the SubSeq section,

with the terminal trinucleotide sequence of SubSeq donated to the target gene. Agarose gel electrophoresis was used to separate excised SubSeq from linearized DNA corresponding to the host plasmid containing the target gene. Extract the latter from the gel as outlined in Subheading 3.2, **step 2**.

9. Recircularize the linear plasmid DNA (~50–100 ng) using T4 DNA ligase as outlined in Subheading 3.2, **step 3**.
10. Transform the resulting library as outlined in Subheading 3.2, **step 4**.
11. The library can now be screened to identify protein variants with the required properties.

### 3.4 Construction of a Domain Insertion Library

The MuDel library constructed as outlined in Subheading 3.1 is the basis for generating the domain insert library. At this stage, all the variants present are assumed to have MuDel inserted within the target gene.

1. Construct the DNA cassette comprising the protein segment/domain insert by PCR using the Phusion high-fidelity DNA polymerase system (New England Biolabs). The general strategy for constructing the domain insert is outlined in Fig. 3. Design terminal oligonucleotides to incorporate linker sequence and open reading frame (ORF) requirements. With regard to the latter, design different ORF versions of the DNA insert cassette to take into account cross-codon trinucleotide deletions inherent on MuDel excision. This increases DNA cassette insertion position diversity within the target gene (*see* **Note 22**).
2. Phosphorylate the different ORF DNA insertion cassettes as outlined in Subheading 3.3, **step 1**.
3. Remove MuDel from the pooled plasmid DNA containing the target gene and dephosphorylate as outlined in Subheading 3.3, **step 2**.
4. Ligate the DNA insert cassette into linear plasmid DNA using T4 DNA ligase for 30 min at room temperature as described in Subheading 3.3, **step 4**. Perform individual ligations for each ORF version of the DNA insert cassette.
5. Transform the ligation reactions into *E. coli* cells by electroporation. Plate transformed cells on LB agar supplemented with 100 μg/ml ampicillin and incubate at 37 °C for at least 16 h.
6. Assess DNA cassette insertion efficiency by screening colonies by colony PCR using the GoTaq DNA polymerase (Promega) system and primers flanking the target gene. Positive clones will produce an amplified DNA product corresponding to the size of the original target gene plus the DNA cassette insert.

7. Further analyze positive clones by PCR to assess frequency of correct DNA cassette insertion orientation and diversity of the insertion position. Use a primer pair comprising a target gene-flanking primer and a DNA cassette insert construction primer from Subheading 3.4, **step 1** (*see* **Note 23**).
8. The libraries are now ready to be screened to identify domain/segment insertion variants with desired properties.

## 4 Notes

1. Plasmids pNOM [10] and pNOM-XP3 [15] are bespoke plasmids available from the authors for use with MuDel and host the target gene. pNOM is the original vector and is based on pUC18 backbone with pNOM-XP3 being an inducible target gene overexpression version of pNOM. pNOM (*p*lasmid *NO M*lyI) was constructed to remove all MlyI sites present in the original pUC18 plasmid (−1 to 1,979 bp region) and to incorporate a useful multiple cloning site with the common restrictions sites NdeI and XhoI at either end. Removal of MlyI sites notably changed 2 bp in the origin of replication and introduced a silent mutation in the *bla* ampicillin resistance gene; neither appeared to be detrimental to the use of the plasmid. pNOM-XP3 is a derivative of pNOM containing the inducible T7 promoter upstream of the target gene. The original T7 promoter was mutated to remove an MlyI recognition sequence. The NdeI recognition sequence is present downstream from the T7 promoter to aid in-frame initiation codon cloning.
2. Versions of pNOM-XP3 are available with NcoI in place of NdeI.
3. Target protein expression can be leaky when using pNOM-XP3. We assume this is due to the mutated T7 promoter.
4. MuDel was generated from the commercially available Cam$^{R}$ containing Mu Entranceposon present within the pEntranceposon plasmid (http://www.thermoscientificbio.com/mutagenesis/entranceposons/). MuDel was recloned back into the pEntranceposon plasmid within the BglII site. MuDel was removed from the plasmid for use in the mutagenesis procedure using BglII and separated from the plasmid backbone by agarose gel electrophoresis followed by gel excision and purification using the Qiagen QIAquick gel extraction kit.
5. We have tried using PCR amplified forms of MuDel but have found that BglII-digested transposon from plasmid gives by far the best results.
6. The SubSeq$^{NNN}$ DNA cassette was constructed by PCR using the Phusion™ high- fidelity DNA polymerase system (New

England Biolabs). The $Kan^R$ gene region of the Entranceposon™ ($kan^R$-3) (Finnzymes) was amplified as two separate fragments to remove an MlyI site in the middle of the gene and introduce MlyI sites at strategic positions toward the termini of the cassette. The nature of the SubSeq used was determined by a terminal primer 5′[Phos]-**NNN**GAAAGGACTCAGTGTGTCG GCGGCCGCGGATCCT-3′ (MlyI site underlined) where the 5′ end 3 bp sequence (NNN) that is ultimately donated to the target gene can be determined by the user.

7. It is important that any MlyI sites are removed in the target gene. This can be achieved through the use of silent mutations.

8. Although manufacturers state that in vitro transposition into linear DNA with MuA is feasible, we have had no success with this approach so we have always used circular plasmid DNA as the target for MuDel insertion.

9. Chemically competent cells with high transformation efficiencies ($>10^9$ cfu/µg DNA) should be used at this stage due to the relatively high concentrations of DNA.

10. We recommend that the calculated number of colonies should be ~2–3 times the size of the target plasmid (plasmid backbone plus target gene) to ensure maximal transposon insertion coverage.

11. Colony PCR (using primers that flank the target gene) can be performed on randomly selected colonies as an alternative way to calculate the percentage of the library containing MuDel insertions within the target gene. Analysis of amplified products by agarose gel electrophoresis should give two amplified products corresponding to (a) the target gene alone and (b) the target gene plus MuDel (1,307 bp). The percentage of analyzed colonies with the larger size band is representative of the percentage of variants in the library with MuDel inserted within the target gene region of the plasmid.

12. As an alternative option, the remaining cells may be spread onto 150 mm diameter LB agar plates supplemented with 20 µg/ml chloramphenicol. Collecting cells from single colonies of roughly equal size straight from plates can help reduce bias which may result from different growth rates in liquid media. Ensure single colony resolution by pipetting 150–200 µl per plate and dispersing cells with a glass spreader. Following overnight incubation at 37 °C, isolate the MuDel library by adding LB broth (5 ml) to each plate and gently shaking the plates for 10–15 min to suspend cells. Collect the LB broth and purify the MuDel insertion plasmid DNA library from liquid culture using Qiagen Plasmid Midi Kit and store at −20 °C.

13. We recommend that the calculated number of colonies should be >3 times the size of the target plasmid (plasmid backbone plus target gene) to ensure maximal transposon insertion coverage.
14. An additional step may be included at this point to confirm that MuDel has inserted randomly and uniformly across the length of the target gene. The pool of plasmid DNA can be digested with MlyI and one of the restriction endonucleases originally used to clone the target gene into the host plasmid (e.g., NdeI or XhoI). Analysis of the digestion mixture by agarose gel electrophoresis should produce a distinctive band at ~1,310 bp corresponding to excised MuDel and two smears: one ranging from the size of the host plasmid (e.g., 2,146 bp for pNOM-XP3) to the size of the original plasmid plus the target gene and another smear ranging from 0 bp to the size of the target gene. We generally use XhoI to clone our target gene into pNOM or derivative. Thus we normally analyze MuDel insertion diversity with XhoI and MlyI. XhoI serves to linearize the plasmid DNA; the resulting variable length DNA fragments are generated by MlyI digestion and depend on the placement of MuDel within the target.
15. An alternate selection strategy to Subheading 3.1, **steps 5–9** can be used if a negative selection or screen is feasible. All the transformed *E. coli* cells can be plated on LB agar supplemented with chloramphenicol to select for clones containing MuDel within the host plasmid which are subsequently screened for target protein activity. Those colonies that *do* display target protein activity are deemed not to have a disrupted gene due to MuDel insertion. Those colonies that *do not* display target protein activity are deemed to contain MuDel inserted within the gene due to disruption of the protein-coding region.
16. It is important to use relatively fresh MlyI at this stage as in some rare instances we have observed downstream problems when using older stocks of the restriction endonuclease.
17. DNA ligated using the Quick Ligation™ Kit protocol must be column-purified prior to transformation by electroporation. PEG in the Quick Ligation™ buffer inhibits transformation by electroporation reducing transformation efficiency and in some cases causes arcing in the electroporation cuvettes. We have also successfully used standard T4 ligase approaches with extended incubation times.
18. If using the pNOM-XP3 plasmid system, we recommend using *E. coli* strains compatible with T7 promoter systems such as BL21-Gold (DE3) or Tuner™ (DE3).

19. We have used different T4 DNA ligase systems, including the Quick Ligation™ kit by New England Biolabs and the Fast-Link™ ligation kits by Epicentre. All have been successful.
20. It is important to maximize the efficiency of ligation and the resulting transformation so as to maintain high molecular diversity. The SubSeq-plasmid DNA is a blunt-end ligation so efficiency may be lower compared to ligation involving base overhangs.
21. At this stage it is wise to isolate plasmid DNA using a "mini-prep" approach to prevent total loss of the library DNA should large-scale "maxi or midiprep" purification fail. Any unused culture can be divided into aliquots, frozen, and stored until required.
22. To facilitate downstream selection for DNA cassette insertion within the target gene, a $Kan^R$ (kanamycin) gene can be inserted within the DNA cassette between convenient restriction sites. We have used this approach successfully [14] although it does add extra steps to the process.
23. An additional step using the DNA cassette insert with $Kan^R$ placed within the coding sequence (Subheading 3.4, **step 1**) has also been used to select solely for clones containing the DNA cassette insert. The additional step involves transforming *E. coli* with the ligation products of Subheading 3.4, **step 4**. The transformed cells were used to inoculate 50 ml LB broth supplemented with 25 μg/ml kanamycin and incubated at 37 °C in a shaking incubator for ~16 h. Plasmid DNA was isolated from the cultures and the $Kan^R$ cassette removed by digestion with the appropriate restriction endonuclease. Restriction digests were separated by agarose gel electrophoresis and DNA bands corresponding to linear library plasmid DNA isolated and purified by gel extraction. Purified linear plasmid library DNA was recircularized by intramolecular ligation (Subheading 3.2, **step 3**) producing the domain/segment insert libraries, which was then used to transform *E. coli* Tuner™ (DE3) cells by electroporation. The transformed cells were grown on LB agar supplemented with 100 μg/ml ampicillin and 150 μM IPTG (if using the pNOM-XP3 plasmid) and incubated for 24 h at 37 °C.

## Acknowledgments

The authors would like to thank the BBSRC (BB/H003746, BB/E001084, BB/FOF/263, BB/E007384), MRC DPFS (G0900868), Merck KGaA, Wellcome (084542/Z/07/Z), and Nuffield Foundation for supporting this work.

## References


1. Cobb RE, Si T, Zhao H (2012) Directed evolution: an evolving and enabling synthetic biology tool. Curr Opin Chem Biol 16:285–291
2. Jackel C, Kast P, Hilvert D (2008) Protein design by directed evolution. Annu Rev Biophys 37:153–173
3. Lutz S, Patrick WM (2004) Novel methods for directed evolution of enzymes: quality, not quantity. Curr Opin Biotechnol 15:291–297
4. Neylon C (2004) Chemical and biochemical strategies for the randomization of protein encoding DNA sequences: library construction methods for directed evolution. Nucleic Acids Res 32:1448–1459
5. Brustad EM, Arnold FH (2011) Optimizing non-natural protein function with directed evolution. Curr Opin Chem Biol 15:201–210
6. Koide S (2009) Generation of new protein functions by nonhomologous combinations and rearrangements of domains and modules. Curr Opin Biotechnol 20:398–404
7. Chothia C, Gough J, Vogel C, Teichmann SA (2003) Evolution of the protein repertoire. Science 300:1701–1703
8. Taylor MS, Ponting CP, Copley RR (2004) Occurrence and consequences of coding sequence insertions and deletions in Mammalian genomes. Genome Res 14:555–566
9. Shortle D, Sondek J (1995) The emerging role of insertions and deletions in protein engineering. Curr Opin Biotechnol 6:387–393
10. Jones DD (2005) Triplet nucleotide removal at random positions in a target gene: the tolerance of TEM-1 β-lactamase to an amino acid deletion. Nucleic Acids Res 33:e80
11. Simm AM, Baldwin AJ, Busse K, Jones DD (2007) Investigating protein structural plasticity by surveying the consequence of an amino acid deletion from TEM-1 β-lactamase. FEBS Lett 581:3904–3908
12. Baldwin AJ, Arpino JA, Edwards WR, Tippmann EM, Jones DD (2009) Expanded chemical diversity sampling through whole protein evolution. Mol Biosyst 5:764–766
13. Baldwin AJ, Busse K, Simm AM, Jones DD (2008) Expanded molecular diversity generation during directed evolution by trinucleotide exchange (TriNEx). Nucleic Acids Res 36:e77
14. Arpino JA, Czapinska H, Piasecka A, Edwards WR, Barker P, Gajda MJ, Bochtler M, Jones DD (2012) Structural basis for efficient chromophore communication and energy transfer in a constructed didomain protein scaffold. J Am Chem Soc 134:13632–13640
15. Edwards WR, Busse K, Allemann RK, Jones DD (2008) Linking the functions of unrelated proteins using a novel directed evolution domain insertion method. Nucleic Acids Res 36:e78. doi:10.1093/nar/gkn363
16. Edwards WR, Williams AJ, Morris JL, Baldwin AJ, Allemann RK, Jones DD (2010) Regulation of β-lactamase activity by remote binding of haem: functional coupling of unrelated proteins through domain insertion. Biochemistry 49:6541–6549
17. Haapa S, Taira S, Heikkinen E, Savilahti H (1999) An efficient and accurate integration of mini-Mu transposons in vitro: a general methodology for functional genetic analysis and molecular biology applications. Nucleic Acids Res 27:2777–2784
18. Fastrez J (2009) Engineering allosteric regulation into biological catalysts. Chembiochem 10:2824–2835
19. Ferraz RM, Vera A, Aris A, Villaverde A (2006) Insertional protein engineering for analytical molecular sensing. Microb Cell Fact 5:15
20. Ostermeier M (2005) Engineering allosteric protein switches by domain insertion. Protein Eng Des Sel 18:359–364

# Part II

## Recombinatorial Methods

# Chapter 12

# Restriction Enzyme-Mediated DNA Family Shuffling

**James B.Y.H. Behrendorff, Wayne A. Johnston, and Elizabeth M.J. Gillam**

## Abstract

DNA shuffling is an established recombinatorial method that was originally developed to increase the speed of directed evolution experiments beyond what could be accomplished using error-prone PCR alone. To achieve this, mutated copies of a protein-coding sequence are fragmented with DNase I and the fragments are then reassembled in a PCR without primers. The fragments anneal where there is sufficient sequence identity, resulting in full-length variants of the original gene that have inherited mutations from multiple templates. Subsequent studies demonstrated that directed evolution could be further accelerated by shuffling similar native protein-coding sequences from the same gene family, rather than mutated variants of a single gene. Generally at least 65–75 % global identity between parental sequences is required in DNA family shuffling, with recombination mostly occurring at sites with at least five consecutive nucleotides of local identity. Since DNA shuffling was originally developed, many variations on the method have been published. In particular, the use of restriction enzymes in the fragmentation step allows for greater customization of fragment lengths than DNase I digestion and avoids the risk that parental sequences may be over-digested into unusable very small fragments. Restriction enzyme-mediated fragmentation also reduces the occurrence of undigested parental sequences that would otherwise reduce the number of unique variants in the resulting library. In the current chapter, we provide a brief overview of the alternative methods currently available for DNA shuffling as well as a protocol presented here that improves on several previous implementations of restriction enzyme-mediated DNA family shuffling, in particular with regard to purification of DNA fragments for reassembly.

**Key words** DNA shuffling, Mutant libraries, Protein engineering, Restriction enzyme

## 1 Introduction

DNA shuffling [1, 2] was introduced 20 years ago as a method by which to easily recombine different genes by template switching in a polymerase chain reaction (PCR). DNA shuffling mimics natural evolution in allowing recombination as well as point mutagenesis, a process described as "molecular breeding." In the basic method, presented in detail in the first edition of this book [3], two different genes are fragmented using DNAse I, separated by

Elizabeth M.J. Gillam et al. (eds.), *Directed Evolution Library Creation: Methods and Protocols*, Methods in Molecular Biology, vol. 1179, DOI 10.1007/978-1-4939-1053-3_12, 

gel electrophoresis, purified, and then allowed to reassemble in a primerless PCR reaction. The fragments hybridize to each other in regions of shared homology, priming extension by a thermostable polymerase. The iterative PCR process allows repeated template switching such that full-length sequences are ultimately generated that represent chimeras (or mosaics) of the original parental templates. DNA shuffling also allowed the addition of synthetic oligonucleotides to enable enrichment of defined regions of the sequence with particular mutations [2].

DNA family shuffling [4] is a modification of the original DNA shuffling method [1], whereby multiple homologous genes sourced from nature are recombined to produce novel mosaic sequences, as distinct from shuffling a single native sequence diversified by error-prone PCR. In DNA family shuffling, the starting sequences (parental genes) are fragmented, separated by gel electrophoresis, gel purified, and then reassembled in a PCR in which fragments from different parental genes can anneal to one another where there is sufficient sequence identity. In the original example, four cephalosporinase genes with 59–82 % nucleotide sequence identity were shuffled, producing mutants with up to 270- to 540-fold improvement in moxolactamase activity, as compared to an eightfold improvement obtained by shuffling any of the four native genes individually [4]. This marked difference could be attributed to the greater sequence space sampled by shuffling divergent genes as compared to the limited point mutations introduced by mutagenic PCR. Furthermore, shuffling preexisting functional genes, rather than genes mutated in vitro, was thought to reduce the frequency of mutants rendered nonfunctional by deleterious point mutations, commonly introduced by random mutagenesis but purged from the extant genes during natural evolution.

Despite the proliferation of alternative random, recombinatorial, and structure-guided approaches developed for directed evolution in the past two decades, methodologies based on DNA family shuffling remain an efficient method for creating libraries in which a high proportion of mutants are properly folded and possess functional properties distinct from the starting material. They provide a reasonable compromise between reagent and labor cost on the one hand and flexibility in customizing the fragmentation step on the other. Importantly, shuffling methods require no understanding of protein structure but can create libraries that contain a high proportion of functional proteins. However, these methods are limited by the requirement for significant (usually >70 %) sequence identity in the parental stock.

Since the publication of the original DNA shuffling and family shuffling methods [1, 4], a host of other variations on this basic technique have emerged [5–10], generally developed to address two issues: the desire to recombine templates with greater sequence diversity and the requirement to control the fragmentation process both in terms of extent and location. Other improvements have been made to the methods used to separate and purify fragments following cleavage [11, 12]. In this chapter, the modifications introduced in each of the variant shuffling techniques and their respective advantages and disadvantages will be reviewed briefly before a detailed protocol is provided for one such variation, restriction enzyme-mediated DNA shuffling, chosen for its technical simplicity and efficiency in creating high-quality libraries. The scope of our overview will be limited to those methods that are related to DNA shuffling by the use of a PCR-based reassembly step to recombine multiple fragments of double-stranded DNA.

### 1.1 Restriction Enzyme-Mediated DNA Family Shuffling

The DNAse I-mediated fragmentation step in DNA shuffling can be difficult to control [9], due to the high activity of this endonuclease. Thus, one of the first variations on DNA family shuffling was the use of restriction enzymes to perform the cleavage reaction [5]. Two or more sets of restriction enzymes are used to cleave the parental genes into fragments of a defined maximal size determined by the cleavage patterns of the enzymes chosen. While fragmentation is not random, the sequence specificity of restriction enzymes means that the fragmentation can be treated as an end point rather than a kinetic assay, markedly reducing the need for optimization of this step and improving reproducibility. We have found that in practice this yields highly diverse libraries with satisfactory structural integrity when using parental genes that share ~80 % identity or more [11, 13, 14]. Judicious choice of restriction enzymes can allow limited control over fragment size. While early exemplifications of this method involved isolation of all short fragments by gel purification [5] or filtration [11, 13], recent work has shown that reassembly efficiency and library quality are both improved by reducing the bias towards purification of very short fragments and increasing the efficiency of recovery of fragments from agarose gels [12]. Choice of restriction enzymes and analysis of recombination patterns in the resultant mutants are both facilitated by specialized purpose-designed software (*see* below). Importantly, restriction enzyme-mediated family shuffling results in libraries with less contamination by parental sequences [5]. The optimized protocol for restriction enzyme-mediated DNA family shuffling is presented in detail below.

### 1.2 Methods Based on dUTP Incorporation into Templates

Nucleotide exchange and excision technology (NExT) shuffling [8] was also developed to simplify the preparation of reassembly templates. In this method, templates are prepared by PCR in the presence of deoxyuridine triphosphate (dUTP). The uracil is later excised enzymatically with uracil DNA glycosylase creating an abasic site at which the DNA backbone is then cleaved chemically with piperidine or sodium hydroxide. The fragments so produced are then reassembled as in the basic DNA shuffling protocol. The key advantages of NExT are that the mean fragment length can be adjusted by changing the dUTP–dTTP ratio and the DNA fragments can be used directly in a reassembly PCR with minimal purification with standard commercial PCR cleanup kits, rather than electrophoretic separation. A detailed method for NExT shuffling has been provided in a previous protocol [15].

T4 endonuclease V can be substituted for chemical treatment in the backbone fragmentation step using templates amplified in the presence of dUTP and pretreated with uracil DNA glycosylase [9]. However, the endonuclease cleavage reaction is thought to yield an unfavorable terminus for extension by the thermostable polymerase, leading to poorer fragmentation than with piperidine or sodium hydroxide [8].

### 1.3 Methods Based on Generation of Single-Stranded Templates

Where parental sequences share less sequence identity, contamination of the mutant library with recombined parental sequences can be a significant problem. Kikuchi et al. [6] addressed this issue by using heteroduplexes formed from hybridizing complementary single-stranded DNA prepared from different templates. In this original method, parental genes were subcloned into a phagemid vector and ssDNA was generated using helper phage. Complementary single-stranded templates from different parents were then hybridized prior to fragmentation with DNAse I and reassembly. Zha et al. [7] streamlined this technique as described in detail in the first edition of this text, by instead amplifying parental templates using primer pairs in which one primer was phosphorylated at the 5′ end. The strand containing a 5′ phosphate was then selectively degraded using lambda exonuclease.

### 1.4 Methods Based on Reassembly of PCR-Amplified Fragments

Degenerate oligonucleotide gene shuffling (DOGS) [10] represents one of the earliest methods to engineer recombination between templates at specific positions. In DOGS, complementary primers with a central nondegenerate clamp flanked by degeneracy at both ends are used to amplify specific segments of parental templates that are then combined in an overlap extension PCR. Although effective, this approach is much more labor intensive than methods involving fragmentation and depends on the presence and position of regions of high sequence identity for the choice of recombination points.

Kolkmann and Stemmer advocated designing recombination points at the naturally occurring exon boundaries in exon shuffling [16]. Here, rather than using degenerate oligonucleotides to amplify modules that could recombine with neighboring modules from multiple parents, the individual segments of parental genes were amplified with multiple primers that each encoded a different recombination event. That is, if there were three different parents, each segment from each parent was amplified by three different forward and reverse primers, the ends of which hybridized exactly with the termini of neighboring segments from each individual parent. The exact specification of recombination sites in multiple nondegenerate primers, while expensive in terms of cost of oligonucleotides, enabled recombinations to be engineered anywhere within the sequence independently of sequence homology, thereby effectively removing the lower limit of sequence identity required for shuffling.

### 1.5 Other Related Recombinatorial Methods

Methods such as Golden Gate shuffling [17], which enables the recombination of fragments based on a limited number of regions of sequence homology distributed throughout the parental sequence using site-selective ligation rather than a PCR-based reassembly step, are beyond the scope of the present chapter. Similarly, assembly of designed oligonucleotides (ADO, explained in detail in the following chapter in this volume) is comparable to shuffling methods in utilizing a PCR-based assembly step, but the mutants are assembled entirely from synthetic oligonucleotides. Likewise, in synthetic shuffling [18] a mixture of degenerate and nondegenerate oligonucleotides that encode all the variations in a set of homologous genes is reassembled by PCR to produce a library of mutant sequences.

In summary, a number of methods based on DNA (family) shuffling are now available in which different aspects of the basic process have been improved. As a general purpose technique, restriction enzyme-mediated DNA family shuffling offers technical advantages in that the fragmentation step is simple, reproducible and can be customized to some degree, and that the method is relatively cost-effective and quick. This protocol below describes an updated and optimized method for restriction enzyme-mediated DNA family shuffling, incorporating improvements from recent publications.

## 2 Materials

### 2.1 Materials for DNA Fragmentation

1. The desired parental coding sequences for shuffling (*see* **Note 1**) cloned into the same backbone vector. Standard alkaline lysis preparations of plasmid DNA [19] are of sufficient quality for DNA shuffling.
2. Restriction enzymes and manufacturer-recommended buffers.

### 2.2 Materials for Agarose Gel-Based Fragment Selection

1. SYBR® Safe DNA gel stain (Invitrogen, Carlsbad, CA, USA).
2. 1 % agarose in 1× TAE buffer: 40 mM Tris–acetate, 1 mM EDTA pH 8.0.
3. 10× DNA loading buffer: 40 mM Tris–acetate, 10 mM EDTA, 30 % glycerol (v/v), pH 8.0.

4. Quick-Load 100-bp DNA ladder (New England Biolabs, Ipswich, MA, USA) or other equivalent molecular size marker.
5. Blue light source (*see* **Note 2**) and amber light filter (*see* **Note 3**).
6. QIAquick Gel Extraction kit (Qiagen, Doncaster, VIC, Australia).

### 2.3 Materials for Selection of Fragments with PCR Filter Units

1. Montage PCR filter units (Millipore, Billerica, MA, USA).

### 2.4 Materials for Reassembly, Amplification, and Cloning of Library Sequences

1. General PCR supplies including a commercially available *Taq* polymerase with provided buffer and deoxynucleotide triphosphates.
2. QIAquick PCR cleanup kit (Qiagen, Doncaster, VIC, Australia) or equivalent.
3. MinElute gel extraction kit (Qiagen, Doncaster, VIC, Australia) or equivalent.
4. SYBR® Safe DNA gel stain (Invitrogen, Carlsbad, CA, USA).
5. 1 % (w/v) agarose in 1× TAE buffer: 40 mM Tris–acetate, 1 mM EDTA pH 8.0.
6. 10× DNA loading buffer: 40 mM Tris–acetate, 10 mM EDTA, 30 % glycerol (v/v), pH 8.0.
7. Restriction enzymes for cloning into the plasmid vector of interest.
8. T4 DNA ligase and manufacturer-supplied buffer.
9. Ultracompetent DH5αF'IQ™ strain *E. coli* (Invitrogen, Carlsbad, CA) or other suitable microbial host for shuffled libraries (*see* **Note 4**).
10. SOC medium: 20 g/L bactotryptone, 5 g/L yeast extract, 10 mM NaCl, 2.5 mM KCl, 10 mM $MgCl_2$, 10 mM $MgSO_4$. Sterilize by autoclaving, then add glucose to 20 mM when cool.
11. LB agar containing appropriate antibiotics to select for the plasmid encoding DNA shuffling products: 10 g/L bactotryptone, 5 g/L yeast extract, 10 g/L NaCl, 15 g/L agar, adjust to pH 7 with 10 N NaOH and autoclave to sterilize.
12. 2× LB medium: 20 g/L bactotryptone, 10 g/L yeast extract, 20 g/L NaCl, adjust to pH 7 with 10 N NaOH and autoclave to sterilize.
13. 60 % (w/v) glycerol, autoclaved (*see* **Note 5**) or filter sterilized.
14. X-Pierce sealing membranes (Excel Scientific, Victorville, CA, USA).

## 3 Methods

The general procedure for creating a library of shuffled enzymes involves the fragmentation of homologous coding sequences using two different sets of restriction enzymes, separation of the desired fragments from undigested material, reassembly of the fragments into mosaic sequences in a primerless PCR, amplification of the shuffled sequences via a conventional PCR, and cloning of the novel sequences into a suitable vector for expression in the desired host organism.

### 3.1 Fragmentation of Parental cDNAs

Two or more coding sequences with >65 % nucleotide sequence identity should be selected (*see* **Note 6**). Each plasmid carrying a different parental coding sequence should be fragmented with two different combinations of restriction enzymes, in order to facilitate hybridization of overlapping fragments.

1. Identify two different sets of restriction enzymes that can be used to fragment the parental coding sequences into segments that are generally between 100 and 600 bp in length (*see* **Notes 7** and **8**). Useful free tools that can aid in this planning step are NEBcutter v2.0 (available at http://tools.neb.com/NEBcutter2) or RE Cut (developed by Dr. Weiliang Huang in the laboratory of Prof. Elizabeth Gillam and available for download at http://www.scmb.uq.edu.au/staff/elizabeth-gillam).
2. In duplicate sterile 0.6-mL tubes, combine parental plasmids in equal quantities so that the total mass of DNA per tube is 50 μg (e.g., 25 μg of each plasmid when using two parental sequences or 12.5 μg of each plasmid when using four parents). One duplicate will be digested with each combination of restriction enzymes.
3. To each tube add the desired restriction enzymes plus buffers to achieve reaction conditions in accordance with the enzyme manufacturer's recommendations, in a total reaction volume of 50 μL.
4. Mix gently and incubate for 4 h at the recommended temperature for the relevant restriction enzymes.

### 3.2 Selection of Fragments by Agarose Gel Electrophoresis with In-Gel Fragment Selection and Concentration Under Reversed Polarity

Following fragmentation with restriction enzymes, large DNA fragments, including undigested plasmid and any full-length parental coding sequences, must be removed to avoid contamination of the shuffled library with native sequences. Two principal methods have been used in this laboratory to accomplish this: filtration through PCR filter units [11] (*see* Subheading 3.2, **step 2**) and physical separation of DNA fragments by an agarose gel-based technique outlined below ([12]; *see* Subheading 3.2, **step 1**, Fig. 1). The gel-based method is recommended, based on a greater ability

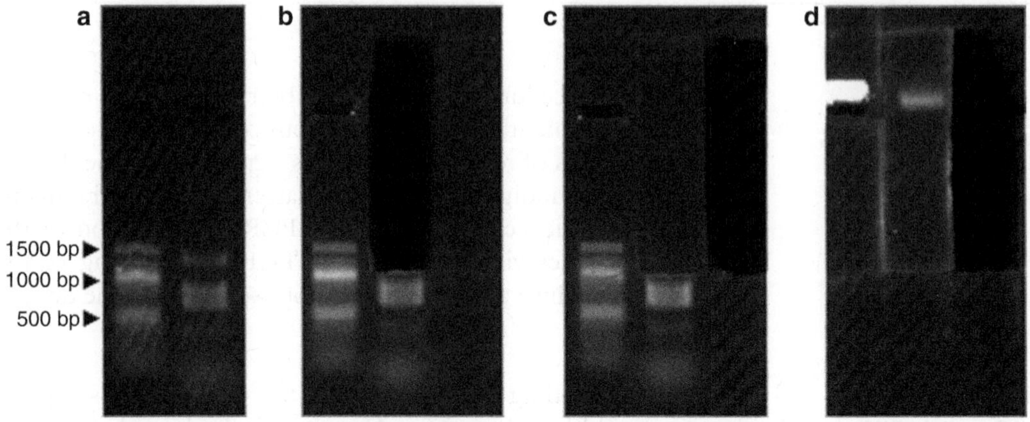


**Fig. 1** Separation, selection, and reconcentration of DNA fragments. (**a**) Electrophoresis of a 100-bp DNA marker ladder (*left*, New England Biolabs Quick-Load 100-bp DNA ladder) and a digested plasmid (*right*, pCW'/NF14 [25] digested with NdeI and NciI). (**b**) Agarose gel containing pCW'/NF14 DNA fragments greater than 1,000 bp was excised from the gel. (**c**) A fresh block of agarose was inserted. (**d**) DNA electrophoresis was reversed to reconcentrate the separated bands into a single band

to customize the fragment range selected and better resultant library quality (*see* **Note 9**).

In a modification of the method by Kadokami and Lewis [20], fragments can be isolated efficiently from agarose gels within a user-defined size range from a complex mixture by excising unwanted fragments and repeating the electrophoresis under reversed polarity to concentrate the fragments of interest prior to extraction from the gel. The steps in this procedure are summarized in Fig. 1 and have been described elsewhere [12, 21].

1. Prepare a 1 % (w/v) agarose gel in TAE buffer containing SYBR Safe dye at 1× concentration. Position the comb ~3 cm from the top end of the gel (*see* **Note 10**).
2. Add 5 μL of 10× DNA loading dye to each 50 μL of fragment mixture. Load the DNA marker ladder and each fragmentation mixture, ensuring at least one lane is left empty next to each sample. Begin electrophoresis. During electrophoresis the DNA fragments can be visualized using a blue light and amber filter.
3. When bands in the DNA ladder are sufficiently resolved (e.g., Fig. 1a), turn off the current and cut the gel at the desired molecular weight cutoff (e.g., 1,000 bp in this example) and around the unwanted fragments as indicated in Fig. 1b. Discard the gel fragment containing the unwanted large fragments and the sample-loading wells (Fig. 1b). Replace the excised section with a similarly sized block of agarose gel containing no DNA from a neighboring, empty lane (Fig. 1c).
4. Reverse the direction of the electric field and rerun the samples for the same length of time as in the initial electrophoresis,

such that the bands containing DNA fragments <1,000 bp converge to form a single band (*see* Fig. 1d and **Note 11**).

5. Excise the bands containing the two different sets of DNA fragments and extract the DNA using the QIAquick Gel Extraction kit in accordance with the manufacturer's instructions.
6. Quantify DNA by measuring absorbance at 260 nm.

### 3.3 Fragment Selection with PCR Filter Units

Filter units designed for extracting products from PCR reactions can be used to recover fragments for DNA shuffling experiments using a modification to the manufacturer's instructions [22].

1. Place PCR filter unit sample reservoirs into collection tubes.
2. Dilute samples to a total volume of 400 μL in sterile Milli-Q water. Load samples into sample reservoirs and centrifuge at 1,000 × *g* for 15 min. Discard sample reservoirs.
3. Precipitate DNA using 0.1 volumes of 3 M sodium acetate and 2.5 volumes of 100 % ethanol for 1 h at −20 °C. Pellet the DNA by centrifugation, wash in 70 % ethanol, and evaporate to dryness. Resuspend the DNA pellet in sterile Milli-Q water.
4. Quantify DNA by measuring absorbance at 260 nm.

### 3.4 Reassembly of Fragments by Primerless PCR

1. Combine equal quantities of DNA fragments isolated from each type of restriction enzyme digestion in a 50-μL PCR mix with routine concentrations of buffer, $MgCl_2$, *Taq* polymerase, and dNTPs (e.g., 50 mM KCl, 0.2 mM dNTPs, 1.5 mM $MgCl_2$, and 0.5 U/μL *Taq* polymerase in 20 mM Tris–Cl (pH 8.4) but without primers). The amount of DNA required for reassembly depends on the fragmentation pattern and on sequence identity. We recommend testing several quantities between 0.1 and 2 μg total DNA and including controls that contain DNA fragments isolated from a single type of restriction enzyme digestion.
2. Amplify using the following cycling conditions adapted from Abecassis et al. [22]. Samples are held at 94 °C while *Taq* polymerase is added, then subjected to 35 cycles of denaturation at 94 °C for 30 s, annealing for 13.5 min over which time the temperature is decreased from 65 °C to 41 °C, and extension at 72 °C for 90 s. This is followed by a final extension at 72 °C for 7 min to ensure complete extension of any remaining incomplete amplifications. The annealing step can also be performed in 25 one-min steps rather than as a 13.5-min smooth gradient.
3. Analyze reassembly products by agarose gel electrophoresis. Reassembly products should appear as a smear, indicating the successful assembly of polynucleotides of various lengths (*see* **Note 12**).

### 3.5 Amplification of Shuffled Products and Cloning into E. coli

The products of the reassembly reaction are used as the template in a conventional PCR with flanking primers to permit efficient isolation of products that are the same length as the full-length parental sequences. The primers used should anneal to the sequence upstream and downstream of the parental open reading frames and contain suitable terminal sequences to facilitate cloning into the vector of interest (e.g., encoding suitable 5′ and 3′ restriction sites to match cloning sites in the vector).

1. Prepare a 50-μL PCR using 5 μL of reassembly products as template DNA and 2.5 μM each of forward and reverse primers. Use the buffer and $MgCl_2$ conditions that are recommended by the manufacturer of the *Taq* polymerase.
2. Amplify the shuffled products using cycling conditions appropriate for amplification of a full-length parental gene with the chosen flanking primers.
3. Clean up reactions using a QIAquick PCR cleanup kit in accordance with the manufacturer's instructions.
4. Digest the purified PCR product and 3 μg of the desired target vector (separately) in 50 μL with the appropriate pair of restriction enzymes using conditions recommended by the restriction enzyme manufacturer.
5. Purify the restriction digest products via agarose gel electrophoresis. To the restriction digest mixtures, add 5 μL of 10× DNA loading buffer and load onto a 1 % (w/v) agarose gel containing 1× SYBR Safe DNA gel stain. Separate by electrophoresis and excise the products corresponding to the anticipated digested full-length coding sequence (*see* **Note 13**) and the double-digested vector backbone.
6. Extract DNA using the MinElute gel extraction kit according to the manufacturer's instructions. Quantify DNA by measuring absorbance at 260 nm.
7. Combine digested PCR products and vector backbone in a 3:1 molar ratio (e.g., 0.03 pmol PCR products and 0.01 pmol vector backbone) in a 20-μL reaction with 1 μL T4 DNA ligase and manufacturer-recommended buffering conditions. Incubate for 12 h at 16 °C.
8. Transform a portion of ligation mixture into the desired microbial host as appropriate (e.g., 2.5 μL ligation mixture into 100 μL chemically competent *E. coli* DH5α).
9. Pick colonies of transformants to inoculate 100 μL of LB medium containing appropriate selection antibiotics in 96-well microtiter plates. Incubate overnight at 37 °C with shaking. Prepare glycerol stocks by adding to each well 100 μL of double-concentration (2×) LB medium containing 30 % (v/v) glycerol. Seal the plate with an X-Pierce sealing membrane (*see* **Note 14**) and mix thoroughly then store at −80 °C.

### 3.6 Verification of Shuffled Products

Plasmid DNA from a random selection of mutants should be prepared and sequenced, in order to confirm that shuffling of the parental genes has occurred. Crossover events and spontaneous mutations can be identified by comparing shuffled mutants with the parental sequences in a multiple sequence alignment or by using the CrossoverX application, which produces a graphical representation of shuffled mutant sequences (software written by Dr. Weiliang Huang in the Gillam laboratory, available for download at http://dl.dropbox.com/u/50010004/CrossoverXSetup.msi).

Archived glycerol stocks of shuffled mutants can be used directly to inoculate cultures for protein expression and screening purposes.

## 4 Notes

1. Although shuffling is possible using PCR-generated templates, the parental sequences must have identical sequence up- and downstream of the coding region in order to facilitate the cloning of shuffled products into a suitable expression vector. The process can be simplified by using parental genes that are cloned in identical format within the same source plasmid.
2. A commercially available blue light transilluminator (Invitrogen Safe Imager, Invitrogen, Carlsbad, CA, USA) or a blue light filter for use with a UV transilluminator may be used for excitation of the dye, but we have successfully used a simple blue multi-LED light (20 LED MR16 lamp, available from most retail electronics suppliers). Such blue LED-based lights show broad emission spectra around 500 nm, suitable for exciting SYBR dyes (SYBR DNA stains are maximally excited at 494 nm (blue) and emit at 521 nm (green), with secondary excitation at 284 nm and 382 nm).
3. Clear visualization of the SYBR Safe-stained sample under blue light requires an appropriate light filter. For ease of use during gel manipulations, Safe Imager™ viewing glasses or equivalent is recommended. Safe Imager viewing glasses can be purchased separately from the manufacturer (Invitrogen, Carlsbad, CA, USA).
4. The use of the cloning strain *E. coli* DH5αF'IQ™ aids in stable maintenance of shuffled-library sequences. If some other strain or organism is preferred for library expression and assay, the shuffled library should be transformed directly into that host rather than into *E. coli* DH5αF'IQ. Sufficiently competent *E. coli* cells can be prepared by the method of Inoue et al. [23].
5. Sterilize glycerol by autoclaving for 20 min at 15 psi (1.05 kg/$cm^2$) on liquid cycle.
6. DNA family shuffling has previously been used to shuffle sequences with as little as 50 % sequence identity, but library

members contain very few or no crossovers where there are so few regions of sufficient identity for cross hybridization [24]. For libraries with frequent recombination, parental sequences with at least 65 % sequence identity to one another are recommended.

7. It is not critical that all fragments lie within the 100- to 600-bp size range. However, increasing fragment length decreases the recombination frequency [12, 24].
8. Within each fragmentation reaction, combinations of enzymes that have compatible buffer and temperature conditions should be used. If these criteria cannot be met, sequential digestions with adjustments to buffer and temperature may be required.
9. Early publications in this area cite the use of PCR filter units to remove large undigested fragments [11, 13, 22]. However, recent work in this laboratory has found that this approach, although technically simple, produces libraries of inferior quality, since it strongly favors the recovery of DNA fragments <100 bp in length. Larger quantities of fragments are required to obtain full-length sequences, due to the bias in fragment size, and because the frequency of mis-hybridization events and spontaneous mutations (generated in the reassembly PCR) is greater when there is a larger proportion of shorter fragments.
10. A suitable gel comb should be used such that wells can accommodate 50 μL of sample.
11. Concentrating the desired DNA fragments into a single band facilitates more efficient extraction of the DNA from the agarose [20].
12. The mean length of reassembly products typically increases with the quantity of DNA fragments added to the reassembly reaction. If the reassembly smear does not contain products of length equal to a full-length parental coding sequence, a greater quantity of DNA fragments should be used in the reassembly reaction.
13. Because the template DNA is a complex mixture, there may be several products of varying lengths.
14. X-Pierce sealing membranes feature a precut crossover of each well to facilitate rapid sampling from 96-well plates.

## References


1. Stemmer WP (1994) Rapid evolution of a protein in vitro by DNA shuffling. Nature 370(6488):389–391
2. Stemmer WPC (1994) DNA shuffling by random fragmentation and reassembly: in vitro recombination for molecular evolution. Proc Natl Acad Sci U S A 91(22):10747–10751
3. Joern JM (2003) DNA shuffling. In: Arnold FH, Georgiou G (eds) Directed evolution library creation, Methods mol biol, vol 231. Humana Press, Totowa, NJ, pp 85–89
4. Crameri A, Raillard SA, Bermudez E, Stemmer WPC (1998) DNA shuffling of a family of

genes from diverse species accelerates directed evolution. Nature 391:288–291

5. Kikuchi M, Ohnishi K, Harayama S (1999) Novel family shuffling methods for the in vitro evolution of enzymes. Gene 236:159–167
6. Kikuchi M, Ohnishi K, Harayama S (2000) An effective family shuffling method using single-stranded DNA. Gene 243(1–2): 133–137
7. Zha W, Zhu T, Zhao H (2003) Family shuffling with single-stranded DNA. In: Arnold FH, Georgiou G (eds) Directed evolution library creation: methods and protocols, vol 231, Methods in molecular biology. Humana Press, Totowa, NJ, pp 91–97
8. Muller KM, Stebel SC, Knall S, Zipf G, Bernauer HS, Arndt KM (2005) Nucleotide exchange and excision technology (NExT) DNA shuffling: a robust method for DNA fragmentation and directed evolution. Nucleic Acids Res 33(13):e117
9. Miyazaki K (2002) Random DNA fragmentation with endonuclease V: application to DNA shuffling. Nucleic Acids Res 30(24)
10. Gibbs MD, Nevalainen KMH, Bergquist PL (2001) Degenerate oligonucleotide gene shuffling (DOGS): a method for enhancing the frequency of recombination with family shuffling. Gene 271(1):13–20
11. Huang W, Johnston WA, Hayes MA, De Voss JJ, Gillam EMJ (2007) A shuffled CYP2C library with a high degree of structural integrity and functional versatility. Arch Biochem Biophys 467:193–205
12. Behrendorff J, Moore CD, Kim KH, Kim DH, Smith CA, Johnston WA, Yun CH, Yost GS, Gillam EMJ (2012) Directed evolution reveals requisite sequence elements in the functional expression of P450 2F1 in Escherichia coli. Chem Res Toxicol 25(9):1964–1974
13. Johnston WA, Huang W, De Voss JJ, Hayes MA, Gillam EMJ (2007) A shuffled CYP1A library shows both structural integrity and functional diversity. Drug Metab Dispos 35(12):2177–2185
14. Hunter DJB, Behrendorff JBYH, Johnston WA, Hayes PY, Huang W, Bonn B, Hayes MA, De Voss JJ, Gillam EMJ (2011) Facile production of minor metabolites for drug development using a CYP3A shuffled library. Metab Eng 13(6):682–693
15. Speck J, Stebel SC, Arndt KM, Muller KM (2011) Nucleotide exchange and excision technology DNA shuffling and directed evolution. In: Park DJ (ed) PCR protocols, vol 687, 3rd edn, Methods in molecular biology. Humana Press, Totowa, NJ, pp 333–344
16. Kolkman JA, Stemmer WPC (2001) Directed evolution of proteins by exon shuffling. Nat Biotechnol 19(5):423–428
17. Engler C, Gruetzner R, Kandzia R, Marillonnet S (2009) Golden gate shuffling: a one-pot DNA shuffling method based on type IIs restriction enzymes. PLoS One 4(5):e5553
18. Ness JE, Kim S, Gottman A, Pak R, Krebber A, Borchert TV, Govindarajan S, Mundorff EC, Minshull J (2002) Synthetic shuffling expands functional protein diversity by allowing amino acids to recombine independently. Nat Biotechnol 20(12):1251–1255
19. Sambrook J, Russell DW (2001) Molecular cloning: a laboratory manual, 3rd edn. Cold Spring Harbor Laboratory Press, Cold Spring Harbor, NY
20. Kadokami Y, Lewis RV (1995) Reverse electrophoresis to concentrate DNA fractions. Anal Biochem 226:193–195
21. Behrendorff JBYH, Johnston WA, Gillam EMJ (2013) DNA shuffling of cytochrome P450 enzymes. In: Phillips IR, Shephard EA, Ortiz de Montellano PR (eds) P450 protocols, vol 987, Methods in molecular biology. Springer, New York, pp 177–188
22. Abécassis V, Pompon D, Truan G (2000) High efficiency family shuffling based on multi-step PCR and in vivo DNA recombination in yeast: statistical and functional analysis of a combinatorial library between human cytochrome P450 1A1 and 1A2. Nucleic Acids Res 28(20):e88
23. Inoue H, Nojima H, Okayama H (1990) High efficiency transformation of Escherichia coli with plasmids. Gene 96:23–28
24. Moore GL, Maranas CD, Lutz S, Benkovic SJ (2001) Predicting crossover generation in DNA shuffling. Proc Natl Acad Sci U S A 98(6):3226–3231
25. Gillam EMJ, Baba T, Kim B-R, Ohmori S, Guengerich FP (1993) Expression of modified human cytochrome P450 3A4 in Escherichia coli and purification and reconstitution of the enzyme. Arch Biochem Biophys 305:123–131

# Chapter 13

# Assembly of Designed Oligonucleotides: A Useful Tool in Synthetic Biology for Creating High-Quality Combinatorial DNA Libraries

**Carlos G. Acevedo-Rocha and Manfred T. Reetz**


## Abstract

The method dubbed Assembly of Designed Oligonucleotides (ADO) is a powerful tool in synthetic biology to create combinatorial DNA libraries for gene, protein, metabolic, and genome engineering. In directed evolution of proteins, ADO benefits from using reduced amino acid alphabets for saturation mutagenesis and/or DNA shuffling, but all 20 canonical amino acids can be also used as building blocks. ADO is performed in a two-step reaction. The first involves a primer-free, polymerase cycling assembly or overlap extension PCR step using carefully designed overlapping oligonucleotides. The second step is a PCR amplification using the outer primers, resulting in a high-quality and bias-free double-stranded DNA library that can be assembled with other gene fragments and/or cloned into a suitable plasmid subsequently. The protocol can be performed in a few hours. In theory, neither the length of the DNA library nor the number of DNA changes has any limits. Furthermore, with the costs of synthetic DNA dropping every year, after an initial investment is made in the oligonucleotides, these can be exchanged for alternative ones with different sequences at any point in the process, fully exploiting the potential of creating highly diverse combinatorial libraries. In the example chosen here, we show the construction of a high-quality combinatorial ADO library targeting sixteen different codons simultaneously with nonredundant degenerate codons encoding various reduced alphabets of four amino acids along the heme region of the monooxygenase P450-BM3.


**Key words** Synthetic biology, Directed evolution, Intellectual property, DNA shuffling, Synthetic shuffling, Degenerate homoduplex recombination, Gene assembly mutagenesis, Saturation mutagenesis, Combinatorial libraries, Reduced amino acid alphabets

## 1 Introduction

A decade ago, we introduced the Assembly of Designed Oligonucleotides (ADO) method as an alternative to classical DNA shuffling for application in directed evolution [1]. The main reason behind the development of ADO was the necessity of having a patent-free substitute method owing to the commercial protection of most available strategies for creating combinatorial libraries [2]. At that time, the most widely used method of directed evolution

Elizabeth M.J. Gillam et al. (eds.), *Directed Evolution Library Creation: Methods and Protocols*, Methods in Molecular Biology, vol. 1179, DOI 10.1007/978-1-4939-1053-3_13, © Springer Science+Business Media New York 2014

was error-prone Polymerase Chain Reaction (epPCR), followed by DNA shuffling, a combinatorial method that was introduced in 1994 by Stemmer [3]. Based on gene recombination of variants of a single parental gene, DNA shuffling or "sexual PCR" [4] was demonstrated to be more successful than epPCR. Later on, it was shown that recombining variants of a family of homologous genes was likewise efficient in providing functional diversity [5]. DNA shuffling of variants of a single gene or a family of genes is based on the random fragmentation of the parent gene variants with DNAse I, followed by their reassembly and amplification in a self-priming PCR reaction. The main disadvantages of classical DNA shuffling, however, are the inability to recombine variant sequences that are very different from each other and the difficulty in shuffling mutations that are very close to each other (resolution of recombination). To overcome some of these limitations, in silico approaches [6] as well as the incremental truncation for the creation of hybrid enzymes (ITCHY) method [7] were developed, but the combinatorial libraries created by these methods still suffered from bias [8]. An alternative that solved completely the above limitations is the random chimeragenesis on transient templates (RACHITT) [9] method, but this protocol is very complicated to perform [8].

Four much simpler methods based on synthetic oligonucleotides that increase the resolution of recombination while diminishing the bias in regions with highly homologous sequences were reported in parallel [1, 10–12]; two were published in the same journal issue. Firstly, Ness and colleagues developed the "synthetic shuffling" (SS) approach in which several homologous overlapping oligonucleotides of equal length coding for the desired degeneracy are reassembled in a PCR reaction [10]. Secondly, Coco et al. reported the "degenerate homoduplex recombination" (DHR) method in which a series of oligonucleotides preferentially encoding the degeneracy in nonoverlapping regions as well as lacking the 5′ phosphate group (to avoid ligation) and harboring a modification at the 3′ amino group (to hamper polymerase extension) are used as scaffolds for the overlap of homologous sequences that are then extended, ligated, and PCR-amplified [11]. Whereas SS relies on thermocycling and polymerase overlap extension (POE), DHR is based on the hybridization and ligation of oligonucleotides followed by PCR, the advantage of DHR versus SS being the creation of higher-quality libraries due to the reduction of divergent sequences between the overlapping regions [8]. Shortly after these studies were published, ADO also appeared, the advantage being a more simplified protocol. ADO comprises a first step based on POE involving a primer-less thermocycling reaction of overlapping oligonucleotides designed so that either the degeneracy does not overlap (i.e., with gaps) or that the mutations to be introduced are specified by dedicated oligonucleotides (without gaps). This is followed by a second step of PCR amplification using the outermost primers of both 5′ ends [1]. Thus, ADO is similar to SS but additionally incorporates the advantages of DHR for the creation of high-quality combinatorial libraries for DNA shuffling. In all cases, the libraries must be subsequently cloned into a suitable vector (*see* below). In the same year in which ADO appeared, a very similar method dubbed "Gene Assembly Mutagenesis" was published, which exemplified the potential of this general approach for saturation mutagenesis [12].

As mentioned earlier, ADO was first developed as an alternative to DNA shuffling, the method being illustrated by the directed evolution of stereoselective lipases, but the desired enantioselectivity was still far from perfect [1]. This prompted us to abandon gene-recombination methods and pursue other approaches [13]. The strategy chosen was saturation mutagenesis (SM), which constitutes a powerful tool with which to introduce mutations in a controlled manner along a gene sequence using mixed oligonucleotides [14]. The advantage of SM- versus DNA-shuffling-based strategies is that in SM the probed sequence space can be varied from a single base change up to several dozens or even hundreds of base changes depending on the method. Furthermore, in some cases, SM has been more successful than DNA shuffling in terms of catalyst improvement [15]. In addition, while some DNA-shuffling strategies are restricted to variants of parental genes or gene families found only in Nature, SM enables an essentially unlimited chemical space to be probed, and the space can be further expanded using unnatural amino acids to design enzymes not found in Nature [16], an emergent research area in directed evolution and synthetic biology [17]. Last but not least, another advantage of SM over DNA shuffling is its simplicity for the efficient and rapid evolution of functional enzymes when it is used in an iterative manner [18], namely, iterative saturation mutagenesis (ISM) [19]. ISM simulates Darwinian evolution in vitro as it involves systematically performing several rounds of variation (library creation), selection (library screening), and amplification (choosing a template for the next mutagenesis round). For more details, *see* Chapter 7 on iterative saturation mutagenesis in this book.

In SM, the introduction of changes at the DNA level usually relies on the use of degenerate ("doped" or "spiked") oligonucleotides; synthetic primers encoding the desired degeneracy (e.g., all 20 canonical amino acids encoded by the 32 codons provided by NNK or NNS; N = A/T/G/C; K = T/G and S = G/C) are usually bought from commercial suppliers for subsequent PCR. However, some libraries created this way often suffer from bias due to several factors including gene size and GC content, position and sequence of the target codon in the gene, melting and annealing temperature of template–primer duplexes, as well as primer length, synthesis quality, and proportion of degenerate bases [20]. In addition, due to the inherent redundancy of the genetic code, the traditional use of NNN or NNK/S degenerate primers has severe disadvantages: some amino acids are encoded by up to six codons (Arg, Leu, Ser), whereas others only by one (Met and Trp). This results in the creation of biased amino acid libraries. For example, the theoretical ratio of Met to Leu obtained from randomizing a given amino acid position with NNN or NNK/S codons is 1:6 or 1:3, respectively, rather than 1:1 as usually desired. Thus, the twenty canonical amino acids (CAAs) should be encoded not by 64 (NNN) or 32 (NNK/S) codons but ideally by 20 [21] or at most 22 [22] codons. Since the simultaneous randomization of more than three residues to all twenty CAAs still results in highly complex libraries [22], limited sets of amino acids are a great aid in reducing the complexity of the protein sequence space to be probed [23].

Reduced amino acid alphabets can be encoded in one or more degenerate and/or nondegenerate codons with some differences: whereas one or more codons can be encoded in a single degenerate oligo, if nondegenerate oligonucleotides are used, the price increases proportionally to the number of codons per encoded amino acid. For example, instead of ordering 12 oligonucleotides encoding 12 chemically balanced amino acids (Phe, Leu, Ile, Val, Tyr, His, Asn, Asp, Cys, Arg, Ser, Gly), a single degenerate codon NDT (N = A/T/G/C; D = A/T/G) suffices for the abovementioned amino acids. Similarly, if two of the target codons are located very close to each other, 144 (12 × 12) oligonucleotides would be required if the codons are nondegenerate, but only a single degenerate primer is necessary with two NDT codons. The advantage of using NDT has been shown in many practical examples [24–26]. In addition, several degenerate codons can be used to encode amino acids with specific features [27]: VRK (V = N − T; R = A/G) for hydrophilicity (12 codons for 8 amino acids), NYC (Y = C/T) for hydrophobicity (8 for 8, respectively), KST for small size (4 for 4, respectively), or RRK for charge at near neutral pH (8 for 7, respectively), for example. However, if there is no degenerate codon encoding the desired amino acids, nondegenerate oligonucleotides need to be purchased separately and then mixed in equimolar amounts. Similarly, if there are amino acids partially encoded in both degenerate and nondegenerate oligonucleotides, these can be mixed. At the end, the desired amino acids dictate the strategy for creating the minimal alphabet, and extreme reduction of the encoded alphabet has been shown to be successful in directed evolution [28–31]. It is nevertheless advisable to estimate the screening effort of the designed combinatorial library based on the reduced amino acid alphabet chosen.

Overall, ADO combines the advantages of using reduced amino acid alphabets—although redundant and nonredundant degenerate and nondegenerate codons encoding the 20 CAAs can also be employed—together with a careful design of overlapping oligonucleotides for the creation of high-quality and bias-free combinatorial libraries. When creating highly combinatorial libraries, at first glance, it sounds reasonable to use established methods such as QuikChange [32] or OmniChange [33] for the respective saturation of multiple codons located contiguously or far away from each other. However, ADO is advantageous in that higher-quality libraries can be created when the templates are difficult to randomize [34], and it is also wise to make an initial investment in the oligonucleotides for the assembly. This is simply because the oligonucleotides can be individually exchanged at any point in the evolution experiment, as is often the case in ISM when creating several libraries for more than one round of SM. Therefore, there is no limit for the creation of many different combinatorial libraries based on ADO.

As mentioned earlier, ADO is performed in two steps. The first one involves the so-called polymerase cycling assembly (PCA; basically a form of POE), whereas the second one is a PCR reaction. Originally described by Stemmer in 1995 [35], PCA is similar to PCR, but without a pair of primers present in excess compared to the template. Since 1995, the power of PCA and related methodologies has been realized for the assembly of synthetic genes and plasmids [35], culminating with the construction of metabolic pathways and even genomes starting from oligonucleotides [36]. However, the creation of combinatorial ADO libraries has not been thoroughly exploited so far, with a few notable exceptions [37, 38]. With the costs of synthetic DNA dropping every year [39], together with an increasing number of methods for the efficient assembly of DNA [40], we believe that ADO represents a powerful tool in synthetic biology for the construction of highly combinatorial DNA libraries. Applications include not only the directed evolution of enzymes [13], as well as metabolic [41] and genome [39] engineering, but also the modulation of gene expression by targeting regulatory regions such as promoters [42].

The most critical factor for the successful creation of a combinatorial ADO library is the design of appropriate overlapping homologous sequences, which may or may not have gaps between oligonucleotides (*see* Fig. 1). Whereas designing overlapping sequences without gaps keeps the oligonucleotides both short in size and stringently hybridized, including gaps between oligonucleotides may slow down the optimization due to the potential formation of mispriming. Experience has nevertheless shown that both strategies are successful [12, 37, 38]. ADO results in the formation of a linear double-stranded DNA library, which can be then either assembled with other DNA fragments or introduced into a suitable vector via three general types of strategy.

1. Ligation- and sequence-dependent methods: Restriction digestion [43] and ligation [44], as well as single-site [45] or multisite [46] GATEWAY cloning based on homologous recombination

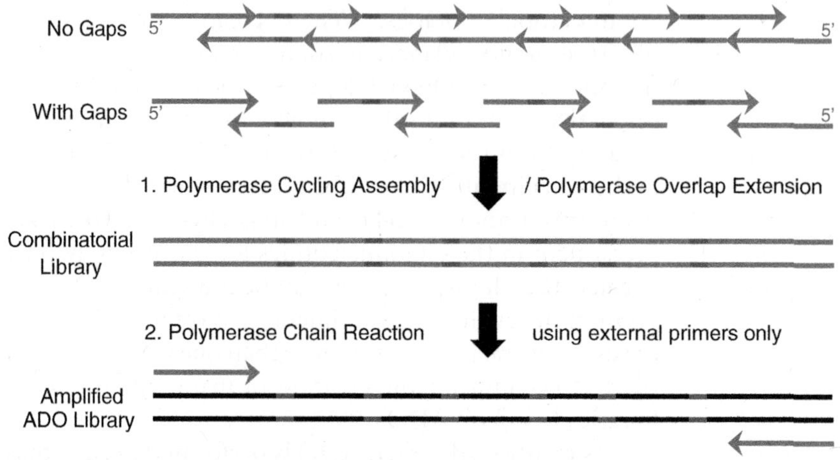


**Fig. 1** ADO scheme. Combinatorial libraries can be designed with or without gaps. The fewer and shorter the gaps, the more primers will be generally required. One or more degenerate and/or nondegenerate codons (*red*) can be incorporated in one or more synthesized oligonucleotides. The first ADO step involves the assembly of oligonucleotides via polymerase cycling assembly (PCA) or polymerase overlap extension (POE) that performs basically the same reaction, followed by a second step of PCR using appropriate external primers at the 5′ ends. The resulting double-stranded ADO library is subsequently cloned into a suitable vector or assembled with one or more other gene fragments. The advantage of ADO over other related methodologies is the ability to reuse individual primers that can be exchanged between different stages of the directed evolution strategy on the same gene, gene fragment, or plasmid and that ADO can be used to generate high-quality libraries without limits to the length of the DNA library or the number of DNA changes (Color figure online)

2. Ligation- and sequence-independent methods: Ligase-independent cloning (LIC)-PCR [47], uracil DNA glycosylase (UDG) cloning [48], and sequence- and ligase-independent cloning (SLIC) [49], as well as site-specific recombination-based tandem assembly (SSRTA) [50]
3. Polymerase-dependent methods: Overlap extension cloning (OEC) [51], overlap extension PCR cloning (OE-PCR-C) [52], and circular polymerase extension cloning (CPEC) [53] that is also available in the form of a detailed protocol [54]

Thus far, the cloning of combinatorial ADO libraries has been performed using only restriction digestion and ligation [1, 37, 38]. The method presented herein demonstrates the potential of ADO for the creation of a highly combinatorial DNA library targeting 16 different codons with nonredundant degenerate codons encoding various reduced alphabets of four amino acids. As a model enzyme, we use the difficult-to-amplify gene template [34] coding for the heme region of the monooxygenase $P450_{BM3}$ from *Bacillus megaterium*, a highly efficient biocatalyst for the oxidative activation of C–H bonds and epoxidation of a wide range of substrates [55].

## 2 Materials

### 2.1 Library Design In Silico

1. The gene or protein sequence(s) of interest obtained from any digital source or a suitable database, for instance, the Universal Protein Resource (UniProt) [56] or GenBank [57].
2. Suitable software to design the oligonucleotides, for example, DNAWorks [58, 59]: http://helixweb.nih.gov/dnaworks/.
3. A tool for selecting the (reduced) amino acid alphabet (optional):
   (a) CASTER: http://www.kofo.mpg.de/media/2/D1108347/0987095526/ISM_tools.zip.
   (b) AA-Calculator: http://guinevere.otago.ac.nz/cgi-bin/aef/AA-Calculator.pl.
4. An algorithm to calculate screening effort (optional):
   (a) TopLib: http://stat.haifa.ac.il/~yuval/toplib/.
   (b) GLUE-IT: http://guinevere.otago.ac.nz/cgi-bin/aef/glue-IT.pl.

### 2.2 Library Creation In Vitro

In general, all reagents should be of analytical grade. All solutions should be prepared using aseptic technique and sterile, deionized water.

1. Oligonucleotides of different size preferentially obtained in suspended form at a defined concentration, for example, 100 μM stock solution in water.
2. KOD Polymerase Buffer 10× (Novagen, Merck Chemicals, Darmstadt, Germany).
3. Deoxynucleotide triphosphates (dNTPs) at 2 mM each (Novagen).
4. 25 mM $MgSO_4$.
5. KOD Hot Start DNA Polymerase (1 U/μL; Novagen).
6. Thermocycler for PCR.
7. Equipment and supplies for agarose gel electrophoresis.
8. QIAquick PCR Purification Kit (QIAGEN, Hilden, Germany).
9. Automated DNA sequencing service.

## 3 Methods

### 3.1 Oligonucleotide Design

#### 3.1.1 Determination of the Number of Overlapping Oligonucleotides

The number of overlapping oligonucleotides depends mainly on the number and location of target codons, i.e., library length (LL), as well as the budget that the researcher is willing to invest. The library can be synthesized as a short linear fragment containing the target codons for subsequent cloning/assembly or as an entire plasmid or extended DNA fragment depending on the budget of the project. Libraries of various lengths have been assembled using

44 (LL, 900 bp) [12], 16 (LL, 500 bp) [38], as well as 5 long (LL, 500 bp) [1] or short (LL, 200 bp) [37] oligonucleotides. Although the construction of full plasmids containing a given combinatorial ADO library has, to our knowledge, not been reported to date, this should be possible because a plasmid of 2.7 kbp [35] and even a genome of 16.3 kbp [36] have been assembled from single oligonucleotides. In principle, there is no restriction on the number of oligonucleotides to be assembled via ADO (*see* **Note 1**). Nevertheless, the total number of oligonucleotides also depends on the length of the overlapping homologous sequences, which in turn dictates the presence or absence of gaps between them. At this stage, it is important to know the method for the cloning/assembly of the double-stranded DNA library because appropriate restriction/homologous sequences will have to be added at both extremes of the gene fragment.

### *3.1.2 Design of Appropriate Overlapping Homologous Sequences*

Despite the number and length of oligonucleotides, the most crucial step for the successful creation of a combinatorial ADO library is the design of appropriate overlapping homologous sequences. As shown in Fig. 1, homologous sequences can be designed with or without gaps. The presence or absence of gaps as well as oligonucleotide length may be arbitrarily chosen (*see* **Note 2**), but there are least two important factors to consider. The first is the overall cost, as more primers are usually required to close any gaps (*see* **Note 3**), whereas the second concerns the optimal assembly conditions. Since every gene sequence is unique, as are its thermodynamic properties, it is advisable to explore first the optimal assembly parameters with and without gaps for the gene in question, regardless of its length. Even if a given gene fragment is very short so that a few primers can be arbitrarily chosen, it is strongly recommended to perform a thermodynamic analysis for assessing oligonucleotide mispriming. Omitting this step may result in failure of the assembly.

Thermodynamic analyses such as these can be easily performed in silico with programs such as "DNAWorks" (DNA-W) [58]. DNA-W was originally developed by Hoover and Lubkowski to optimize the construction of synthetic genes for heterologous expression [59], but it can also be exploited to create overlapping oligonucleotides that can be later supplied with one or more degenerate or nondegenerate codons, depending on the desired amino acid alphabet for randomization. The DNA-W interface not only is easy to use but also allows the user to specify limits for the oligonucleotide annealing temperature and size as well as the presence or absence of gaps for their assembly. Lastly, DNA-W also calculates these thermodynamic parameters with various concentrations of ions depending on the experimental conditions and provides the option for inclusion of restriction sites.

### *3.1.3 Determination of the Amino Acid Alphabet to be Used*

The choice of the (reduced) amino acid alphabet (i.e., which amino acids to incorporate at a given position) depends basically on the effort that has to be invested in screening (or selecting) a given library when more than two codons are randomized simultaneously. The decision also depends on the available information (in terms of sequence alignment or fitness) of a given amino acid residue or more at a given protein.

#### Screening Effort (SE)

SE refers to the number of samples required to partially or almost completely analyze part of a given protein sequence space. Whereas the SE may be minimized by using selection strategies in the directed evolution of more active or fluorescent enzymes [60], in many cases, the number of mutants that need to be screened is unduly large. This so-called numbers problem constitutes the biggest bottleneck in the directed evolution of enzymes [24]. There are two methods available to assess SE. One can calculate the probability of screening 95 % of a given library [61]. Alternatively, it is possible to estimate the necessary samples to be screened for finding with 95 % confidence at least one or more of the fittest mutants [62]. For example, if two codons were simultaneously targeted with NDT degeneracy ($12 \times 12 = 144$ variants), it would be necessary to screen 430 or 143 samples in order to cover 95 % of the library or to find at least one out of the three best mutants with 95 % probability, respectively. While the first method is useful to explore almost completely a given library for fitness landscapes [63, 64], the second is more practical for the rapid identification of hits (C.G. Acevedo-Rocha et al., unpublished; this principle is expanded upon by Yuval Nov in Chapter 18 of this volume). However, even if fit mutants can be found with a reasonable SE, the complexity of the combinatorial libraries rises dramatically as the number of target codons increases, even when using a reduced amino alphabet such as NDT (*see* Table 1).

Therefore, the further reduction of protein sequence space with more compact amino acid alphabets represents an attractive strategy, e.g., using only four amino acids as building blocks (*see* Table 1). However, how can the chemically diversity offered by the genetic code be maximized while minimizing the number of amino acids? This is a case-dependent question depending on the enzyme under study and the information available regarding its structure–function relationships.

#### Available Information (AI)

The AI can be in the form of amino acid residues that are either conserved across protein families or whose fitness parameters are known. One strategy, "Combinatorial Consensus Mutagenesis" (CCM), involves using a multiple sequence alignment (MSA) for the identification of "consensus-like" mutations across protein families based on the assumption that substitutions not occurring naturally are more

**Table 1**
**Two different types of screening effort necessary for the randomization of one or more amino acid residues simultaneously with the NDT degenerate codon and an amino acid alphabet of four members (4-AAA)**

| Residue number | Screening effort required according to randomization method | | | |
|---|---|---|---|---|
| | 95 % probability of library coverage [61] | | 95 % probability of finding at least one out of the three best mutants [62] | |
| | NDT | 4-AAA | NDT | 4-AAA |
| 1 | 35 | 11 | 11 | 3 |
| 2 | 430 | 47 | 143 | 15 |
| 3 | 5,176 | 95 | 1,726 | 63 |
| 4 | 62,119 | 766 | 20,707 | 255 |
| 5 | $7.454\times10^{5}$ | 3,067 | 248,500 | 1,023 |
| 6 | $8.945\times10^{6}$ | 12,270 | $2.982\times10^{6}$ | 4,090 |
| 7 | $1.073\times10^{8}$ | 49,081 | $3.587\times10^{7}$ | 16,361 |
| 8 | $1.288\times10^{9}$ | $1.963\times10^{5}$ | $4.294\times10^{8}$ | 65,443 |
| 9 | $1.546\times10^{10}$ | $7.853\times10^{5}$ | $5.152\times10^{9}$ | 261,800 |
| 10 | $1.855\times10^{11}$ | $3.141\times10^{6}$ | $6.183\times10^{10}$ | $1.047\times10^{6}$ |

likely to be deleterious [65]. A second strategy, described here as "combinatorial fitness mutagenesis" (CFM) and also referred to as "data-driven protein engineering" [66], involves using the accumulated knowledge of "hot spots" or mutations, sometimes in the form of databases [67], contributing to known fitness parameters like activity, stability, and/or selectivity. Directed evolution via CCM using various extremely reduced amino acid alphabets (between 2 and 9 different residues) has been shown in beta-lactamases [65], penicillin G acylases [68], glucose dehydrogenases [69], monooxygenases [28], esterases [70, 29], haloalkane dehalogenases [30], and lipases [31]. There are fewer examples where CFM has been used in directed evolution (C.G. Acevedo-Rocha et al.; unpublished). Of interest in the CFM approach are extremely reduced nonredundant degenerate codons such as KST or NTT that, respectively, code for small (Ser, Ala, Cys, Gly) or large hydrophobic (Phe, Leu, Ile, Val) amino acid residues [27] depending on the desired physicochemical property to be included. Finally, the quest for the minimal amino acid alphabet that maximizes chemical diversity is a research area of great current interest [71–73].

#### 3.1.4 Case Study: Construction of an ADO Library for $P450_{BM3}$

The monooxygenase $P450_{BM3}$ from *B. megaterium* belongs to the large cytochrome P450 (CYP) protein superfamily. P450 proteins contain a heme cofactor and require a protein partner to deliver one or more electrons to reduce the iron and molecular oxygen necessary for the catalytic oxidation of substrates. Unlike most CYP superfamily members, in $P450_{BM3}$, the heme region (55 kDa) is naturally fused to its redox partner (65 kDa), resulting in a protein of approx. 120 kDA (1,049 amino acid residues) that catalyzes a wide range of biotransformations [55]. The $P450_{BM3}$ gene is 3,147 bp long, of which the first 1,416 bp correspond to the heme domain and the remainder to the reductase domain.

Although we developed a PCR-based method for the creation of combinatorial libraries on difficult-to-randomize templates such as $P450_{BM3}$ [34], the simultaneous saturation of three or more amino acids residues often requires more than one round of mutagenesis, even without the problems encountered due to repetitive and AT-rich sequences. This prompted us to construct an ADO library of the heme region (1,416 bp), the sequence of which, together with appropriate homologous sequences at both ends for subsequent gene assembly (total gene length about 1,500 bp), was queried in DNA-W choosing gap-free (due to degenerate codons; *see* below) overlapping oligonucleotides of various lengths (50- to 60-mers) and melting temperatures (Tm) of 55–65 °C (*see* **Note 4**). The program offered several sets of primers with various optimal lengths and Tm, from which one containing 64 primers was selected (*see* **Note 5**). The primers were copied to a Microsoft Excel database and the codons of 16 target residues for saturation mutagenesis were changed to various minimal alphabets of four amino acids encoded in a single degenerate codon (*see* **Note 6**). The reduced amino acid alphabets included KKG (K = T/G), RWG (R = A/G; W = A/T), RKC, GNC (N = A/T/G/C), and KKC. The corresponding spiked oligonucleotides were ordered (*see* **Note 7**).

### 3.2 Oligonucleotide Assembly

1. Dilute each primer to 10 μM in water.
2. In a new tube, pool equimolar amounts of each primer, e.g., for 64 primers, mix 1 μL of each, to give a total of 64 μL.
3. Reaction without flanking primers (*see* **Note 8**): In a PCR tube, mix 4 μL of the equimolar mixture of oligonucleotides with 5 μL of KOD buffer, 5 μL of 2 mM dNTPs, 3.5 μL of 25 mM $MgSO_4$, 1 μL of KOD polymerase, and 31.5 μL of $H_2O$ (*see* **Note 9**).
4. Close the tube and place it in the thermocycler.
5. First ADO step: Perform 20 cycles of 95 °C for 20 s, 50 °C for 30 s, and 70 °C for 1 min and then cool at 4 °C (*see* **Note 10**).
6. Reaction with flanking primers (*see* **Note 11**): Take 1 μL of the resulting assembly reaction to a new PCR tube and mix it

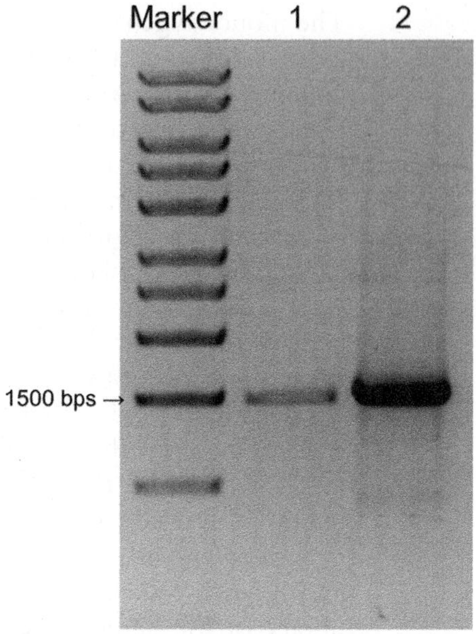


**Fig. 2** Agarose gel electrophoresis of ADO library amplification products created from $P450_{BM3}$. Lane 1 shows the traditional ADO protocol after the second PCR step. The expected size of the heme region of $P450_{BM3}$ with appropriate homologous sequences is about 1,500 base pairs (bps). Lane 2 shows also the expected gene length but using a modification of ADO where the outer primers (sense #1 and antisense #64) were added at tenfold the molar ratio of all other primers in the first ADO step, followed by PCR amplification using the same flanking primers in the second step (*see* **Note 8**)

with 5 μL of KOD buffer, 5 μL of 2 mM dNTPs, 3.5 μL of 25 mM $MgSO_4$, 1 μL of 10 μM sense primer (#1 in the $P450_{BM3}$ example), 1 μL of 10 μM antisense primer (#64 in the $P450_{BM3}$ example), 1 μL of KOD polymerase, and 32.5 μL of $H_2O$ (*see* **Note 12**).

7. Close the tube and place it in the thermocycler.
8. Second ADO step: Heat the sample at 95 °C for 2 min; perform 25 cycles of 95 °C for 20 s, 50 °C for 30 sec, and 70 °C for 1 min, followed by a step of 70 °C for 5 min; and then cool at 4 °C (*see* **Note 10**).
9. Verify the final product by agarose gel electrophoresis (*see* Fig. 2).
10. Purify the PCR product with the QIAquick PCR Purification Kit according to the manufacturer's instructions.
11. Using a sense primer (*see* **Note 13**), check the quality of the library (*see* **Note 14**) by sequencing.
12. If the library has the desired degeneracy (*see* Fig. 3), clone/assemble it in the appropriate plasmid/gene fragment(s).

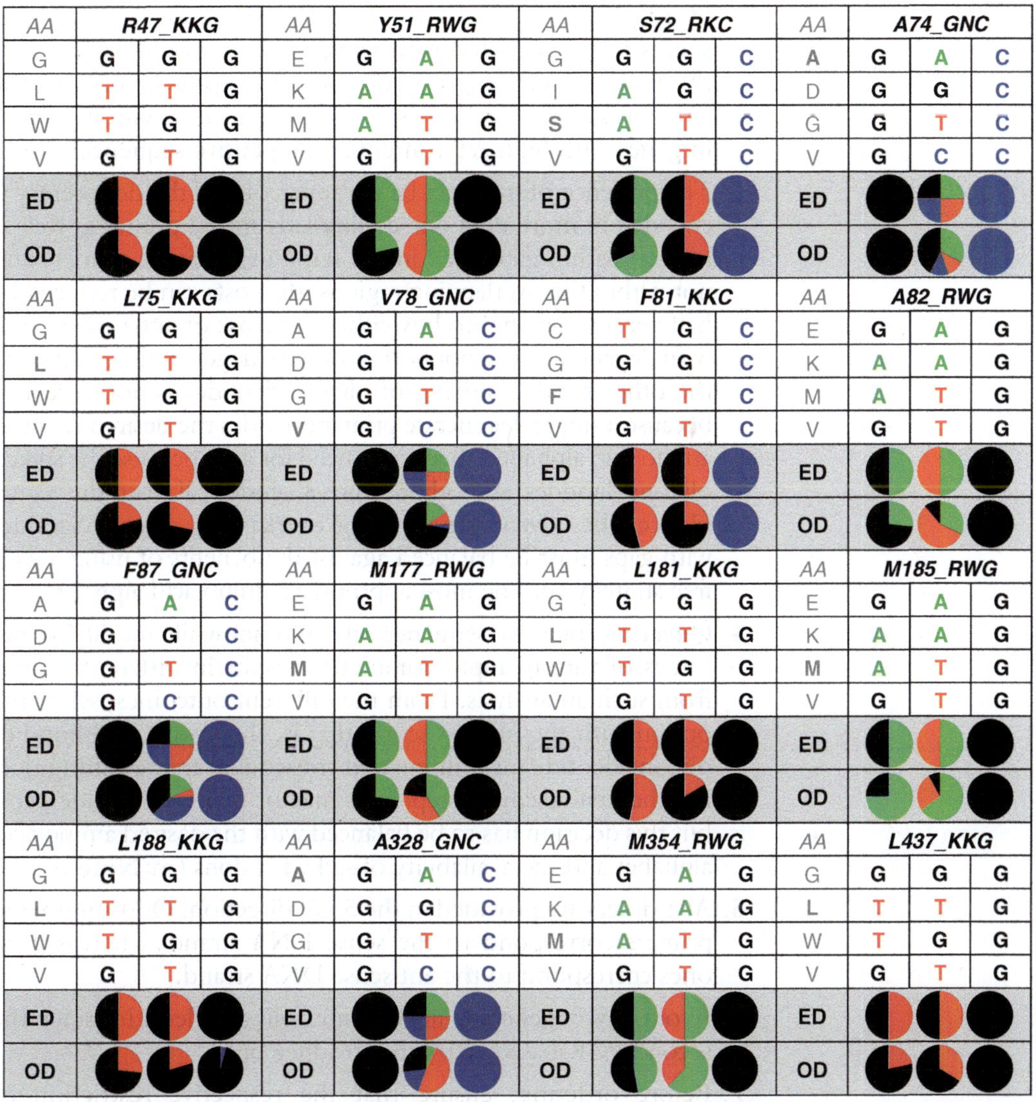

| AA | R47_KKG | | | AA | Y51_RWG | | | AA | S72_RKC | | | AA | A74_GNC | | |
|---|---|---|---|---|---|---|---|---|---|---|---|---|---|---|---|
| G | G | G | G | E | G | A | G | G | G | G | C | A | G | A | C |
| L | T | T | G | K | A | A | G | I | A | G | C | D | G | G | C |
| W | T | G | G | M | A | T | G | S | A | T | C | G | G | T | C |
| V | G | T | G | V | G | T | G | V | G | T | C | V | G | C | C |
| ED | | | | ED | | | | ED | | | | ED | | | |
| OD | | | | OD | | | | OD | | | | OD | | | |
| AA | L75_KKG | | | AA | V78_GNC | | | AA | F81_KKC | | | AA | A82_RWG | | |
| G | G | G | G | A | G | A | C | C | T | G | C | E | G | A | G |
| L | T | T | G | D | G | G | C | G | G | G | C | K | A | A | G |
| W | T | G | G | G | G | T | C | F | T | T | C | M | A | T | G |
| V | G | T | G | V | G | C | C | V | G | T | C | V | G | T | G |
| ED | | | | ED | | | | ED | | | | ED | | | |
| OD | | | | OD | | | | OD | | | | OD | | | |
| AA | F87_GNC | | | AA | M177_RWG | | | AA | L181_KKG | | | AA | M185_RWG | | |
| A | G | A | C | E | G | A | G | G | G | G | G | E | G | A | G |
| D | G | G | C | K | A | A | G | L | T | T | G | K | A | A | G |
| G | G | T | C | M | A | T | G | W | T | G | G | M | A | T | G |
| V | G | C | C | V | G | T | G | V | G | T | G | V | G | T | G |
| ED | | | | ED | | | | ED | | | | ED | | | |
| OD | | | | OD | | | | OD | | | | OD | | | |
| AA | L188_KKG | | | AA | A328_GNC | | | AA | M354_RWG | | | AA | L437_KKG | | |
| G | G | G | G | A | G | A | C | E | G | A | G | G | G | G | G |
| L | T | T | G | D | G | G | C | K | A | A | G | L | T | T | G |
| W | T | G | G | G | G | T | C | M | A | T | G | W | T | G | G |
| V | G | T | G | V | G | C | C | V | G | T | G | V | G | T | G |
| ED | | | | ED | | | | ED | | | | ED | | | |
| OD | | | | OD | | | | OD | | | | OD | | | |

**Fig. 3** Quality control of ADO combinatorial library of the heme region of $P450_{BM3}$ where 16 residues were targeted for saturation mutagenesis using various reduced alphabets of four amino acids (AA). In general, the expected degeneracy (ED) at the DNA level by each alphabet is in good agreement with the observed degeneracy (OD) upon sequencing (*see* **Note 15**)

## 4 Notes

1. The assembly of large metabolic pathways and full genomes is hierarchical. That is, smaller fragments are first assembled and then these fragments are subsequently assembled to construct larger fragments and so on.
2. If the gap-free strategy is chosen, a gene or gene fragment can be rationally divided in sections of equal length of, for example, 40- [12, 35] or 60-mers [36]. While oligonucleotides

of 40 [37], 60 [38], or even up to 140 bases [1] have been successfully assembled with gaps, priority should be given to the thermodynamic properties of the oligonucleotides of the gene in question; otherwise, mispriming can occur, especially in genes with high AT content and repetitive sequences.

3. The presence of gaps can be advantageous. If the degeneracy is encoded in more than one equimolarly mixed primer as well as located within gaps (i.e., in the sense or antisense primer) and not within the overlapping regions, the costs can be reduced by half because one has to buy either the sense or antisense primer with degeneracy and not both groups of degenerate primers. On the other hand, if mixing of oligonucleotides is not necessary because a single degenerate primer provides the desired reduced amino acid alphabet, there is no need for gaps because the spiked oligonucleotides at both 5′ and 3′ ends will cost the same. Hence, the advantages of using overlapping oligonucleotides with gaps must be balanced against the benefits of using a gap-free strategy and the most appropriate amino acid alphabet.
4. Querying the gene sequence several times while modifying the values of various input parameters results in different output from such an analysis. From the different outcomes, select the option with the highest score, that is, the lowest likelihood of oligonucleotide mispriming and consequent frameshifting during the first assembly step. This may or may not include gaps, but this decision has to be balanced with the desired amino acid alphabet and the availability of spiked codons (*see* **Note 3**).
5. All primers are provided in the 5′–3′ direction. Odd-numbered primers correspond to the sense DNA strand, whereas even ones correspond to the antisense DNA strand.
6. Avoid any degeneracy in the outer oligonucleotides since the second PCR-based step can introduce bias.
7. Before ordering, ensure that the respective restriction or homologous sequences have been incorporated at the extremes of the most outer primers. This depends on the method for cloning or assembling the combinatorial ADO library. Primers should be obtained at good purity (e.g., HPLC-purified) to avoid mispriming and to improve the quality of the library.
8. The flanking primers can be also added to this reaction in tenfold molar excess relative to all the primers to increase the yield of the subsequent PCR step (*see lane 2* in Fig. 2). Depending on the gene length and composition, the ratio between template and outer primers should be optimized [74].
9. If the amount of reagents is limited, it is advisable to do this reaction in a smaller volume, for instance, 25 μL or less, because only 1 μL is needed for the next PCR step.

10. The number of cycles can vary from 20 up to 30 or more, depending on the gene length and composition.
11. Although separate flanking primers can be synthesized, the outermost ones from the initial step can be used as primers in this step. In this example, primers #1 (sense) and #64 (antisense) were employed.
12. Make more than one reaction in small volumes so as to have sufficient DNA for downstream processing.
13. Sequence with a sense primer that allows the DNA polymerase to "sit" and "slide" well along the gene; in the provided example, instead of sequencing with the sense oligonucleotide #1, oligonucleotide #3 was used, resulting in a large improvement of the quality of the electropherograms (data not shown).
14. This is perhaps the most important step since not having the expected diversity will result in a futile screening effort and incorrect interpretation of results, wasting time and resources.
15. AT-rich sequences within the gene may impose constraints on the degeneracy that can be achieved. In addition, the possibility of having other percentages of the bases than the expected ones in the spiked oligonucleotides is a common phenomenon due to limitations in the fidelity of DNA synthesis [20]. This is the reason why it is seldom possible to obtain libraries that exactly reflect those that were designed; for a further discussion of this issue, see [22]. In the example chosen (Fig. 3), the ADO library can be considered of high quality, except at positions V78, F87, and A328 where there is likelihood of missing the codons for Val, Gly, and Ala, respectively (*see* the observed degeneracy at the middle base). In such cases as this, the ADO library could be optimized by obtaining the underrepresented codons in a separate primer and by mixing this with the degenerate primer in an appropriate ratio. Lastly, in other cases such as A82, M177, and M185, a frameshift in the second base appears to occur with different intensities (guanine is present unexpectedly in the observed degeneracy), very likely resulting from mispriming due to the A/T sequence, but this can be considered negligible because it does not appear to affect the subsequent bases along the gene sequence.

## Acknowledgments

We thank Yuval Nov for providing statistical advice and for updating his server. We also thank the editors for their excellent editorial work. This work was supported by the Max-Planck Society and the Arthur C. Cope Foundation.

## References


1. Zha D, Eipper A, Reetz MT (2003) Assembly of designed oligonucleotides as an efficient method for gene recombination: a new tool in directed evolution. Chembiochem 4(1):34–39
2. Neylon C (2004) Chemical and biochemical strategies for the randomization of protein encoding DNA sequences: library construction methods for directed evolution. Nucleic Acids Res 32(4):1448–1459
3. Stemmer WP (1994) Rapid evolution of a protein in vitro by DNA shuffling. Nature 370(6488):389–391
4. Smith GP (1994) Applied evolution. The progeny of sexual PCR. Nature 370(6488):324–325
5. Crameri A, Raillard SA, Bermudez E, Stemmer WP (1998) DNA shuffling of a family of genes from diverse species accelerates directed evolution. Nature 391(6664):288–291
6. Moore GL, Maranas CD, Lutz S, Benkovic SJ (2001) Predicting crossover generation in DNA shuffling. Proc Natl Acad Sci U S A 98(6):3226–3231
7. Ostermeier M, Shim JH, Benkovic SJ (1999) A combinatorial approach to hybrid enzymes independent of DNA homology. Nat Biotechnol 17(12):1205–1209
8. Ostermeier M (2003) Synthetic gene libraries: in search of the optimal diversity. Trends Biotechnol 21(6):244–247
9. Coco WM, Levinson WE, Crist MJ, Hektor HJ, Darzins A, Pienkos PT, Squires CH, Monticello DJ (2001) DNA shuffling method for generating highly recombined genes and evolved enzymes. Nat Biotechnol 19(4):354–359
10. Ness JE, Kim S, Gottman A, Pak R, Krebber A, Borchert TV, Govindarajan S, Mundorff EC, Minshull J (2002) Synthetic shuffling expands functional protein diversity by allowing amino acids to recombine independently. Nat Biotechnol 20(12):1251–1255
11. Coco WM, Encell LP, Levinson WE, Crist MJ, Loomis AK, Licato LL, Arensdorf JJ, Sica N, Pienkos PT, Monticello DJ (2002) Growth factor engineering by degenerate homoduplex gene family recombination. Nat Biotechnol 20(12):1246–1250
12. Bessette PH, Mena MA, Nguyen AW, Daugherty PS (2003) Construction of designed protein libraries using gene assembly mutagenesis. Methods Mol Biol 231: 29–37
13. Reetz MT (2011) Laboratory evolution of stereoselective enzymes: a prolific source of catalysts for asymmetric reactions. Angew Chem Int Ed Engl 50(1):138–174
14. Derbyshire KM, Salvo JJ, Grindley ND (1986) A simple and efficient procedure for saturation mutagenesis using mixed oligodeoxynucleotides. Gene 46(2–3):145–152
15. Parikh MR, Matsumura I (2005) Site-saturation mutagenesis is more efficient than DNA shuffling for the directed evolution of beta-fucosidase from beta-galactosidase. J Mol Biol 352(3):621–628
16. Hoesl MG, Budisa N (2011) In vivo incorporation of multiple noncanonical amino acids into proteins. Angew Chem Int Ed Engl 50(13):2896–2902
17. Brustad EM, Arnold FH (2011) Optimizing non-natural protein function with directed evolution. Curr Opin Chem Biol 15(2):201–210
18. Reetz MT, Prasad S, Carballeira JD, Gumulya Y, Bocola M (2010) Iterative saturation mutagenesis accelerates laboratory evolution of enzyme stereoselectivity: rigorous comparison with traditional methods. J Am Chem Soc 132(26):9144–9152
19. Reetz MT, Carballeira JD (2007) Iterative saturation mutagenesis (ISM) for rapid directed evolution of functional enzymes. Nat Protoc 2(4):891–903
20. Airaksinen A, Hovi T (1998) Modified base compositions at degenerate positions of a mutagenic oligonucleotide enhance randomness in site-saturation mutagenesis. Nucleic Acids Res 26(2):576–581
21. Tang L, Gao H, Zhu X, Wang X, Zhou M, Jiang R (2012) Construction of "small-intelligent" focused mutagenesis libraries using well-designed combinatorial degenerate primers. Biotechniques 52(3):149–158
22. Kille S, Acevedo-Rocha CG, Parra LP, Zhang ZG, Opperman DJ, Reetz MT, Acevedo JP (2013) Reducing codon redundancy and screening effort of combinatorial protein libraries created by saturation mutagenesis. ACS Synth Biol 2:83–92. http://pubs.acs.org/doi/abs/10.1021/sb300037w
23. Li T, Fan K, Wang J, Wang W (2003) Reduction of protein sequence complexity by residue grouping. Protein Eng 16(5):323–330
24. Reetz MT, Kahakeaw D, Lohmer R (2008) Addressing the numbers problem in directed evolution. Chembiochem 9(11):1797–1804
25. Liu J, Cropp TA (2012) A method for multi-codon scanning mutagenesis of proteins based on asymmetric transposons. Protein Eng Des Sel 25(2):67–72
26. Bonnet J, Subsoontorn P, Endy D (2012) Rewritable digital data storage in live cells via

engineered control of recombination directionality. Proc Natl Acad Sci U S A 109(23): 8884–8889
27. Balint RF, Larrick JW (1993) Antibody engineering by parsimonious mutagenesis. Gene 137(1):109–118
28. Reetz MT, Wu S (2008) Greatly reduced amino acid alphabets in directed evolution: making the right choice for saturation mutagenesis at homologous enzyme positions. Chem Commun (Camb) 43:5499–5501
29. Jochens H, Bornscheuer UT (2010) Natural diversity to guide focused directed evolution. Chembiochem 11(13):1861–1866
30. van Leeuwen JG, Wijma HJ, Floor RJ, van der Laan JM, Janssen DB (2012) Directed evolution strategies for enantiocomplementary haloalkane dehalogenases: from chemical waste to enantiopure building blocks. Chembiochem 13(1):137–148
31. Sandstrom AG, Wikmark Y, Engstrom K, Nyhlen J, Backvall JE (2012) Combinatorial reshaping of the Candida antarctica lipase A substrate pocket for enantioselectivity using an extremely condensed library. Proc Natl Acad Sci U S A 109(1):78–83
32. Papworth C, Bauer JC, Braman J, Wright DA (1996) Site-directed mutagenesis in one day with >80 % efficiency. Strategies 9(3):3–4
33. Dennig A, Shivange AV, Marienhagen J, Schwaneberg U (2011) OmniChange: the sequence independent method for simultaneous site-saturation of five codons. PLoS One 6(10):e26222
34. Sanchis J, Fernandez L, Carballeira JD, Drone J, Gumulya Y, Hobenreich H, Kahakeaw D, Kille S, Lohmer R, Peyralans JJ, Podtetenieff J, Prasad S, Soni P, Taglieber A, Wu S, Zilly FE, Reetz MT (2008) Improved PCR method for the creation of saturation mutagenesis libraries in directed evolution: application to difficult-to-amplify templates. Appl Microbiol Biotechnol 81(2):387–397
35. Stemmer WP, Crameri A, Ha KD, Brennan TM, Heyneker HL (1995) Single-step assembly of a gene and entire plasmid from large numbers of oligodeoxyribonucleotides. Gene 164(1):49–53
36. Gibson DG, Smith HO, Hutchison CA 3rd, Venter JC, Merryman C (2010) Chemical synthesis of the mouse mitochondrial genome. Nat Methods 7(11):901–903
37. Lendrihas T, Hunter GA, Ferreira GC (2010) Targeting the active site gate to yield hyperactive variants of 5-aminolevulinate synthase. J Biol Chem 285(18):13704–13711
38. Hokanson CA, Cappuccilli G, Odineca T, Bozic M, Behnke CA, Mendez M, Coleman WJ, Crea R (2011) Engineering highly thermostable xylanase variants using an enhanced combinatorial library method. Protein Eng Des Sel 24(8):597–605
39. Carr PA, Church GM (2009) Genome engineering. Nat Biotechnol 27(12):1151–1162
40. Ellis T, Adie T, Baldwin GS (2011) DNA assembly for synthetic biology: from parts to pathways and beyond. Integr Biol (Camb) 3(2):109–118
41. Keasling JD (2012) Synthetic biology and the development of tools for metabolic engineering. Metab Eng 14(3):189–195
42. Cox RS 3rd, Surette MG, Elowitz MB (2007) Programming gene expression with combinatorial promoters. Mol Syst Biol 3:145
43. Cohen SN, Chang AC, Boyer HW, Helling RB (1973) Construction of biologically functional bacterial plasmids in vitro. Proc Natl Acad Sci U S A 70(11):3240–3244
44. Scharf SJ, Horn GT, Erlich HA (1986) Direct cloning and sequence analysis of enzymatically amplified genomic sequences. Science 233(4768):1076–1078
45. Hartley JL, Temple GF, Brasch MA (2000) DNA cloning using in vitro site-specific recombination. Genome Res 10(11):1788–1795
46. Petersen LK, Stowers RS (2011) A Gateway MultiSite recombination cloning toolkit. PLoS One 6(9):e24531
47. Aslanidis C, de Jong PJ (1990) Ligation-independent cloning of PCR products (LIC-PCR). Nucleic Acids Res 18(20):6069–6074
48. Nisson PE, Rashtchian A, Watkins PC (1991) Rapid and efficient cloning of Alu-PCR products using uracil DNA glycosylase. PCR Methods Appl 1(2):120–123
49. Li MZ, Elledge SJ (2007) Harnessing homologous recombination in vitro to generate recombinant DNA via SLIC. Nat Methods 4(3):251–256
50. Zhang L, Zhao G, Ding X (2011) Tandem assembly of the epothilone biosynthetic gene cluster by in vitro site-specific recombination. Sci Rep 1:141
51. Garces C, Laborda J (1995) Single-step, ligase-free cloning of polymerase chain reaction products into any restriction site of any DNA plasmid. Anal Biochem 230(1):178–180
52. Bryksin AV, Matsumura I (2010) Overlap extension PCR cloning: a simple and reliable way to create recombinant plasmids. Biotechniques 48(6):463–465

53. Quan J, Tian J (2009) Circular polymerase extension cloning of complex gene libraries and pathways. PLoS One 4(7):e6441
54. Quan J, Tian J (2011) Circular polymerase extension cloning for high-throughput cloning of complex and combinatorial DNA libraries. Nat Protoc 6(2):242–251
55. Whitehouse CJ, Bell SG, Wong LL (2012) P450(BM3) (CYP102A1): connecting the dots. Chem Soc Rev 41(3):1218–1260
56. Consortium U (2010) The Universal Protein Resource (UniProt) in 2010. Nucleic Acids Res 38(Database issue):D142–D148
57. Benson DA, Karsch-Mizrachi I, Lipman DJ, Ostell J, Wheeler DL (2008) GenBank. Nucleic Acids Res 36(Database issue):D25–D30
58. Hoover DM, Lubkowski J (2002) DNAWorks: an automated method for designing oligonucleotides for PCR-based gene synthesis. Nucleic Acids Res 30(10):e43
59. Hoover D (2012) Using DNAWorks in designing oligonucleotides for PCR-based gene synthesis. Methods Mol Biol 852:215–223
60. Aharoni A, Griffiths AD, Tawfik DS (2005) High-throughput screens and selections of enzyme-encoding genes. Curr Opin Chem Biol 9(2):210–216
61. Patrick WM, Firth AE (2005) Strategies and computational tools for improving randomized protein libraries. Biomol Eng 22(4):105–112
62. Nov Y (2012) When second best is good enough: another probabilistic look at saturation mutagenesis. Appl Environ Microbiol 78(1):258–262
63. Reetz MT, Sanchis J (2008) Constructing and analyzing the fitness landscape of an experimental evolutionary process. Chembiochem 9(14):2260–2267
64. Gumulya Y, Sanchis J, Reetz MT (2012) Many pathways in laboratory evolution can lead to improved enzymes: how to escape from local minima. Chembiochem 13(7):1060–1066
65. Amin N, Liu AD, Ramer S, Aehle W, Meijer D, Metin M, Wong S, Gualfetti P, Schellenberger V (2004) Construction of stabilized proteins by combinatorial consensus mutagenesis. Protein Eng Des Sel 17(11):787–793
66. Chaparro-Riggers JF, Polizzi KM, Bommarius AS (2007) Better library design: data-driven protein engineering. Biotechnol J 2(2):180–191
67. Braun A, Halwachs B, Geier M, Weinhandl K, Guggemos M, Marienhagen J, Ruff AJ, Schwaneberg U, Rabin V, Torres Pazmino DE, Thallinger GG, Glieder A (2012) MuteinDB: the mutein database linking substrates, products and enzymatic reactions directly with genetic variants of enzymes. Database (Oxford) 2012:bas028
68. Polizzi KM, Chaparro-Riggers JF, Vazquez-Figueroa E, Bommarius AS (2006) Structure-guided consensus approach to create a more thermostable penicillin G acylase. Biotechnol J 1(5):531–536
69. Vazquez-Figueroa E, Chaparro-Riggers J, Bommarius AS (2007) Development of a thermostable glucose dehydrogenase by a structure-guided consensus concept. Chembiochem 8(18):2295–2301
70. Jochens H, Aerts D, Bornscheuer UT (2010) Thermostabilization of an esterase by alignment-guided focussed directed evolution. Protein Eng Des Sel 23(12):903–909
71. Etchebest C, Benros C, Bornot A, Camproux AC, de Brevern AG (2007) A reduced amino acid alphabet for understanding and designing protein adaptation to mutation. Eur Biophys J 36(8):1059–1069
72. Munoz E, Deem MW (2008) Amino acid alphabet size in protein evolution experiments: better to search a small library thoroughly or a large library sparsely? Protein Eng Des Sel 21(5):311–317
73. Pape S, Hoffgaard F, Dur M, Hamacher K (2013) Distance dependency and minimum amino acid alphabets for decoy scoring potentials. J Comput Chem 34(1):10–20
74. Prodromou C, Pearl LH (1992) Recursive PCR: a novel technique for total gene synthesis. Protein Eng 5(8):827–829

# Chapter 14

# One-Pot Simple Methodology for Cassette Randomization and Recombination for Focused Directed Evolution (OSCARR)

**Aurelio Hidalgo, Anna Schließmann, and Uwe T. Bornscheuer**

## Abstract

The OSCARR methodology (*O*ne-pot *S*imple methodology for *CA*ssette *R*andomization and *R*ecombination) bridges the gap between site-directed mutagenesis and full randomization by making use of carefully designed mutagenic cassettes and an optimized one-pot megaprimer PCR. The method is especially suited to construct libraries of up to ten randomized codons for focused directed evolution, exhibits up to 97 % efficiency in the amplification of mutated over wild-type products, and is sufficiently versatile to allow mutagenesis and recombination of several cassettes within the same gene.

**Key words** Focused directed evolution, Megaprimer PCR, Cassette mutagenesis, Gene randomization, Gene recombination

## 1 Introduction

There is a considerable gap in the spectrum of mutagenesis methodologies between randomizing techniques such as epPCR, the mutation rate of which is not high enough to focus mutations within a 30 bp cassette, and site-directed mutagenesis, in which only 1–6 bp can be mutated. Within this gap lies "focused randomization" mutagenesis, an approach that is instrumental to "focused directed evolution," in which determined regions of a gene (generally closer to the active site [1, 2]) are targeted for mutagenesis, in order to maximize library quality and minimize library size. This approach is supported by the increasing availability of genomic and structural information, such that it is becoming more and more usual to start a protein engineering campaign having some information about the target [3]. Even when a crystal structure is not available, computational methods to calculate a structural model are increasingly available through user-friendly servers for homology modeling, protein threading [4], and even de novo design [5, 6].

Elizabeth M.J. Gillam et al. (eds.), *Directed Evolution Library Creation: Methods and Protocols*, Methods in Molecular Biology, vol. 1179, DOI 10.1007/978-1-4939-1053-3_14, 

The OSCARR methodology (*O*ne-pot *S*imple methodology for *CA*ssette *R*andomization and *R*ecombination) presented here was established with the aim of bridging the abovementioned gap [7]. It was developed and optimized making use of the *Pseudomonas fluorescens* esterase I (PFEI, accession number U12537) as a model, because of the biocatalytic potential of this protein and the relatively limited success of previous error-prone methods in improving its activity (unpublished data). In our particular case and in congruence with the "closer is often better" principle of focused directed evolution [1], we chose target regions for mutagenesis based exclusively on their proximity (<10 Å) to the active site.

## 2 Materials

### 2.1 Biological and Chemical Materials

1. Chemically competent *Escherichia coli* DH5α strain.
2. Ultrapure water.
3. Proofreading polymerase (such as Pfu Plus polymerase, EURx, Poland) and corresponding 10× reaction buffer.
4. PCR and gel purification kits (e.g., Qiagen, Hilden, Germany).
5. Two plasmids containing the gene to be mutagenized flanked by different upstream and downstream sequences (to achieve this, in our study the *pfei* gene was cloned into the plasmids pGASTON-PFEI and pET22-PFEI).
6. Oligonucleotide primers that anneal downstream of the target gene in one of the two plasmids (reverse primer) and upstream in the other plasmid (forward primer).
7. A mutagenic oligonucleotide containing up to ten randomized codons flanked by ten perfectly matched bases on each side.
8. dNTP solution: 10 mM of each dNTP (Carl Roth, Karlsruhe, Germany). Store in small aliquots at −20 °C.
9. Appropriate restriction enzymes and digestion buffers.
10. T4 DNA ligase and ligase buffer (e.g., Promega, Mannheim, Germany).

### 2.2 Equipment

1. Microcentrifuge (e.g., MiniSpin Plus, Eppendorf, Germany).
2. Thermocycler (e.g., MasterCycler Personal, Eppendorf, Germany).
3. Orbital incubator.
4. Agarose gel electrophoresis equipment and supplies.
5. Thermoshaker (e.g., Thermomixer Comfort, Eppendorf, Germany).

## 3 Methods

### *3.1 Plasmid and External Primer Design*

To use the OSCARR method, two constructs are needed in which the gene of interest (in our particular case, *pfei*) is cloned using the same restriction enzymes, but the flanking upstream and downstream sequences are different in each plasmid (in our particular case, plasmid 1 is pGASTON-PFEI and plasmid 2 is pET22-PFEI).

A pair of external primers is required, which meets the following conditions:

1. The forward primer must hybridize on one flanking region of the first construct.
2. The reverse primer must hybridize on the opposite flanking region of the second construct (e.g., primer RhaP_forward hybridizing upstream of the *pfei* gene on pGASTON-PFEI and primer t7_reverse hybridizing downstream of the *pfei* gene on pET22-PFEI).

By fulfilling these two requirements, amplification of the mutagenized product over the wild-type template is favored (*see* **Note 1**). The process is schematically illustrated in Fig. 1.

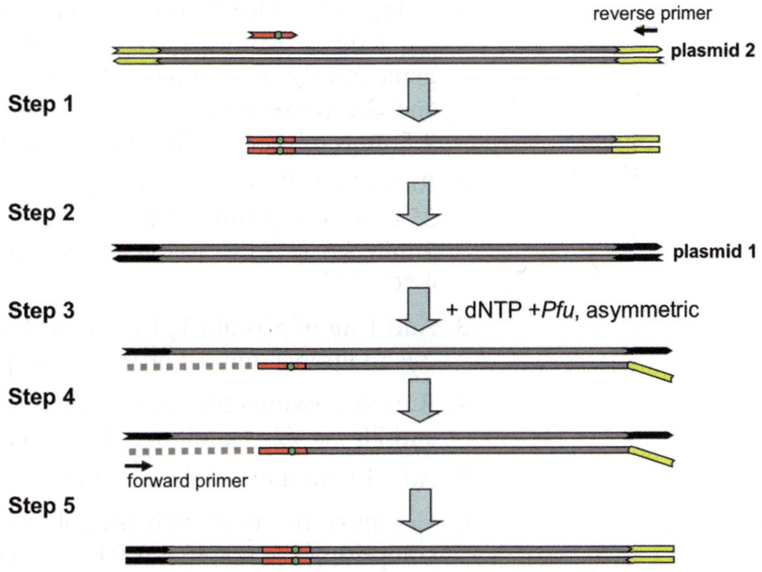


**Fig. 1** Schematic representation of the optimized OSCARR procedure. Step 1, megaprimers are generated from the plasmid 2 template by PCR using an internal mutagenic oligonucleotide (*red*, mutation site represented by *green circle*) and a reverse primer that binds within a unique plasmid 2 region. Steps 2 and 3, the megaprimers are extended asymmetrically into a full-length product using plasmid 1 as template. Steps 4 and 5, the extended megaprimers are amplified using a forward primer that binds within a unique plasmid 1 region and the reverse primer from step 1. The use of external primers that are unique to the different plasmid templates ensures amplification of the mutagenic construct only

### 3.2 Design and Synthesis of Mutagenic Primer

A degenerate oligonucleotide should be designed to mutate all target residues as desired, e.g., via incorporation of NNN or NNK degeneracies at each target codon. In our previous study [7], we used an algorithm to calculate the appropriate proportion of each nucleotide in each position [8]. This algorithm requires a target population of amino acids as input and returns nucleotide composition per position as output; in our particular case, we had no restrictions for the mutant population. Alternative methods for identifying target residues are described in Chapter 20 by Sebestova et al. in this book. Irrespective of how many codon positions are targeted and what degeneracies are employed, the mutated codons should be flanked by ten bases of unmodified (i.e., wild-type) sequence on each side.

### 3.3 Library Construction by One-Pot Megaprimer PCR

In a first stage (step 1, Fig. 1), this modified PCR generates megaprimers from the mutagenic oligonucleotides. In a second stage (steps 2 and 3, Fig. 1), these megaprimers are extended asymmetrically into a full-length product, and finally (steps 4 and 5, Fig. 1), the extended megaprimers are amplified, making use of a second external primer by means of a one-pot reaction, with virtually no contamination from wild-type amplicons.

1. Using thin-walled PCR tubes, set up a 100 μl reaction containing 1 ng of plasmid 2, 1× polymerase buffer, 25 pmol mutagenic primer, and 50 pmol of the reverse primer (*see* **Notes 2** and **3**), a final concentration of 100 μM of each dNTP and 2.5 units of proofreading DNA polymerase.
2. Run the PCR with the following parameters: 3 min at 94 °C; 25 cycles comprising each of 45 s at 94 °C, 1 min at 53 °C, and 1 min/kb at 72 °C; followed by 10 min at 72 °C and a hold at 4 °C.
3. Add 1 ng of plasmid 1, 1 μl 10 mM dNTP stock solution, and 2.5 additional units of proofreading polymerase.
4. Run five asymmetric cycles consisting each of 45 s at 94 °C and 75 s at 72 °C, followed by a hold at 4 °C (*see* **Note 4**).
5. Add 50 pmol of the forward primer (*see* **Note 5**).
6. Complete the PCR with the following parameters: 25 cycles comprising each of 45 s at 94 °C, 1 min at 66 °C, and 1 min/kb at 72 °C; followed by 10 min at 72 °C and a hold at 4 °C.
7. After completion, run the total volume of the reaction in an agarose gel and excise the band corresponding to the full-length gene. Ideally, only this band should be present, but unincorporated megaprimer may also be found depending on the amount of mutagenic primer used.
8. Gel-purify the mutagenized, reconstituted full-length gene using the PCR purification kit as per the manufacturer's protocol.

9. Digest the reaction product and the desired cloning plasmid with the appropriate restriction enzymes to create compatible sticky ends in both insert and vector.
10. Ligate the compatible sticky-end fragments overnight at 16 °C according to the manufacturer's protocol for T4 DNA ligase.
11. Transform into suitable strain for library screening.

## 4 Notes

1. Having the forward and reverse external primers anneal on different templates reduces the risk of having amplification of the wild-type gene [9]. This is particularly true when lower amounts of mutagenic primer are used (*see* **Note 2**).
2. The amount of megaprimer obtained can be controlled by lowering the amount of mutagenic primer relative to the reverse primer. This is relevant in order to avoid having leftover megaprimer in the product, as a result of an inefficient use of the (usually expensive) mutagenic primer.
3. It is possible to simultaneously combine mutagenic primers for at least two sites, giving rise to a population of diverse megaprimers that include mutations in either or both sites. The ratio of these megaprimers is controlled by the ratio of the mutagenic primers and their proximity to the reverse primer [7].
4. By asymmetrically generating only single-stranded mutated megaprimer first, the reverse primer has a higher likelihood of annealing to a mutated strand rather than amplifying the wild-type sequence [10].
5. The external primers should differ significantly in $T_m$. The first primer to be used should have a lower $T_m$. In this way, using a higher $T_a$ for the final stage compared to the megaprimer generation stage, no purification is needed, and the first primer may remain in the reaction mix, since it will not anneal [11].

### References


1. Morley KL, Kazlauskas RJ (2005) Improving enzyme properties: when are closer mutations better? Trends Biotechnol 23:231–237
2. Park S, Morley KL, Horsman GP et al (2005) Focusing mutations into the *P. fluorescens* esterase binding site increases enantioselectivity more effectively than distant mutations. Chem Biol 12:45–54
3. Bornscheuer UT, Huisman GW, Kazlauskas RJ et al (2012) Engineering the third wave of biocatalysis. Nature 485:185–194
4. Kelley LA, Sternberg MJE (2009) Protein structure prediction on the Web: a case study using the Phyre server. Nat Protoc 4:363–371
5. Jiang L, Althoff EA, Clemente FR et al (2008) De novo computational design of retro-aldol enzymes. Science 319:1387–1391
6. Röthlisberger D, Khersonsky O, Wollacott AM et al (2008) Kemp elimination catalysts by computational enzyme design. Nature 453:190–195
7. Hidalgo A, Schließmann A, Molina R et al (2008) A one-pot, simple methodology for

cassette randomization and recombination for focused directed evolution (OSCARR). Protein Eng Des Sel 21:567–576

8. Jensen LJ, Andersen KV, Svendsen A et al (1998) Scoring functions for computational algorithms applicable to the design of spiked oligonucleotides. Nucleic Acids Res 26:697–702
9. Urban A, Neukirchen S, Jaeger K-E (1997) A rapid and efficient method for site-directed mutagenesis using one-step overlap extension PCR. Nucleic Acids Res 25:2227–2228
10. Datta AK (1995) Efficient amplification using "megaprimer" by asymmetric polymerase chain reaction. Nucleic Acids Res 23:4530–4531
11. Ke S-H, Madison EL (2005) Rapid and efficient site-directed mutagenesis by single-tube "megaprimer" PCR method. Nucleic Acids Res 25:3371–3372

# Chapter 15

# USER Friendly DNA Recombination (USERec): Gene Library Construction Requiring Minimal Sequence Homology

**Benoit Villiers and Florian Hollfelder**


## Abstract

USER friendly DNA recombination (USERec) is based on near homology-independent recombination of DNA fragments (~40–400 bp) that are efficiently reassembled into full-length genes, proven for up to ten fragments. USERec requires a minimal crossover sequence (a 5′-$AN_{4-8}$T-3′ motif) of the fragments that can be implemented wherever structural or functional comparisons suggest a fragment boundary. The greatly reduced sequence constraints of this method facilitate directional assembly of gene fragments for applications such as exon or domain shuffling, loop grafting, reassembly of natural modular biosynthetic assembly lines, and rearrangement of structurally (but not sequence) homologous proteins.


**Key words** Directed evolution, DNA library, Exon shuffling, Homology-independent DNA recombination, USER enzyme, ITCHY, SCRATCHY, SHIPREC, SISDC

## 1 Introduction

DNA libraries made by the most popular methods of site saturation mutagenesis and error-prone PCR generate variants that are relatively similar to the original gene and typically only afford minor structural rearrangements (often corresponding to relatively small improvements in function). Attempts to cover greater distances of protein sequence space are attractive, because of the prospect of further-reaching evolutionary innovation, resulting in larger structural and functional changes. Recombination strategies that combine the functional properties of two or more genes by swapping individual residues, secondary structure elements, motifs, domains, and whole proteins are therefore attractive alternatives to harvest information from a more diverse pool of variants.

Homologous recombination, e.g., by DNA shuffling [1, 2] and StEP PCR [3], typically requires >70 % sequence identity to be effective. Homology-independent recombination offers a means to recombine genes with lower sequence identities. It is likely that such recombination strategies will still rely on structural homology

Elizabeth M.J. Gillam et al. (eds.), *Directed Evolution Library Creation: Methods and Protocols*, Methods in Molecular Biology, vol. 1179, DOI 10.1007/978-1-4939-1053-3_15, © Springer Science+Business Media New York 2014

to a certain extent, in order to increase the likelihood of generating stable proteins as library members, but additional degrees of freedom in recombining protein fragments will enable a more encompassing exploration of sequence space. For us, the immediate targets for this approach will be the structurally homologous but functionally diverse TIM barrels and modular biosynthetic systems of secondary metabolism (such as polyketide or non-ribosomal peptide synthetases) that promise to generate new natural product analogues, where controlled recombination of defined structural motifs can be based on existing knowledge of these enzymes. Creation of chimeric proteins containing crossovers will also mimic events thought to have played a central role in evolution, namely, gene fusion and domain recruitment and exon shuffling.

A number of methods address the need for libraries made by homology-independent recombination. For example, ITCHY [4], SCRATCHY [5], and SHIPREC [6] are based on truncation of distantly related genes. Using these methods, two genes can be recombined at a time, resulting in relatively few crossovers (typically one or two). More than two parental sequences can be simultaneously recombined by overlap extension PCR [1, 7] (requiring ~15 nucleotide overlaps between fragments and successive assembly PCRs) and by random multi-recombinant PCR (RM-PCR) [8] (which does not require homologous sequences at the recombination sites, but the accessible number and size of the recombined fragments is very limited due to the reliance on synthetic DNA sequences). In sequence-independent site-directed chimeragenesis (SISDC), type IIb restriction enzymes that cut outside their recognition site have been used [9–11]. Here, DNA fragments flanked by suitable restriction sites are first prepared by PCR and then reassembled into their parent genes by overlap extension PCR.

This protocol outlines a nearly homology-independent DNA recombination method in which the assembly of DNA fragments is achieved with USER enzyme [12–14] and T4 DNA ligase (Fig. 1). USER enzyme is a commercially available enzyme mixture composed of uracil DNA glycosylase (UDG) and endonuclease VIII. In combination, the two enzymes catalyze the excision of uracil residues from DNA thereby generating a single base pair gap. If a uracil residue is introduced close to the 5′ end by PCR using uracil-containing oligonucleotides, the ends of a linear DNA fragment can be endowed with unique single-stranded 3′ extensions (Fig. 1, step 2). These can subsequently direct the assembly of two DNA fragments with complementary overlapping ends (Fig. 1, step 3). The main advantages of this method over previous recombination strategies based on restriction digestion and ligation are ease of handling, directionality, flexibility, and minimal sequence requirements: only an adenine and a thymidine spaced apart by several nucleotides (i.e., a consensus of 5′-$AN_{4-8}T$-3′) are needed to ensure efficient dissociation of the excised regions [12–14].

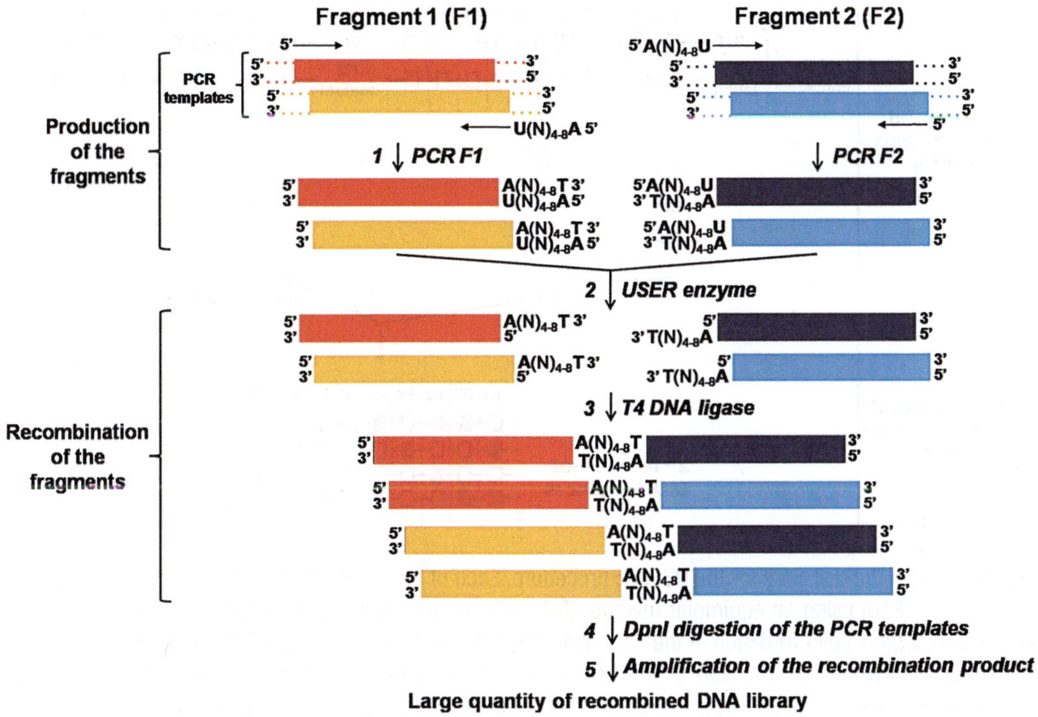


**Fig. 1** Experimental steps in USER friendly DNA recombination (USERec). The two principal experimental stages, namely, the production of the fragments (*see* Subheading 3.2) and the recombination of the fragments (*see* Subheading 3.3), are illustrated for two fragments. Fragments are produced by standard PCR using uracil-containing primers and *PfuTurbo* $C_x$ hotstart DNA polymerase (a mutant of *Pfu* DNA polymerase able to amplify uracil-containing templates) (step 1). The genetic diversity is introduced for each fragment (F1, F2) by using a mixture of DNA templates (shown with different shadings) leading to different fragment variants. For recombination between F1 and F2, the reverse uracil-containing primer of F1 and the forward uracil-containing primer of F2 are designed to be complementary for ~8 bases [13] at the 5′ end, starting with A/U and finishing with U/A (5′-AN$_{4-8}$U-3′). The PCR-amplified fragments are mixed and incubated with USER enzyme to generate a single nucleotide gap at the location of the uracil (U), creating complementary 3′ single-stranded extensions (step 2). T4 DNA ligase is used to ligate the different fragment variants simultaneously, leading to the recombination of the genetic diversity (step 3). The DpnI digestion allows removal of the PCR templates that were used to produce the fragments (step 4). Finally, the correct recombination product is amplified by PCR with external oligonucleotides and *PfuTurbo* DNA polymerase, generating a large quantity of the recombined DNA library (step 5). The diagram was adapted from reference [15] with permission

We have extended this original method (which had initially been used for gene synthesis) to a DNA recombination tool [15]. The recombination process was exemplified with the gene coding for the A-domain of the tyrocidine synthetase 1 (TycA) [16–18]. For this purpose, the TycA A-domain was arbitrarily divided into ten DNA fragments of varying sizes ranging from ~40 to 400 bp and each fragment was marked by a diagnostic single alanine mutation in one of the ten DNA fragment regions, so that recombination fidelity could be assessed (Fig. 2) [15]. Large libraries are

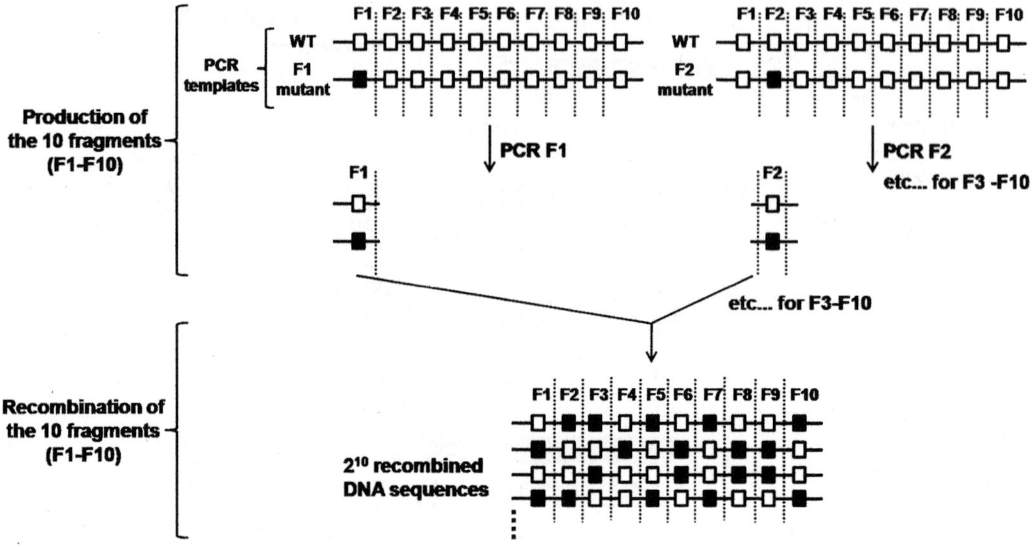


**Fig. 2** Outline of the test case for the USERec procedure. Each of the ten fragments (F1–F10) was produced by PCR (PCR F1–F10) using an equimolar mixture of the WT and a corresponding alanine mutant plasmid template (*filled square*: point mutation in the corresponding amplified fragment region) of the model gene encoding tyrocidine synthetase 1 (TycA) [16]. The ten fragments were recombined simultaneously using the procedure detailed in Fig. 1, leading to a theoretical library size of $2^{10}$ (=1,024) recombined variants

accessible with our method, due to the large amount of recombined DNA library produced, e.g., ~20 μg of purified recombined DNA library of 1,600 bp in length can be easily obtained, representing ~$10^{13}$ copies. In our TycA study, ten DNA fragments were recombined simultaneously with a low error rate (25 %) and 90 % of sequences in frame [15].

Thus far, the value of this method has only been assessed at the DNA level and a selection of proteins from templates assembled by USERec is still to be shown. However, we hope that USERec will find applications in gene engineering where several genetically diverse DNA fragments need to be assembled. The size of the DNA fragments that were recombined in our work [15] was ~40–400 bp and matches the domain size (30–130 amino acids) that is observed in natural modularly assembled proteins [1], suggesting that USERec is suitable for exon shuffling. USERec is set up to allow facile design of experiments that seek to take advantage of the ability to combine protein "building blocks" at will. The generation of near homology-independent recombination libraries composed of multiple DNA fragments in one step, within a day and with minimal sequence constraints, should make this method a valuable addition to the toolkit of gene library construction.

## 2 Materials

### 2.1 Reagents

1. Uracil-containing oligonucleotides as PCR primers to produce the fragments and standard oligonucleotides to amplify the recombined DNA library (in our study, these were supplied salt-free by Eurofins Genomics, Ebersberg, Germany). PHUSER software can be used for the design of the uracil-containing primers (http://www.cbs.dtu.dk/services/PHUSER/) [19].
2. Plasmid DNA templates (source of genetic diversity) produced in *Escherichia coli*, so that they can subsequently be removed via digestion with the restriction enzyme DpnI after the ligation of fragments.
3. A high-fidelity thermostable DNA polymerase that is compatible with DNA templates that contain uracil residues, such as *PfuTurbo Cx* hotstart DNA polymerase (available with 10× *PfuTurbo Cx* reaction buffer, Agilent Technologies, Santa Clara, CA, USA), to amplify the uracil-containing fragments.
4. A standard high-fidelity thermostable DNA polymerase, such as *PfuTurbo* DNA polymerase (available with 10× *PfuTurbo* reaction buffer, Agilent Technologies, Santa Clara, CA, USA), to amplify the recombination product.
5. 10 mM dNTPs (containing 2.5 mM of each dNTP).
6. DNA molecular weight markers to analyze small (e.g., 25–500 bp, HyperLadder 25 bp, Bioline, London, UK) and large (e.g., 200–10,037 bp, HyperLadder 1 kb, Bioline, London, UK) PCR products.
7. Spin column kit for purifying DNA from PCR reactions (e.g., QIAquick PCR purification kit, Qiagen, Venlo, Netherlands).
8. Spin column kit for purifying recombined DNA from agarose gels (e.g., QIAquick gel extraction kit, Qiagen, Venlo, Netherlands).
9. Spin column kit for purifying plasmid DNA from *E. coli* cells (e.g., QIAprep spin miniprep kit, Qiagen, Venlo, Netherlands).
10. Sterile $ddH_2O$ for DNA elution.
11. USER enzyme (New England Biolabs, Ipswich, MA, USA) to generate a single nucleotide gap at the location of the uracil and to create complementary 3′ single-stranded extensions.
12. T4 DNA ligase and 10× T4 DNA ligase buffer (e.g., New England Biolabs, Ipswich, MA, USA) to ligate the fragments.
13. DpnI restriction enzyme and 10× NEBuffer 4 (New England Biolabs, Ipswich, MA, USA) to remove PCR templates.

14. Reagents and plasmid vector to clone the recombined DNA library.
15. Competent cells (e.g., XL1-Blue, Agilent Technologies, Santa Clara, CA, USA) and reagents for their transformation, growth, and selection to amplify the recombined DNA library.
16. Reagents for standard agarose gel electrophoresis.

### 2.2 Equipment

1. PCR thermocycler (e.g., Mastercycler nexus, Eppendorf, Hamburg, Germany).
2. Apparatus for standard agarose gel electrophoresis.
3. Spectrophotometer for determination of DNA concentration (e.g., NanoDrop 2000, Thermo Fisher Scientific, Waltham, MA, USA).
4. Water bath or electroporator for *E. coli* transformation.
5. Incubator for bacterial growth.

## 3 Methods

### 3.1 Construction of Mutant Templates

The templates to be recombined can be mutant templates. For exemplification purposes, ten alanine-labeled mutant templates were produced by site-directed mutagenesis based on the QuikChange method (Agilent Technologies, Santa Clara, CA, USA). Details of this standard procedure can be found elsewhere [15] and are not elaborated upon here, because the template construction will differ according to every system that is to be recombined.

### 3.2 PCR Amplification of Fragments as Substrates for Recombination (Fig. 1, Step 1)

1. Amplify the recombination substrates (DNA fragments) from appropriate proportions of mutant templates in each PCR. In our case [15], ten recombination substrates (F1–F10, Fig. 2) were prepared by PCR, each using an equimolar mixture of the wild-type (WT) TycA [16] and the corresponding alanine mutant plasmid template (i.e., in total, ten mutant plasmids, each containing one alanine point mutation, were used) (*see* **Note 1**). Primer sequences should be designed to ensure an efficient and directional subsequent recombination of the fragments (*see* **Note 2**). For a reaction of 50 μl, prepare a PCR mix on ice containing 5 ng of DNA template (*see* **Note 3**) (e.g., in our study, 2.5 ng of plasmid pSU18-*tycA-PheATE-His* and 2.5 ng of the corresponding alanine mutant plasmid [15]), 0.2 mM of each dNTP, 0.3 μM of the forward and reverse oligonucleotides, and 2.5 U of the high-fidelity uracil-compatible polymerase in its appropriate reaction buffer. Place reaction tubes into a thermocycler and use the following cycling protocol: initial denaturation for 2 min at 95 °C; 30 cycles of

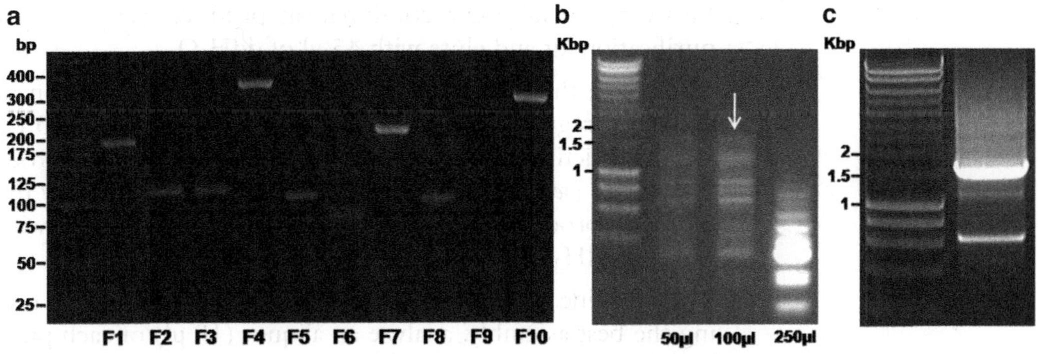


**Fig. 3** Agarose gels at three stages of USERec: (**a**) production of the fragments (corresponding to Fig. 1, step 1), (**b**) recombination of the fragments (Fig. 1, steps 2–4), and (**c**) amplification of the recombination product (Fig. 1, step 5). *Panel a* depicts a 3.5 % agarose gel showing the ten DNA fragments (F1, 191 bp; F2, 113 bp; F3, 114 bp; F4, 365 bp; F5, 106 bp; F6, 85 bp; F7, 218 bp; F8, 104 bp; F9, 42 bp; F10, 316 bp). *Panel b* depicts a 1 % agarose gel showing the recombination products obtained from 50 μl, 100 μl, or 250 μl of fragment mixture. The largest quantity of correct recombination product (1,582 bp) was obtained from the 100 μl mixture (*white arrow*). This recombination product was used as DNA template for the following step, depicted in *panel c*: the amplification of the recombination product with the external oligonucleotides F1_forward and F10_reverse (1 % agarose gel). The major band obtained from this amplification (the recombined DNA library) had the correct size (1,582 bp) and was purified by gel extraction

denaturation for 30 s at 95 °C, annealing for 30 s at 55 °C, and polymerization for 1 min/kbp at 72 °C; and final polymerization for 10 min at 72 °C.

2. To check PCR amplification specificity, analyze an aliquot (10 μl) of each PCR product by agarose gel electrophoresis. For small PCR products (e.g., 25–500 bp), you can use 3.5 % TAE-agarose gel electrophoresis with 5 μl of HyperLadder 25 bp (Bioline) as molecular weight marker (Fig. 3a).
3. Mix all the PCR products to make the fragment mixture (e.g., for ten fragments, you will obtain a total volume of 500 μl; *see* **Note 4**). To find the most appropriate fragment mixture concentration for their subsequent recombination (*see* **Note 5**), purify different aliquots (e.g., 50 μl, 100 μl, or 250 μl) of this fragment mixture using the PCR purification kit and elute each preparation into 40 μl of dd$H_2O$ (*see* **Note 6**) (*see* **Note 7** for the timing of the procedure).

### *3.3 Recombination of the Fragments (Fig. 1, Steps 2–4)*

1. For a 50 μl recombination reaction, prepare a mix containing 40 μl of purified fragment mixture, 5 μl of 10× T4 DNA ligase buffer, and 5 μl (5 U) of USER enzyme. Place the tube in a PCR thermocycler and incubate for 20 min at 37 °C and then for a further 20 min at 25 °C.
2. Add 5 μl of T4 DNA ligase (2,000 U) to the mixture (final volume: 55 μl) and incubate for 15 min at 25 °C.

3. Purify the resulting recombination product using a PCR purification kit and elute with 45 µl of $ddH_2O$.
4. To remove the plasmid templates used for the production of the DNA fragments, add 5 µl 10× NEBuffer 4 and 1 µl (20 U) DpnI restriction enzyme to the purified recombination product (45 µl) (*see* **Note 8**). Incubate for 1 h at 37 °C. Purify the digestion product with a PCR purification kit and elute with 30 µl of $ddH_2O$.
5. To determine the purified fragment mixture concentration giving the best assembly, analyze an aliquot (10 µl) of each purified DpnI-digested recombination product by agarose gel electrophoresis. For large PCR products (e.g., 200–10,000 bp), you can use 1 % TAE-agarose gel electrophoresis with 5 µl of HyperLadder 1 kb as molecular weight marker (Fig. 3b). In our case [15], the best assembly was obtained for the purified fragment mixture concentration corresponding to 100 µl (Fig. 3b, white arrow), so this purified DpnI-digested recombination product was used for the later steps.

### 3.4 PCR Amplification of the Recombination Product (Fig. 1, Step 5)

1. For a 50 µl reaction, prepare a PCR mix containing an aliquot (10 µl) of the purified DpnI-digested recombination product, 0.3 µM of the external oligonucleotides (e.g., F1_forward and F10_reverse, Fig. 1), 0.2 mM of each dNTP, and 2.5 U of the standard (i.e., non-uracil-compatible) DNA polymerase in its appropriate reaction buffer. Place the reaction tube in a PCR thermocycler and use the following cycling protocol: initial denaturation for 2 min at 95 °C; 30 cycles of denaturation for 30 s at 95 °C, annealing for 30 s at 55 °C, and polymerization for 1 min/kbp at 72 °C; and final polymerization for 10 min at 72 °C.
2. Load the entire volume of PCR product (50 µl) on an agarose gel with an appropriate DNA molecular weight marker (Fig. 3c). Under UV light, cut the band of agarose corresponding to the recombined DNA library size (e.g., the 1,582 bp band in Fig. 3c). Purify the DNA from this piece of agarose by gel extraction using the gel extraction kit and elute with 50 µl of $ddH_2O$. Measure the DNA concentration by absorbance at 260 nm (*see* **Note 9**).

### 3.5 Cloning of the Recombined DNA Library

1. USERec allows a wide range of cloning strategies (*see* **Note 10**). If the recombination product is highly homologous to a plasmid template (i.e., differs only in terms of point mutations), a method analogous to the QuikChange method can be used (using the recombined DNA library as megaprimers); this method is described here. For a 50 µl reaction, prepare a PCR mix containing 250 ng of recombined DNA library, 50 ng of plasmid template (e.g., plasmid pSU18-*tycA-PheATE-His* [15]),

0.2 mM of each dNTP, and 2.5 U of the standard high-fidelity DNA polymerase in its appropriate reaction buffer. Place the reaction tube in a PCR thermocycler and use the following cycling protocol: initial denaturation for 30 s at 95 °C, 18 cycles of denaturation for 30 s at 95 °C, annealing for 1 min at 55 °C, and polymerization for 1 min/kbp at 68 °C.

2. Add 1 μl (20 U) of DpnI restriction enzyme to the PCR product (50 μl) and incubate for 1 h at 37 °C to digest the original plasmid template.
3. Transform competent *E. coli* cells (e.g., 100 μl XL1-Blue) with 10 μl of DpnI digestion product and incubate cells overnight in 5 ml 2YT medium containing the appropriate antibiotics.
4. Prepare plasmid DNA from the 5 ml overnight culture with the miniprep kit and evaluate the quality of the recombined DNA library by sequencing random clones (*see* **Note 11**).

## 4 Notes

1. Assuming equal amplification efficiencies and unbiased recombination, each variant then had an anticipated recombination probability of 50 %. This outcome was confirmed by sequencing (*see* **Note 11**).
2. Only minimal sequence requirements are imposed on the crossover sites allowing a range of different DNA fragments to be recombined. Considering the consensus sequence of 5′-$AN_{4–8}T$-3′, crossover sites can easily be identified in the sequence of any gene as the distance between A and T is flexible [13]. PHUSER software can be used for the design of the uracil-containing primers [19].
3. We found that the amount of DNA template in the PCR had a significant impact on the efficiency of the subsequent assembly step, e.g., use of 50 ng DNA template instead of 5 ng decreased the assembly efficiency. The efficiency of producing the correct recombination product should be tested by agarose gel electrophoresis. Approximately 30–40 ng of correct recombination product should be obtainable in a 10 μl aliquot. This significant amount of DNA template limits the number of times the same sequence is amplified during the amplification reaction, thus making the polymerization-induced errors negligible (~1 % of the sequences [15]) (*see* **Note 11**).
4. The demonstrated number of ten DNA fragments that were simultaneously recombined should not be considered an upper limit. However, we anticipate that the number of errors (substitutions and deletions) in the recombined DNA library will increase with the number of fragments (*see* **Note 11**).

5. The assembly efficiency was tested for three purified fragment mixture concentrations (obtained from 50 μl, 100 μl, or 250 μl of fragment mixture), covering a range above and below the recommended amounts of uracil-containing DNA (as defined in the USER enzyme manufacturer's manual).
6. Purifying the fragment mixture prior to the assembly process is necessary to remove the polymerase. It was observed that, in the absence of a purification step, the number of single nucleotide deletions (dA or dT) at the ends of the fragment regions was significantly increased (*see* **Note 11**).
7. Once all primers are designed and mutant templates are available, the procedure leading to a recombined DNA library can be produced within a day: ~2 h for the PCR amplification of fragments as substrates for recombination (*see* Subheading 3.2), ~3 h for the recombination of the fragments (*see* Subheading 3.3), and ~3 h for the PCR amplification of the recombination product (*see* Subheading 3.4).
8. Before the amplification step, a DpnI restriction digestion should be carried out (Fig. 1, step 4) to remove the plasmid that served as template for the amplification of the fragments. Efficient removal of the PCR templates by the DpnI digestion can be tested by sequencing of the recombined library, which should show no wild-type clones.
9. We have typically obtained ~1.25 μg of recombined DNA library from one single 50 μl amplification reaction (or ~19 μg of recombined DNA library from the initial 500 μl mixture of fragments). This value is based on using 10 μl of recombination product (of a total volume of 30 μl) for the amplification reaction and 100 μl of fragment mixture (of a total volume of 500 μl) to produce the 30 μl of recombination product (i.e., 1.25 μg × 3 × 5 = 19 μg). This large quantity of DNA makes USERec suitable for constructing large libraries.
10. USERec allows a wide range of cloning strategies since specially designed external oligonucleotides can be used during the amplification step. For example, the most common cloning strategy based on type II restriction enzymes and a ligase (as described in detail in Chapter 1 of this volume) could be applied by using external oligonucleotides containing a unique restriction enzyme recognition site. Considering the high degree of DNA sequence identity between our recombined DNA library and the WT sequence, we decided to optimize a cloning strategy suitable for libraries with point mutations, since an important potential application of USERec is the construction of complex multiple site-directed mutagenesis libraries [20]. The strategy is inspired by the QuikChange method (Agilent Technologies) in which the complementary mutagenic primers have been replaced by the double-stranded

recombined DNA library (acting as megaprimers). With this method, about $6 \times 10^4$ clones could be obtained from 1.25 μg of the recombined DNA library and lab-made XL1-Blue competent cells with a transformation efficiency of ~$10^6$ cfu/μg pUC19.

11. Sequencing a sample of the recombined library forms the basis of the analysis of the quality of the recombined DNA library. In our study [15], we observed near-ideal distribution of the marker sequence, suggesting that there is no recombination bias. All the sequences contained the ten fragments in the right order, indicating perfect directionality of the method. However, errors were detected in 25 % of the sequenced clones, comprising ~10 % deletions and ~15 % substitutions. These results indicate that ~90 % of sequences were in frame, which is particularly remarkable considering the large number (ten) of DNA fragments that were recombined simultaneously. We suspect the deletions to be due to traces of *PfuTurbo* $C_X$ hotstart DNA polymerase, the 3′ overhangs being particularly sensitive to 3′→5′ exonuclease activity. All the substitutions occurred in oligonucleotide regions, which represent only ~24 % of the sequence, suggesting that substitutions are not induced by polymerization but more likely by initial errors in oligonucleotide sequences. Thus, it appears that the percentage of errors with USERec will rise with increasing number of crossovers (or recombination sites, nine in our case). In other words, more crossovers would lead to more fragments, hence more oligonucleotides and more potential deletion sites.

## Acknowledgments

This research was supported by the Medical Research Council. B. V. was supported by a fellowship from the EU Marie Curie Early-Stage Training Site ChemBioCam. F. H. is an ERC Starting Investigator. We thank David Ackerley for his thorough editing of this article.

### References

1. Kolkman JA, Stemmer WP (2001) Directed evolution of proteins by exon shuffling. Nat Biotechnol 19(5):423–428
2. Stemmer WP (1994) Rapid evolution of a protein in vitro by DNA shuffling. Nature 370:389–391
3. Zhao H, Giver L, Shao Z, Affholter JA, Arnold FH (1998) Molecular evolution by staggered extension process (StEP) in vitro recombination. Nat Biotechnol 16:258–261
4. Ostermeier M, Nixon AE, Shim JH, Benkovic SJ (1999) Combinatorial protein engineering by incremental truncation. Proc Natl Acad Sci U S A 96:3562–3567
5. Lutz S, Ostermeier M, Moore GL, Maranas CD, Benkovic SJ (2001) Creating multiple-crossover

DNA libraries independent of sequence identity. Proc Natl Acad Sci U S A 98:11248–11253
6. Sieber V, Martinez CA, Arnold FH (2001) Libraries of hybrid proteins from distantly related sequences. Nat Biotechnol 19(5): 456–460
7. Horton RM, Hunt HD, Ho SN, Pullen JK, Pease LR (1989) Engineering hybrid genes without the use of restriction enzymes: gene splicing by overlap extension. Gene 77(1): 61–68
8. Tsuji T, Onimaru M, Yanagawa H (2001) Random multi-recombinant PCR for the construction of combinatorial protein libraries. Nucleic Acids Res 29(20):E97
9. Hiraga K, Arnold FH (2003) General method for sequence-independent site-directed chimeragenesis. J Mol Biol 330:287–296
10. Meyer MM, Hiraga K, Arnold FH (2006) Combinatorial recombination of gene fragments to construct a library of chimeras. Curr Protoc Protein Sci Chapter 26, Unit 26.2
11. Meyer MM, Hochrein L, Arnold FH (2006) Structure-guided SCHEMA recombination of distantly related beta-lactamases. Protein Eng Des Sel 19:563–570
12. Geu-Flores F, Nour-Eldin HH, Nielsen MT, Halkier BA (2007) USER fusion: a rapid and efficient method for simultaneous fusion and cloning of multiple PCR products. Nucleic Acids Res 35(7):e55
13. Bitinaite J, Rubino M, Varma KH, Schildkraut I, Vaisvila R, Vaiskunaite R (2007) USER friendly DNA engineering and cloning method by uracil excision. Nucleic Acids Res 35(6): 1992–2002
14. Nour-Eldin HH, Hansen BG, Norholm MH, Jensen JK, Halkier BA (2006) Advancing uracil-excision based cloning towards an ideal technique for cloning PCR fragments. Nucleic Acids Res 34(18):e122
15. Villiers BR, Stein V, Hollfelder F (2010) USER friendly DNA recombination (USERec): a simple and flexible near homology-independent method for gene library construction. Protein Eng Des Sel 23(1):1–8
16. Mootz HD, Marahiel MA (1997) The tyrocidine biosynthesis operon of *Bacillus brevis*: complete nucleotide sequence and biochemical characterization of functional internal adenylation domains. J Bacteriol 179(21):6843–6850
17. Grünewald S, Mootz HD, Stehmeier P, Stachelhaus T (2004) In vivo production of artificial nonribosomal peptide products in the heterologous host *Escherichia coli*. Appl Environ Microbiol 70(6):3282–3291
18. Villiers B, Hollfelder F (2011) Directed evolution of a gatekeeper domain in nonribosomal peptide synthesis. Chem Biol 18(10): 1290–1299
19. Olsen LR, Hansen NB, Bonde MT, Genee HJ, Holm DK, Carlsen S, Hansen BG, Patil KR, Mortensen UH, Wernersson R (2011) PHUSER (Primer Help for USER): a novel tool for USER fusion primer design. Nucleic Acids Res 39(Web server issue):W61–W67
20. Bhat KS (1996) Multiple site-directed mutagenesis. Methods Mol Biol 57:269–277

# Chapter 16

# ITCHY: Incremental Truncation for the Creation of Hybrid Enzymes

**Wayne M. Patrick and Monica L. Gerth**


## Abstract

Incremental Truncation for the Creation of Hybrid enzYmes (ITCHY) is a directed evolution technique for randomly recombining two genes. The chief advantage of ITCHY is that there is no requirement for the two genes to share any sequence similarity. This distinguishes ITCHY from directed evolution methods that are based on homologous recombination, such as DNA shuffling. In ITCHY, *Escherichia coli* exonuclease III is used to incrementally truncate one of the parental genes from its 3′ end and the other from its 5′ end. Ligation of the randomly truncated gene fragments yields a combinatorial library of chimeras. In this chapter, we provide detailed protocols for constructing libraries using both the user-friendly thio-ITCHY method and also time-dependent incremental truncation. We illustrate the protocols with the data that we obtained when we recombined two alcohol dehydrogenase genes that only share 47 % sequence identity.


**Key words** Incremental truncation, Exonuclease III, Nonhomologous recombination, Protein engineering, Thio-ITCHY, Time-dependent truncation

## 1 Introduction

In vitro recombination techniques have yielded many success stories in the field of directed evolution [1]. Most methods for in vitro recombination either rely on a high level of sequence identity between the genes being recombined, or they are underpinned by computational predictions of favored crossover locations; these methods are described elsewhere in this volume.

In contrast, Incremental Truncation for the Creation of Hybrid enzYmes (ITCHY) is a method for randomly recombining any two genes, with no requirement that they share sequence similarity [2]. ITCHY has provided a means to utilize the power of recombination for sampling sequence space and finding unexpected solutions, in cases where the two parental proteins are structural homologues but their genes are <70 % identical. For example, ITCHY was used to recombine two homologous deoxynucleoside

Elizabeth M.J. Gillam et al. (eds.), *Directed Evolution Library Creation: Methods and Protocols*, Methods in Molecular Biology, vol. 1179, DOI 10.1007/978-1-4939-1053-3_16, © Springer Science+Business Media New York 2014

kinases with 48 % sequence identity, yielding a chimeric enzyme that could phosphorylate the nucleoside analogue prodrug, d4T—an activity that neither parent possessed [3]. ITCHY is also a tractable method for exploring the role of nonhomologous end joining in the (natural) evolution of protein folds and functions [4].

The basic ITCHY protocol involves incrementally truncating one parental gene from its 3′ end, creating a set of sequences in which every single-base truncation is represented. Simultaneously, the second parental gene is incrementally truncated from its 5′ end. In practice, the two parental genes are cloned into a single plasmid, in tandem. The plasmid is linearized with a restriction enzyme that cleaves between the two genes, and then the sets of incrementally truncated sequences are generated by digestion with *Escherichia coli* exonuclease III. In the original ITCHY method [2], time-dependent exonuclease III digestion yielded incrementally truncated genes. An important advance was the development of thio-ITCHY [5]. In this user-friendly method, the linearized plasmid is amplified in a PCR that uses a mixture of dNTPs and α-phosphorothioate dNTPs (αS-dNTPs). The phosphorothioate-containing linkages that are incorporated randomly into the PCR product are resistant to cleavage by exonuclease III, thus enabling sets of incrementally truncated genes to be generated. In each protocol, exonuclease III digestion is followed by the preparation of blunt ends and intramolecular ligation to recircularize the plasmid, thereby fusing together randomly sized fragments of the two parental genes. An overview of the steps in the thio-ITCHY protocol, using an example from our laboratory, is shown in Fig. 1.

A number of variations and alterations to the basic ITCHY and thio-ITCHY methodologies have also been implemented. Hart and coworkers successfully incorporated the time-dependent truncation protocol into their method for finding folded fragments of recalcitrant proteins, ESPRIT (Expression of Soluble Proteins by Random Incremental Truncation) [6]. Rather than cloning the two parental genes into the same plasmid, one of us (MLG) explored the possibility of beginning with each gene residing on a different plasmid (albeit with identical backbones). The full-length thio-ITCHY template (with both parental genes on the same, linear molecule) could then be assembled in an overlap extension PCR [3]. However, this protocol is difficult to optimize, and subsequently we have returned to constructing two-gene templates such as the one shown in Fig. 1. Finally, the pINSALECT vector was designed to remove out-of-frame chimeras from ITCHY libraries [7]. Two-thirds of the variants in an ITCHY library will contain a frameshift at the crossover point. Removing frameshifted variants is particularly important if the ITCHY library is to be used as a starting point for further diversification by DNA shuffling (a method named SCRATCHY [8]) or in ORF-selector ESPRIT, where a single gene is truncated from both termini [9]. The pINSALECT vector contains the gene for the *Saccharomyces cerevisiae* VMA split intein, sandwiched between the Tat signal sequence and

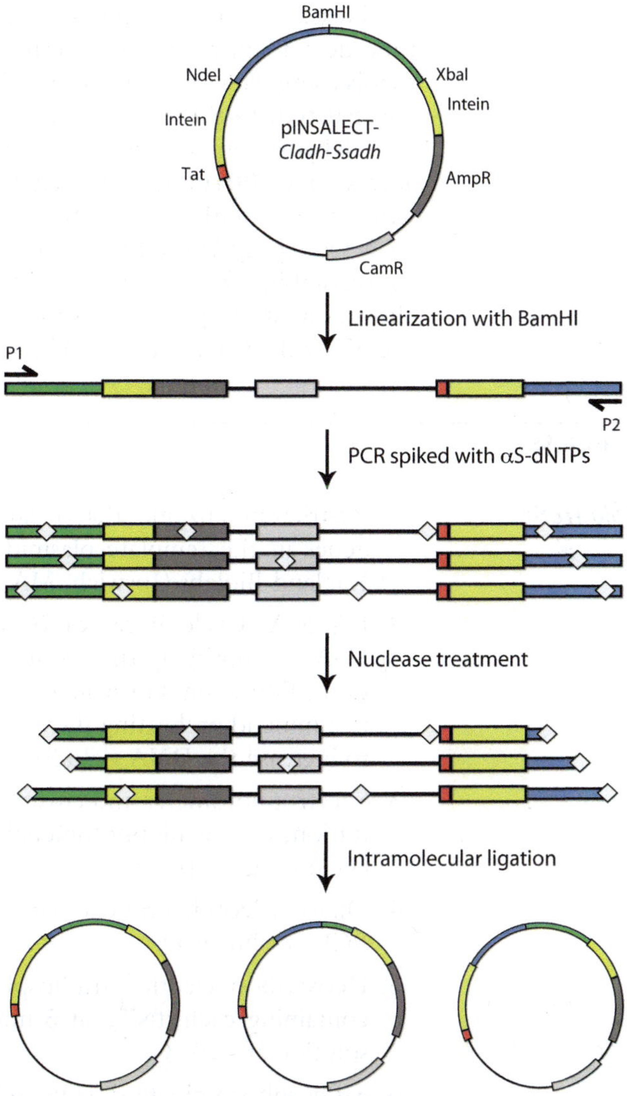


**Fig. 1** Overview of the thio-ITCHY protocol. In this example, the genes for the *C. ljungdahlii* alcohol dehydrogenase (*Cladh*, *blue*) and the *S. solfataricus* alcohol dehydrogenase (*Ssadh*, *green*) were cloned in tandem into the pINSALECT backbone, using the NdeI, BamHI, and XbaI sites indicated. The *adh* genes form part of a continuous open reading frame that begins with the Tat signal sequence (*red*), includes the VMA split intein (*yellow*), and ends with the gene for β-lactamase (AmpR). The AmpR marker is used for reading frame selection. The plasmid also contains a chloramphenicol resistance marker (CamR) for maintenance during library construction and sequence analysis. After the plasmid is linearized, it is amplified using the primers indicated (P1 and P2), with a mixture of dNTPs and αS-dNTPs. The (random) sites of αS-dNTP incorporation are shown as *white diamonds*

the TEM-1 β-lactamase gene (Fig. 1). The two genes to be recombined are cloned into a loop of the intein homing endonuclease domain. Only in-frame ITCHY variants will give rise to full-length polyproteins, in which the intein ligates Tat to β-lactamase and the host cell displays an ampicillin resistance phenotype.

Here, we present our protocols for thio-ITCHY and the time-dependent truncation method. We have illustrated the protocols using data from a thio-ITCHY library that we have recently constructed. In this example, we recombined genes for the alcohol dehydrogenases from *Clostridium ljungdahlii* (GenBank accession CP001666; locus tag CLJU_c24860) and *Sulfolobus solfataricus* (GenBank accession AJ010590). These genes share 47 % sequence identity; therefore, it would have been impossible to recombine them using DNA shuffling. As recommended above, our starting point was the two genes cloned in tandem, in the pINSALECT backbone (Fig. 1).

## 2 Materials

### 2.1 Thio-ITCHY

1. Restriction enzyme that cleaves between the two parental genes in the template plasmid, e.g., BamHI-HF from New England Biolabs (Ipswich, MA, USA) (Fig. 1).
2. E.Z.N.A. Cycle Pure Kit (Omega Bio-Tek, Norcross, GA, USA) for purifying the products of PCRs and digestion reactions. Equivalent kits from other suppliers are also suitable. We recommend preheating the elution buffer to 60 °C before use, to increase the DNA yield from each spin column.
3. Spectrophotometer and cuvettes for measuring DNA concentration, e.g., a Biophotometer and UVettes from Eppendorf (Hamburg, Germany).
4. Oligonucleotide primers that are suitable for the thio-ITCHY PCR (*see* **Note 1**).
5. Deoxyribonucleotide triphosphates (dNTPs): stock solution containing each dNTP at 5 mM. Store aliquots of this stock solution at −20 °C.
6. α-Phosphorothioate dNTPs (αS-dNTPs): stock solution containing each αS-dNTP at 1 mM. Store aliquots of this stock solution at −20 °C. High-quality αS-dNTPs can be purchased from Glen Research Corporation (Sterling, VA, USA).
7. Taq DNA polymerase and its reaction buffer, e.g., Kapa Taq with 10× Kapa Taq Buffer A (Kapa Biosystems, Woburn, MA, USA) (*see* **Note 2**).
8. Thermocycler with a heated lid (e.g., an MJ Mini from Bio-Rad (Hercules, CA, USA)).
9. An agarose gel (0.8 % or 1.0 % w/v agarose), stained with ethidium bromide at 0.5 μg/mL.
10. A suitable DNA ladder and apparatus for agarose gel electrophoresis.

11. *E. coli* exonuclease III (100 U/μL) and its accompanying reaction buffer, NEBuffer 1 (New England Biolabs, Ipswich, MA, USA).
12. Mung bean nuclease (10 U/μL) and 10× mung bean nuclease reaction buffer (New England Biolabs, Ipswich, MA, USA).
13. DNA polymerase I, Klenow fragment (5 U/μL) and its accompanying reaction buffer, NEBuffer 2 (New England Biolabs, Ipswich, MA, USA).
14. Ethylenediaminetetraacetic acid (EDTA) stock solution: 0.5 M, pH 8.0.

### 2.2 Size Selection and Library Construction

1. An agarose gel (0.8 % w/v agarose), stained with 1× SYBR Safe DNA gel stain (Invitrogen, Carlsbad, CA, USA).
2. A size standard, constructed by linearizing a plasmid that has the same backbone as the ITCHY template plasmid, but only one of the two parental gene inserts (*see* **Note 3**).
3. Gel loading dye, a suitable DNA ladder, and apparatus for agarose gel electrophoresis.
4. Safe Imager 2.0 Blue-Light Transilluminator (Invitrogen, Carlsbad, CA, USA).
5. Clean razor blade for excising bands from a gel.
6. GenElute Gel Extraction Kit (Sigma-Aldrich, St. Louis, MO, USA). Equivalent kits from other manufacturers are also suitable.
7. T4 DNA ligase (400 cohesive end units per microliter) and 10× T4 DNA ligase reaction buffer (New England Biolabs, Ipswich, MA, USA). Other T4 DNA ligases are also suitable (*see* **Note 4**).
8. E.Z.N.A. Cycle Pure Kit (Omega Bio-Tek, Norcross, GA, USA), or equivalent, with elution buffer pre-warmed to 60 °C.
9. Aliquots (50 μL) of electrocompetent *E. coli* cells (*see* **Note 5**).
10. Gene Pulser electroporation cuvettes (0.2 cm electrode gap, Bio-Rad, Hercules, CA, USA).
11. MicroPulser electroporator (Bio-Rad, Hercules, CA, USA).
12. Sterile SOC medium: 20 g/L tryptone; 5 g/L yeast extract; 10 mM NaCl; 2.5 mM KCl; 20 mM glucose.
13. LB-agar plates containing the correct antibiotic for selecting plasmid-containing cells.

### 2.3 Sequence Analysis and Scale-Up

1. Thermocycler with a heated lid.
2. Plasmid backbone-specific primers for amplifying the recombined ITCHY inserts.

3. Reagents for a standard PCR screen. While there are many alternate suppliers, we routinely use the 2× Kapa Taq ReadyMix with dye (Kapa Biosystems, Woburn, MA, USA). This contains reaction buffer, dNTPs, and Taq polymerase.
4. Ethidium bromide-stained agarose gels and electrophoresis equipment.
5. E.Z.N.A. Cycle Pure Kit (Omega Bio-Tek, Norcross, GA, USA), or equivalent, with elution buffer pre-warmed to 60 °C.
6. Standard LB-agar plates (circular, 85–90 mm diameter) containing the appropriate antibiotic for maintaining the ITCHY plasmid.
7. Two square bioassay dishes (245 mm × 245 mm) from Corning (Tewksbury, MA, USA) or Nunc (Thermo Fisher Scientific, Waltham, MA, USA). Each bioassay dish holds 200 mL of LB-agar, supplemented with the appropriate antibiotic for selecting plasmid-containing cells.
8. A fresh batch of electrocompetent *E. coli* cells (*see* **Note 5**).
9. Electroporation cuvettes, an electroporator, and SOC medium, as described in Subheading 2.2.
10. LB medium, containing the appropriate antibiotic (20 mL).
11. Sterile glycerol (50 % v/v).
12. Cryogenic vials, suitable for storage at −80 °C.

#### 2.4 ITCHY by Time-Dependent Truncation

1. Materials for ITCHY, as listed in Subheading 2.1: **items 1–3** and **8–14**.
2. NaCl stock solution: 120 mM, sterilized by autoclaving.
3. A laboratory timer (or a smart phone; *see* **Note 6**).

## 3 Methods

#### 3.1 Thio-ITCHY

1. Construct a plasmid containing the two genes that are to be recombined, separated by a sequence that contains a unique restriction site (as shown in Fig. 1).
2. Purify the plasmid DNA and linearize ~3 μg of it by digestion with the restriction enzyme that cleaves between the two parental genes. We typically use BamHI-HF for this purpose (Fig. 1), though other enzymes are also suitable (*see* **Note 7**). After digestion to completion (e.g., 4 h incubation at 37 °C, with 20 U enzyme), purify the DNA with the Cycle Pure Kit. Elute the DNA from the spin column in 50 μL of elution buffer and determine its concentration spectrophotometrically. Use water to dilute an aliquot to 10 ng/μL.
3. In thin-walled 0.2 mL tubes, prepare three PCRs with different ratios of dNTPs to αS-dNTPs, as shown in Table 1 (*see* **Note 8**).

**Table 1**
**Components of thio-ITCHY PCRs**

| Reagent | Control | 1:7 ratio | 1:9 ratio |
|---|---|---|---|
| 10× polymerase reaction buffer | 5.0 μL | 10.0 μL | 10.0 μL |
| dNTP mix (from 5 mM stock) | 2.0 μL | 3.5 μL | 3.6 μL |
| αS-dNTP mix (from 1 mM stock) | 0.0 μL | 2.5 μL | 2.0 μL |
| Forward primer (from 10 μM stock) | 2.0 μL | 4.0 μL | 4.0 μL |
| Reverse primer (from 10 μM stock) | 2.0 μL | 4.0 μL | 4.0 μL |
| Linearized plasmid DNA (from 10 ng/μL stock) | 1.0 μL | 2.0 μL | 2.0 μL |
| Taq DNA polymerase (5 U/μL) | 0.2 μL | 0.4 μL | 0.4 μL |
| Water | 37.8 μL | 73.6 μL | 74.0 μL |

**Table 2**
**Recommended PCR cycling conditions for amplifying long templates**

| | | |
|---|---|---|
| Step 1 | 2 min | 94 °C |
| Step 2 | 10 s | 94 °C |
| Step 3 | 20 s | Annealing temperature for primers (*see* **Note 1**) |
| Step 4 | 1 min/kb | 68 °C |
| Step 5 | Repeat steps 2–4 for an additional 24 cycles | |
| Step 6 | 5 min | 68 °C |
| Step 7 | Hold | 4 °C (for product storage, if necessary) |

4. Mix the reactions well, and then split each of the two 100 μL reactions into 2×50 μL aliquots. Place all five tubes in the thermocycler.
5. Run a PCR program that is suitable for amplifying long templates (*see* **Note 9**). In our experience, the program listed in Table 2 works well.
6. Pool the 2×50 μL aliquots of each αS-dNTP-containing reaction back together.
7. Run 3 μL of each PCR product on an ethidium bromide-stained agarose gel, alongside a DNA ladder. All three reactions ("control," "1:7 ratio," and "1:9 ratio") should yield amplified products, although the presence of αS-dNTPs may decrease yields marginally, particularly when present at their highest concentration (i.e., the 1:7 ratio) (*see* **Note 10**).

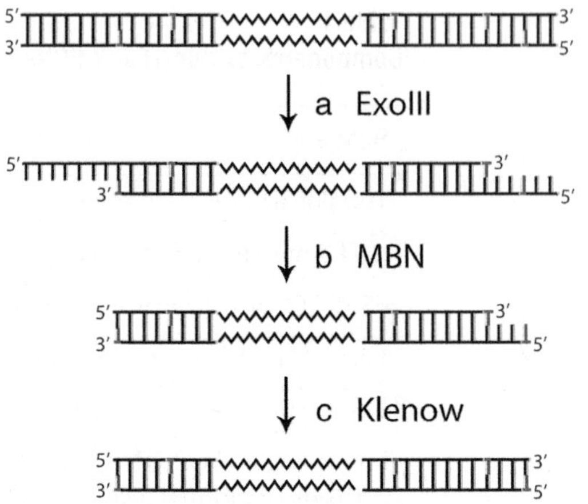


**Fig. 2** Critical steps in the thio-ITCHY protocol. The thio-ITCHY PCR product has randomly incorporated phosphorothioate-containing linkages (shown in *red*), due to the presence of αS-dNTPs in the PCR. (**a**) Treatment of the PCR product with exonuclease III removes nucleotides from the 3′ termini. DNA strand hydrolysis is halted when exonuclease III encounters a phosphorothioate-containing linkage. (**b**) Mung bean nuclease (MBN) removes nucleotides from the single-stranded 5′ overhangs. Like exonuclease III, it halts at phosphorothioate-containing linkages (e.g., right-hand end). (**c**) Treatment with Klenow DNA polymerase fills in the remaining gaps, making blunt ends that are suitable for intramolecular ligation

8. Pool the remainder of the "1:7 ratio" and "1:9 ratio" reaction products and purify them using the Cycle Pure Kit. Elute the DNA from the spin column in 50 μL elution buffer and determine its concentration spectrophotometrically. A typical yield is 4–5 μg of DNA (i.e., 50 μL at 80–100 ng/μL).
9. Set up an exonuclease III digestion reaction containing all of the DNA from **step 8**, 1× NEBuffer 1, and 100 U of exonuclease III per microgram of DNA. A typical total reaction volume is 100 μL. Add the exonuclease III last, mix the reaction components well, and incubate at 37 °C for 30 min. This step removes nucleotides from the 3′ ends of the αS-dNTP-containing PCR products (Fig. 2a).
10. Stop the exonuclease III digestion reaction by adding 5 volumes of buffer CP from the Cycle Pure Kit. Purify the DNA using this kit, and elute from the spin column in 50 μL elution buffer. Determine the DNA concentration using a spectrophotometer. A typical yield at this stage is ~2 μg of purified DNA.
11. Set up a mung bean nuclease digestion reaction containing all of the DNA from step 10, 1× mung bean nuclease reaction buffer, and 1.5 U of mung bean nuclease per microgram of DNA (*see* **Note 11**). A typical total reaction volume is 60 μL. Add the mung bean nuclease last, mix the reaction components well,

and incubate at 30 °C for 30 min. This step removes the single-stranded overhangs at the 5′ ends of the αS-dNTP-containing PCR products, although it is important to note that mung bean nuclease is unable to cleave phosphorothioate-containing linkages on the 5′ overhangs (Fig. 2b).

12. Stop the mung bean nuclease digestion reaction by adding 5 volumes of buffer CP from the Cycle Pure Kit. Use the kit to purify the digested DNA and elute from the spin column in 50 μL elution buffer. Determine the DNA concentration using a spectrophotometer. A typical yield is ~1.5 μg of purified DNA.
13. Set up an end-polishing reaction containing all of the DNA from **step 12**, 1× NEBuffer 2, dNTPs at a final concentration of 100 μM (i.e., a 50-fold dilution of the stock used in **step 3**), and 1 U of Klenow DNA polymerase per microgram of DNA. A typical total reaction volume is 60 μL. Add the Klenow DNA polymerase last, mix the reaction components well, and incubate at 25 °C for 15 min. This step removes any 3′ overhangs, and fills in any 5′ overhangs, that may remain after treatment with mung bean nuclease (Fig. 2c).
14. Stop the end-polishing reaction by adding EDTA to a final concentration of 10 mM and heating at 75 °C for 20 min.
15. Use the Cycle Pure Kit to purify the DNA and elute from the spin column in 50 μL elution buffer. Determine the DNA concentration spectrophotometrically. A typical yield is ~1.1 μg of purified DNA.

### 3.2 Size Selection and Library Construction

1. Mix the purified DNA from the thio-ITCHY protocol (above) with gel loading dye (1× final concentration). Separately, mix ~100 ng of the size standard (*see* **Note 3**) with an aliquot of the same gel loading dye. Run the two samples, together with a DNA ladder, on a 0.8 % w/v agarose gel. A reduced voltage is highly recommended, to improve the resolution of the randomly sized thio-ITCHY DNA fragments. We typically electrophorese at ~13 V/cm (i.e., 80 V for a 6 cm wide mini-gel) for at least 60 min.
2. By comparison with the DNA ladder and the size standard, excise the part of the smear that corresponds to the size standard, ±500 bp (or whatever size range is desired in your library). We strongly recommend using a blue-light transilluminator and SYBR Safe DNA stain, rather than a UV transilluminator and ethidium bromide, for this step (*see* **Note 12**).
3. Purify the DNA from the excised gel band. We use the GenElute Gel Extraction Kit and elute the DNA from the spin column in 50 μL Elution Solution that has been pre-warmed to 60 °C. Use an aliquot (1–2 μL) to determine the DNA concentration spectrophotometrically. A typical yield at this stage is ~600 ng of purified DNA.

4. Set up an intramolecular ligation reaction, using all of the purified DNA from **step 3**. It is critical to use a DNA concentration < 3 ng/μL in the reaction, for high-efficiency intramolecular ligation [10]. We would typically set up a 300 μL ligation reaction that contains ~600 ng DNA, 1×T4 DNA ligase reaction buffer, and T4 DNA ligase (400 cohesive end units; *see* **Note 4**). Incubate the ligation reaction at 16 °C for 12–16 h, and then inactivate the ligase by heating to 65 °C for 10 min.
5. Use the Cycle Pure Kit to purify the DNA from the ligation reaction, and elute from the spin column using 30 μL prewarmed elution buffer.
6. Use a 2 μL aliquot of the ligated DNA to transform a 50 μL aliquot of *E. coli*, by electroporation (*see* **Note 13**).
7. Immediately after electroporation, add 500 μL of SOC medium to the cuvette, mix by pipetting, and transfer the cells to a sterilized, capped test tube or a 15 mL tube (Falcon or equivalent). Allow the cells to recover by incubating them at 37 °C, with shaking, for 1 h.
8. Store the remainder of the purified ligation reaction (from **step 6**, above) at −20 °C.
9. Spread aliquots (5 μL and 50 μL) of the recovery culture on LB-agar plates. Incubate the plates at 37 °C for 12–16 h.
10. Count the number of colonies on each plate. The colony counts can be used to estimate the size of the final, scaled-up library (*see* Subheading 3.3, **step 10**). The final library is likely to be ~1,500 times larger than the number of colonies on the plate on which 5 μL of the recovery culture was spread. Our scaled-up libraries typically contain $10^5$–$10^6$ clones (*see* **Notes 14** and **15**).

### 3.3 Sequence Analysis and Scale-Up (See Note 16)

1. Use sterile toothpicks (or 2 μL pipette tips) to pick ~30 colonies at random from the plates described above (Subheading 3.2, **step 10**). Transfer each colony into a thin-walled, 0.2 mL tube containing 5 μL of sterile water.
2. Lyse the cells by incubating the tubes in a thermocycler, at 95 °C for 5 min.
3. Amplify the ITCHY inserts by PCR. While many variations are possible, we have listed our standard protocol below. We typically set up 25 μL PCRs in thin-walled 0.2 mL tubes, as per Table 3.
4. Mix each sample and place the tubes in the thermocycler. Run an appropriate PCR program, with an extension time that is sufficient to amplify inserts with a maximum size that is equal to the combined lengths of the two parental genes.
5. Run 3 μL aliquots of each PCR product on agarose gels, to confirm successful amplification. We typically observe amplified inserts from 60 to 90 % of the colonies screened.

**Table 3**
**Components for a standard colony PCR**

| Reagent | Volume |
|---|---|
| Water | 9.5 μL |
| 2×Kapa Taq ReadyMix with dye | 12.5 μL |
| Forward primer (from 10 μM stock) | 1.0 μL |
| Reverse primer (from 10 μM stock) | 1.0 μL |
| Cell lysate (from **step 2**, above) | 1.0 μL |

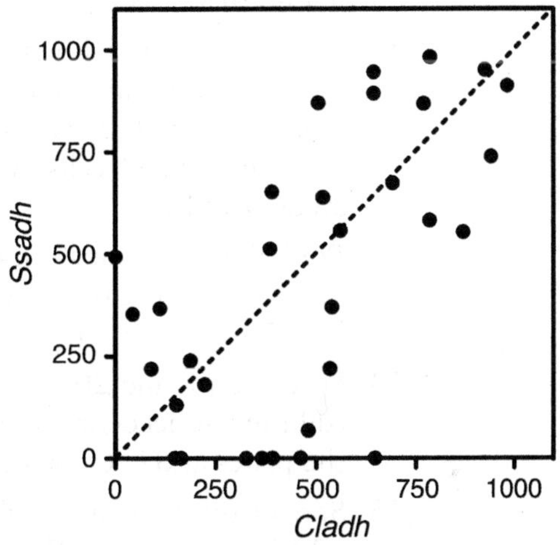


**Fig. 3** Crossover distribution of a library made by recombining *Cladh* and *Ssadh*. Sequences from the library are plotted such that the coordinates (*x*, *y*) represent ("last base from *Cladh*," "first base from *Ssadh*"). The *Cladh* gene is 1,056 bp and the *Ssadh* is 1,050 bp. Therefore, points on the diagonal (*dashed line*) correspond to hybrid genes that are the same length as one parent. Points above the *dashed line* indicate hybrid genes that are shorter than either parent, with points below the *dashed line* indicating longer hybrid genes

6. Purify the remainder of each PCR product using the Cycle Pure Kit. Elute the DNA from each spin column in 30 μL of elution buffer.
7. Sequence each PCR product, using the forward and/or reverse primers from the PCR as the sequencing primer(s).
8. Align the sequence of each PCR product with the sequences of the two parental genes, and determine the location of the crossover in each ITCHY variant. Software applications such as MacVector are useful for this analysis.
9. Assess the ITCHY library for potential bias, by plotting the crossover data as shown in Fig. 3. Sequences from an unbiased

library, with effective size selection (Subheading 3.2, **step 2**), should give points that are scattered along the diagonal, as shown in Fig. 3.

10. If the results from Subheading 3.2 (**step 10**) and Subheading 3.3 (**step 9**) are satisfactory, then proceed with constructing the full-sized library. Begin by preparing the LB-agar plates on which the transformed cells of the library will be spread. For each ITCHY library, we typically use two 245 mm × 245 mm square bioassay dishes (*see* **Note 17**) and two regular Petri dishes.
11. Prepare a fresh batch of electrocompetent *E. coli* cells (*see* **Note 5**). We observe transformation efficiencies for freshly prepared cells that are two- to fourfold higher than cells that have undergone a freeze/thaw cycle.
12. Thaw the stored library ligation mixture (Subheading 3.2, **step 8**). Add 2 μL aliquots of it to 14 × 50 μL aliquots of electrocompetent *E. coli* cells. Transform each aliquot and recover the transformed cells, as described above (Subheading 3.2, **steps 6** and **7**).
13. Pool all of the transformed cells in a sterile 15 mL tube, and mix by inverting the tube 2–3 times. The total volume should be 7.7 mL (14 recovery cultures; 550 μL per culture).
14. Spread 2 μL and 20 μL aliquots on the regular LB-agar plates (diluting as necessary to obtain a spreadable volume). Incubate the plates at 37 °C for 12–16 h.
15. Spread the remainder of the library on the two large plates, ~3.85 mL per plate. We recommend incubating these plates at 30 °C, to avoid the possibility of a confluent lawn.
16. Count the number of colonies on the plates from **step 14** above, and extrapolate to estimate the number of colonies on the large plates.
17. The large library plates should be covered in thousands (or millions) of small colonies. To recover them, pipette 4 mL of LB medium (with the appropriate antibiotic) onto the center of the plate. Use a sterilized glass spreader to scrape the cells off the surface and pool them in one corner.
18. Remove the resuspended cells with a P1000 pipettor and transfer them to a sterile 50 mL tube (Falcon tube or similar).
19. Pipette a second 4 mL aliquot of LB onto the same plate and repeat the scraping step.
20. Repeat **steps 17–19** with the second library plate. Pool all of the recovered cells from both plates in a single 50 mL tube.
21. Mix the pooled cells well, by pipetting up and down repeatedly with a P1000 pipettor.
22. Centrifuge the cells at 3,000 × *g* and 4 °C for 15 min. Use a P1000 pipettor to remove the supernatant.

23. Resuspend the cell pellet in 1 mL of LB medium, with antibiotic. Split the library into 100 μL aliquots and transfer each aliquot to a cryogenic vial. Add 50 μL of sterile glycerol (50 % v/v) to each aliquot and mix well by pipetting.
24. Store the aliquots at −80 °C until you are ready to proceed with downstream screening or selection.

#### *3.4 ITCHY by Time-Dependent Truncation (See Note 18)*

1. If time-dependent truncation is preferred to thio-ITCHY, then we strongly recommend that users determine the exonuclease III truncation rate under the conditions in their own laboratory. To begin, purify the plasmid containing the two genes that are to be recombined.
2. Linearize the plasmid (~5 μg) by digestion with a restriction enzyme that cleaves between the two parental genes, such as BamHI-HF in Fig. 1 (*see* **Note 19**). After digestion to completion, purify the DNA with the Cycle Pure Kit. Elute from the spin column in 50 μL of pre-warmed elution buffer and determine the DNA concentration using a spectrophotometer. A typical yield of DNA is 4 μg (i.e., 50 μL at 80 ng/μL).
3. A standard truncation reaction should contain DNA at 33 ng/μL and 100 U of exonuclease III per microgram of DNA. The rate of truncation can be controlled by varying the temperature and the concentration of NaCl (*see* **Note 20**). We recommend determining the truncation rate empirically, using a 22 °C incubation and 30 mM NaCl. Mix reagents in a thin-walled 0.2 mL tube, as shown in Table 4.
4. Set a thermocycler (or heating block) to incubate at a constant temperature of 22 °C. Add the tube to the thermocycler and allow the reaction to equilibrate for 2–5 min.
5. Dispense 75 μL aliquots of buffer CP from the Cycle Pure Kit into 8 thin-walled 0.2 mL tubes. Label the tubes "A" to "H" and place them on ice.
6. Remove a 15 μL sample ($t = 0$) from the tube that is incubating at 22 °C. Add the sample to tube A. Mix thoroughly, but quickly, by pipetting up and down.

**Table 4**
**Reaction components for exonuclease III truncation of DNA**

| Reagent | Volume |
|---|---|
| Linearized plasmid (80 ng/μL) | 50 μL |
| 10× NEBuffer 1 | 12 μL |
| NaCl (from 120 mM stock) | 30 μL |
| Water | 28 μL |

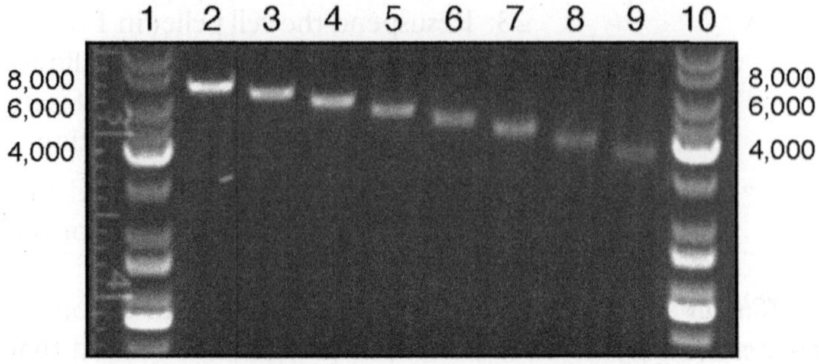


**Fig. 4** Empirical determination of exonuclease III truncation rate. A linearized plasmid (pINSALECT-*Cladh-Ssadh*, 6,683 bp) was truncated as described in the text. Samples were taken at 5 min intervals, quenched, purified, and run on an ethidium bromide-stained gel (0.8 % w/v agarose). Lanes 1 and 10 contain a DNA ladder, with the sizes of key bands indicated in base pairs. Lanes 2–9 contain the samples, from $t=0$ (lane 2) to $t=35$ min (lane 9). The average fragment size decreased from 6,683 bp at $t=0$ to ~3,900 bp at $t=35$ min. Therefore, the truncation rate was estimated to be ~80 bases/min

7. There is now 3.5 μg of DNA remaining in the tube at 22 °C. Add 3.5 μL of exonuclease III (i.e., 350 U) to this tube. Mix quickly and thoroughly by pipetting up and down. Start a timer (*see* **Note 6**). Leave the cap of the tube open, to minimize temperature fluctuations due to handling.
8. After incubation for 5 min at 22 °C, remove a 15 μL sample of the truncation reaction and add it to tube B (on ice). Mix with the CP buffer in tube B, by pipetting.
9. Repeat **step 8** at 5 min intervals, until the final sample is taken at $t=35$ min (and added to tube H, on ice).
10. Purify the DNA in tubes A–H using the Cycle Pure Kit. Elute from each of the eight spin columns in 30 μL of pre-warmed elution buffer.
11. Run 10 μL aliquots of each sample on a 0.8 % w/v agarose gel, alongside a suitable ladder. An example of the expected result is shown in Fig. 4. Use the gel to estimate the truncation rate in bases per minute. For example, from Fig. 4, we estimate the truncation rate under our conditions to be ~80 bases/min. In contrast, the previously reported formula for truncation rate (*see* **Note 20**) would have predicted a truncation rate of ~30 bases/min.
12. With an accurate estimate of the truncation rate in hand, it is now possible to design an optimized strategy for library construction. As with thio-ITCHY, we typically aim to construct hybrids that are the length of one parental gene, ±500 bp (Subheading 3.2, **step 2**). Therefore, in our example (recombining *Cladh* and *Ssadh*, which are 1,050 bp and 1,056 bp,

respectively), we would aim to truncate a total of ~550–1,550 bases from the two genes (which have a combined length of 2,106 bp). A worked example, designed to accomplish this goal, follows (*see* **Note 21**).

13. Prepare more linearized plasmid DNA, as described in **steps 1** and **2** above. Set up a truncation reaction, as described in **steps 3** and **4** above.
14. Dispense 75 μL aliquots of buffer CP from the Cycle Pure Kit into two thin-walled 0.2 mL tubes, labeled “A” and “C.” Dispense a 450 μL aliquot of buffer CP into a 1.5 mL tube, labeled “B.” Place the three tubes on ice.
15. Remove a 15 μL sample ($t = 0$) from the tube that is incubating at 22 °C. Add the sample to tube A, and mix thoroughly by pipetting up and down.
16. Add 3.5 μL of exonuclease III (i.e., 350 U) to the tube that is incubating at 22 °C. Mix quickly and thoroughly by pipetting up and down. Start a timer (*see* **Note 6**). Leave the cap of the tube open, to minimize temperature fluctuations due to handling.
17. After incubation for 30 s at 22 °C, remove a 2.25 μL sample of the truncation reaction and add it to tube B (on ice). Mix with the CP buffer in tube B, by pipetting.
18. Repeat **step 17** at 30 s intervals, adding each successive sample to tube B. Take the final sample for tube B at $t = 20$ min.
19. Immediately after adding the final 2.25 μL aliquot to tube B, take one final sample of 15 μL and add it to tube C.
20. Purify the DNA in tubes A–C using the Cycle Pure Kit. Elute from spin columns A and C in 30 μL of elution buffer. Elute from spin column B in 50 μL of elution buffer.
21. Check that the truncation has worked as desired, by running 10 μL aliquots of samples A and C (i.e., $t = 0$ and $t =$ final) on an ethidium bromide-stained agarose gel.
22. Use a spectrophotometer to determine the DNA concentration in sample B. Proceed with mung bean nuclease and Klenow DNA polymerase treatment, as described in Subheading 3.1, **steps 11–15** (*see* **Note 22**).
23. The gel in **step 21** (above) will confirm whether or not a size selection step is necessary, prior to setting up the intramolecular ligation reaction. In most cases, careful control of the time-dependent truncation will obviate the need for size selection. Therefore, library construction can proceed as described in Subheading 3.2, **steps 4–10** (*see* **Note 23**).
24. Library sequence analysis, scale-up, and storage should proceed exactly as described in Subheading 3.3, above.

## 4 Notes

1. Primers should be designed to effect amplification of the entire plasmid, after it has been linearized. The annealing locations of the forward and reverse primers are indicated in Fig. 1 (P1 and P2, respectively). The two primers should be noncomplementary and should have melting temperatures that are within 5 °C of each other. We estimate melting temperatures with the OligoAnalyzer tool from Integrated DNA Technologies (http://www.idtdna.com/analyzer/applications/oligoanalyzer). The optimal annealing temperature to use in the PCR is typically 3–5 °C cooler than the lower of the two primer melting temperatures.
2. As noted previously [10], it is important to use a polymerase that lacks 3′–5′ exonuclease activity. When strand synthesis is completed, proofreading polymerases (Phusion, Vent, etc.) "idle": they continually remove and resynthesize 3′ ends. However, these polymerases are unable to hydrolyze phosphorothioate-containing linkages. Therefore, in the thio-ITCHY PCR, idling would lead to the accumulation of phosphorothioates at the 3′ ends of the amplified products and result in a library that is biased toward genes with small (or nonexistent) truncations.
3. In most scenarios, the user is likely to be recombining two genes and looking for recombinants that are approximately the same length as one parent or the other. In order to enrich the library for these variants, it is important to carry out a size selection step (Subheading 3.2). For accurate size selection, we construct a standard of the desired size (i.e., a linearized plasmid that contains one of the two parental genes).
4. We have had success with the T4 DNA ligases supplied by New England Biolabs (Ipswich, MA, USA), Fermentas (Vilnius, Lithuania), and Enzymatics Inc. (Beverly, MA, USA). We strongly recommend using a ligase reaction buffer that does not contain crowding agents such as polyethylene glycol (PEG), because these agents favor intermolecular ligation rather than the desired intramolecular ligation events. Note that New England Biolabs uses cohesive end units to quantify ligase activity, while other suppliers use Weiss units. We have recommended using 400 cohesive end units of ligase in the ITCHY ligation. This is equivalent to ~6 Weiss units.
5. Within the constraints of downstream screening or selection steps, strains with high transformation efficiencies (such as *E. coli* MC1061) are preferred. We prepare electrocompetent cells according to standard protocols [11].

6. A regular lab timer or stopwatch can be used to determine when samples should be withdrawn from the exonuclease III digestion reaction. We have also found the VersaTimer iPhone app (Vivendo Software; http://www.vivendosoftware.com), to be useful, because it can be programmed to sound an audible alarm at any given interval.
7. It is important that the plasmid DNA is digested to completion; otherwise, the uncut vector can be carried forward through all successive steps and will contaminate the final library with non-truncated variants. We find the high-fidelity restriction enzymes from New England Biolabs (Ipswich, MA, USA) to be preferable for ensuring complete digestion. If digestion is incomplete, it will be necessary to run the digested DNA on an agarose gel and excise the band that corresponds to the linearized plasmid.
8. Varying the ratio of αS-dNTPs to dNTPs affords some control over the lengths of the truncations in the library. Increasing the concentration of αS-dNTPs over dNTPs leads to shorter truncations. However, we have found that PCR product yields begin to decrease when the ratio of αS-dNTPs to dNTPs is increased above 1:7. Empirically, the mixtures that we recommend (1:7 and 1:9 ratios) give an even distribution of truncation lengths, over a range that is suitable for many applications (truncations from 0 to ~1,500 base pairs).
9. Thio-ITCHY amplicons are typically > 4 kb. Therefore, optimization (in the absence of αS-dNTPs) may be required. In particular, long incubations at elevated temperatures can result in depurination of the template strands [12]. We recommend reducing the extension temperature from 72 °C to 68 °C and minimizing the denaturation time in each cycle.
10. If the dNTP-only control reaction does not yield an amplified product, then the primers and/or cycling conditions will need to be optimized further. Previous protocols have recommended treating this control in parallel with the αS-dNTP-containing products, for the remainder of the thio-ITCHY protocol [10]. This is good practice for first-time users and controls for the activity of exonuclease III. However, experienced users may wish to forego these steps in the interests of expediency and discard the remainder of the "control" reaction after it has been used to verify the success of the PCR amplification.
11. Note that mung bean nuclease is supplied at a concentration of 10 U/μL. Therefore, it may be necessary to dilute it, prior to adding it to the digestion reaction. We have found 1× mung bean nuclease reaction buffer to be a suitable diluent.

12. UV transillumination of ethidium bromide-stained DNA can induce damage, even when exposure is kept to less than 60 s. This damage can decrease the efficiencies of downstream ligation and transformation steps [13]. Constructing a large ITCHY library requires the highest possible quality of DNA. In our hands, the use of a blue-light transilluminator results in libraries that are five- to tenfold larger than those prepared with DNA that is excised from UV-illuminated gels.

13. Thaw the aliquot of electrocompetent cells on ice. Add the aliquot of DNA to the thawed cells, and transfer the mixture to a pre-chilled, sterile Gene Pulser cuvette. We electroporate using a MicroPulser unit, set to 2.5 kV.

14. In an ITCHY library, the total number of possible variants is found by multiplying the lengths of the two parental genes. For example, we recombined the *C. ljungdahlii* alcohol dehydrogenase gene (1,056 bp) and the *S. solfataricus* alcohol dehydrogenase gene (1,050 bp). There are $1{,}056 \times 1{,}050 = 1.1 \times 10^6$ possible variants. Assuming that each variant is equally likely, our library analysis program GLUE [14] can be used to estimate the number of clones that are required to achieve a desired level of sampling. For example, a library containing $\sim 3.3 \times 10^6$ clones is required to sample 95 % of all the possible variants in our alcohol dehydrogenase library.

15. If the test transformation yields an unsuitably low number of colonies, then we recommend scaling up the thio-ITCHY PCRs (and subsequent steps). For example, it is straightforward to set up 200 μL reactions with each ratio of αS-dNTPs to dNTPs (1:7 and 1:9) and split each into 4× 50 μL aliquots for thermocycling (Subheading 3.1, **steps 3** and **4**). We have also had success using drop dialysis, rather than silica-based spin columns, for desalting the ligation reaction prior to electroporation [15].

16. We recommend sequencing randomly chosen ITCHY variants from the initial transformation (Subheading 3.2) to assess the distribution of crossovers in the library. This avoids wasting time and resources on a scaled-up library that contains unwanted biases (e.g., too many clones that are under- or over-truncated). We have summarized our protocol for library analysis in Subheading 3.3, though we note that it has also been published in an earlier volume [16].

17. The large volume (200 mL) and high surface area of each bioassay dish make it liable to "sweat" when incubated at 30–37 °C. This can be minimized by cooling the LB-agar to ~55 °C, before pouring the plates. However, the plates may also need to be pre-warmed (37 °C, 4–6 h) and/or dried in a laminar flow hood or class II biosafety cabinet (10–15 min), before they are dry enough to use.

18. In most cases, we recommend the thio-ITCHY protocol, both for its ease of use and for the unbiased libraries that typically result. However, the original incarnation of ITCHY [2], which uses time-dependent exonuclease III digestions, remains useful in cases where the user desires more control over the truncation process. Our protocol modifies and updates the one that has been described previously [10].
19. Exonuclease III will act on blunt or 3′-recessed ends, but not on 3′ overhangs. Ensure that the selected restriction enzyme does not cleave to leave 3′ overhangs. Common enzymes that are unsuitable include KpnI, PstI, and SphI.
20. It is well known that the activity of exonuclease III is temperature and NaCl dependent [17]. Previously, Ostermeier and Lutz reported a formula for estimating the truncation rate at 22 °C, with 33 ng/μL DNA and 100 U of exonuclease III per microgram of DNA [10]:

$$\text{Rate}\left(\text{bases}/\text{min}\right) = 47.9 \times 10^{(-0.00644N)}$$

where $N$ is the concentration of NaCl in millimoles/L (for concentrations in the range 0–150 mM). However, we often observe truncation rates to be higher than this formula suggests, so we recommend estimating truncation rates empirically, to avoid biasing the library toward over-truncated variants.

21. In designing a sampling strategy, consider the maximum truncation length (~1,550 bases in our worked example), the observed truncation rate (~80 bases/min), and the total volume of the exonuclease III digestion reaction (120 μL). We recommend taking 15 μL $t=0$ and $t=$ final samples for gel analysis. Therefore, in our example, we aimed to sample a total volume of 90 μL, over the course of 20 min (20 min × 80 bases/min = maximum truncation of ~1,600 bases). Sampling every 30 s suggests that each sample should be 2.25 μL. Note that exonuclease III digestion produces a Poisson distribution of truncation lengths [17]. Both our empirical data and a theoretical treatment of the method [18] confirm that uniform distributions of truncation lengths can result, even when the sampling interval (e.g., 30 s) is such that there is an average truncation of more than one base between samples.
22. In theory, the Klenow DNA polymerase step is not necessary in the time-dependent truncation protocol because there are no phosphorothioate-containing linkages to stop mung bean nuclease from digesting the 5′ overhangs (viz., Fig. 2b, c). In practice, we and others [6, 10] have found an end-polishing step to be critical for constructing large libraries.

23. Previous protocols for ITCHY by time-dependent truncation [10] have recommended digestion of the truncated products with a restriction enzyme (e.g., XbaI in Fig. 1), prior to an intermolecular ligation step. The idea was to uncouple the fragments that result from synchronized truncations. However, in our hands, this extra restriction digest leads to ten- to 100-fold decreases in library size. Moreover, we have observed variants with little evidence of synchronization in the truncation lengths for each parental gene. On balance, we recommend omitting the restriction digest.

## Acknowledgments

The authors gratefully acknowledge financial support for this work from the University of Otago Priming Partnerships fund. W.M.P. is the recipient of a Rutherford Discovery Fellowship.

### References

1. Lutz S, Patrick WM (2004) Novel methods for directed evolution of enzymes: quality, not quantity. Curr Opin Biotechnol 15:291–297
2. Ostermeier M, Shim JH, Benkovic SJ (1999) A combinatorial approach to hybrid enzymes independent of DNA homology. Nat Biotechnol 17:1205–1209
3. Gerth ML, Lutz S (2007) Non-homologous recombination of deoxyribonucleoside kinases from human and *Drosophila melanogaster* yields human-like enzymes with novel activities. J Mol Biol 370:742–751
4. Eisenbeis S, Höcker B (2010) Evolutionary mechanism as a template for protein engineering. J Pept Sci 16:538–544
5. Lutz S, Ostermeier M, Benkovic SJ (2001) Rapid generation of incremental truncation libraries for protein engineering using α-phosphothioate nucleotides. Nucleic Acids Res 29:e16
6. Tarendeau F, Boudet J, Guilligay D, Mas PJ, Bougault CM, Boulo S, Baudin F, Ruigrok RW, Daigle N, Ellenberg J, Cusack S, Simorre JP, Hart DJ (2007) Structure and nuclear import function of the C-terminal domain of influenza virus polymerase PB2 subunit. Nat Struct Mol Biol 14:229–233
7. Gerth ML, Patrick WM, Lutz S (2004) A second-generation system for unbiased reading frame selection. Protein Eng Des Sel 17:595–602
8. Lutz S, Ostermeier M (2003) Preparation of SCRATCHY hybrid protein libraries: size- and in-frame selection of nucleic acid sequences. Methods Mol Biol 231:143–151
9. An Y, Yumerefendi H, Mas PJ, Chesneau A, Hart DJ (2011) ORF-selector ESPRIT: a second generation library screen for soluble protein expression employing precise open reading frame selection. J Struct Biol 175:189–197
10. Ostermeier M, Lutz S (2003) The creation of ITCHY hybrid protein libraries. Methods Mol Biol 231:129–141
11. Hanahan D, Jessee J, Bloom FR (1991) Plasmid transformation of *Escherichia coli* and other bacteria. Methods Enzymol 204:63–113
12. Cheng S, Fockler C, Barnes WM, Higuchi R (1994) Effective amplification of long targets from cloned inserts and human genomic DNA. Proc Natl Acad Sci U S A 91:5695–5699
13. Hartman PS (1991) Transillumination can profoundly reduce transformation frequencies. Biotechniques 11:747–748
14. Firth AE, Patrick WM (2005) Statistics of protein library construction. Bioinformatics 21:3314–3315
15. Saraswat M, Grand RS, Patrick WM (2013) Desalting DNA by drop dialysis increases library size upon transformation. Biosci Biotechnol Biochem 77:402–404
16. Hanson-Manful P, Patrick WM (2013) Construction and analysis of randomized protein-encoding libraries using error-prone PCR. Methods Mol Biol 996:251–267
17. Hoheisel JD (1993) On the activities of *Escherichia coli* exonuclease III. Anal Biochem 209:238–246
18. Ostermeier M (2003) Theoretical distribution of truncation lengths in incremental truncation libraries. Biotechnol Bioeng 82:564–577

# Chapter 17

# Generating Random Circular Permutation Libraries

Stefan Lutz, Ashley B. Daugherty, Ying Yu, and Zhen Qian

## Abstract

Protein engineering by random circular permutation is an effective tool for tailoring protein topology with potential functional benefits including improved catalytic activity. This method involves covalently connecting the native protein termini with a peptide linker and cleaving a peptide bond elsewhere in the polypeptide sequence. Termini relocation can impact protein ternary and quaternary structure and translate into functional enhancements due to changes in protein conformation and flexibility. As the effects of new termini in specific protein locations are difficult to predict, the preparation of a library constituting all possible permutation sites is an effective search strategy for identifying variants with novel properties.

**Key words** Protein engineering, Directed evolution, Circular permutation, Flexibility, Protein ternary/quaternary structure, Biocatalysis

## 1 Introduction

Protein engineering by rational and combinatorial methods has proven to be a powerful tool for studying protein structure and function, as well as a key technology for creating new and improved biocatalysts, sensors, and materials [1–4]. In contrast to the many traditional mutagenesis and in vitro recombination methods which modify protein sequences through the substitution of one or more amino acids [5, 6], a less-known strategy in protein engineering is circular permutation (CP). Conceptually, CP of a protein involves covalently linking the native N- and C-termini with a short peptide sequence and reintroducing new termini elsewhere in the protein sequence through the cleavage of a peptide bond (*see* Fig. 1). As such, CP does not alter the amino acid composition of a protein but simply reorganizes the order of amino acids in the polypeptide sequence. As a laboratory technique, CP was initially embraced by the protein folding community to probe protein structures for folding units and domains [7–12]. More recently, the method has also been explored by protein engineers to alter protein modularity [13, 14], as well as catalytic properties, substrate profiles, and stability [15–22].

Elizabeth M.J. Gillam et al. (eds.), *Directed Evolution Library Creation: Methods and Protocols*, Methods in Molecular Biology, vol. 1179, DOI 10.1007/978-1-4939-1053-3_17, © Springer Science+Business Media New York 2014

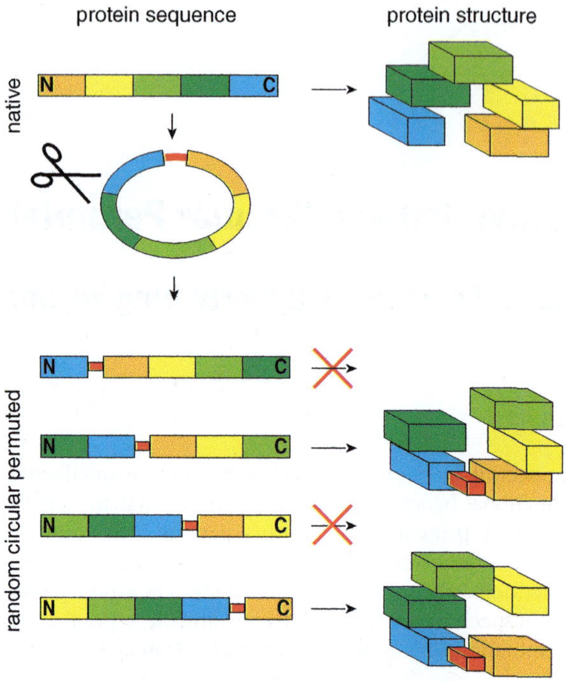


**Fig. 1** Protein sequence reorganization by random circular permutation. The process initially covalently links the native N- and C-termini with the help of a peptide sequence (shown in *red*). Subsequent random cleavage of one of the peptide bonds elsewhere in the primary sequence results in the structured reorganization of the protein. While some permutated sequences will no longer be able to assume a defined tertiary structure, foldable candidates typically adopt native-like conformations with local structural perturbations near the old and new termini. Adopted with permission from ref. [22]; copyright 2006, Nature Publishing Group

These studies have demonstrated that CP can result in substantive improvements in catalytic performance which have been linked to perturbation of rate-limiting conformational changes. In addition, termini relocation can introduce significant changes in ternary and quaternary structure [20].

The two key factors for experimental success are the peptide linker connecting the native termini and the location of the new termini [22]. Firstly, the length and composition of the peptide linker are critical design features. If the linker is too short or rigid, it may perturb protein topology and interfere with proper function. However, if the linker is too long and flexible, it creates a floppy surface loop which can result in poor folding and an overall reduction of protein stability. The simple rules for linker design outlined in Subheading 3.1 should be helpful in guiding experimentalists. Secondly, the structural perturbation resulting from the introduction of new termini through cleavage of the protein backbone is largely a local effect. New termini positions that are far from the active site will likely have minimal effects on protein function, assuming that the protein variant still folds into a native-like conformation. In contrast, cleavage of peptide bonds in the active site will probably be detrimental to function as the increased conformational flexibility will interfere with the correct topology of catalytic residues. Early engineering studies typically focused on a select few CP variants; however, a more effective strategy for identifying the functionally most beneficial locations of new protein termini is the creation of a random CP library encompassing (ideally) all possible variants of a target protein. Such a comprehensive library approach may seem excessive as knowledge about protein structure–function relationships would lead one to favor new termini in surface loop regions and deter from disrupting secondary structure elements. However, experimental studies have demonstrated that proteins tolerate cleavage in structurally well-defined regions which in some cases result in substantive and beneficial functional change [19, 21]. An experimental method for generating random CP libraries was first reported by Schachman and coworkers [15] and, with few modifications, is described in detail in this chapter.

## 2 Materials

1. QIAquick PCR and gel purification kits (Qiagen; Valencia, CA) including buffers PB and EB.
2. *Taq* DNA polymerase (NEB, Ipswich, MA).
3. 10× *Taq* DNA polymerase buffer: 500 mM KCl, 100 mM Tris–HCl, pH 9.0, 1 % Triton X-100.
4. 100 mM $MgCl_2$.
5. 100 μM oligonucleotide primer stock solutions.
6. Deoxyribonucleotide triphosphates (dNTPs) mix: 2.5 mM per dNTP stock solution. Aliquots stored at −20 °C.
7. Agarose gels: NuSieve GTG agarose in 1× TAE buffer (40 mM Tris–acetate, 1 mM EDTA) and 0.5 μg/mL ethidium bromide.
8. 1 kb DNA ladder: New England Biolabs (Ipswich, MA) stored at −20 °C.
9. DNase I (magnesium-free preparation) (Roche Biochemicals, Indianapolis, IN).
10. 10× DNase I buffer: 500 mM Tris–HCl pH 7.5, 10 mM $MnCl_2$.
11. Restriction endonucleases (with reaction buffers) (NEB, Ipswich, MA).
12. 0.5 M EDTA.

13. Exonuclease III (Exo III) (Promega, Madison, WI).
14. 10× Exo III buffer: 660 mM Tris–HCl, pH 8.0, 6.6 mM $MgCl_2$.
15. 100 % ethanol stored at –20 °C.
16. 70 % ethanol (v/v) stored at –20 °C.
17. 3 M sodium acetate.
18. T4 DNA ligase (NEB, Ipswich, MA).
19. 10× T4 DNA ligase buffer: 500 mM Tris–HCl, pH 7.5, 100 mM $MgCl_2$, 100 mM dithiothreitol, 10 mM ATP, 250 μg/mL bovine serum albumin.
20. T4 DNA polymerase (NEB, Ipswich, MA).
21. 10× T4 DNA polymerase buffer: 1 M potassium acetate, 250 mM Tris–acetate, pH 7.7, 100 mM Mg–acetate, 10 mM dithiothreitol.
22. Shrimp alkaline phosphatase (NEB, Ipswich, MA).
23. 10× shrimp alkaline phosphatase buffer: 500 mM Tris–HCl, pH 9.0, 100 mM $MgCl_2$.
24. pET-16b DNA plasmid (EMD Millipore Corp., Billerica, MA).
25. Electrocompetent *E. coli* strain DH5α-E (Life Technologies, Grand Island, NY) stored at –80 °C.
26. SOC medium: per liter, mix 5 g yeast extract, 20 g tryptone, 0.5 g NaCl, and 2.5 mM KCl. Autoclave at 121 °C for 25 min and let cool to ambient temperature. Autoclave each of the following separately: 1 M $MgSO_4$; 1 M $MgCl_2$; and 20 % (w/v) glucose. Using these stocks, add the following to the media: $MgCl_2$ to 10 mM; $MgSO_4$ to 20 mM; and glucose to 20 mM; and store at ambient temperature.
27. Luria–Bertani broth (LB) medium: per liter, mix 5 g yeast extract, 10 g tryptone, and 10 g NaCl. Autoclave at 121 °C for 25 min and let cool to ambient temperature.
28. LB agar medium: per liter, mix 5 g yeast extract, 10 g tryptone, 10 g NaCl, and 15 g agar. Autoclave at 121 °C for 25 min and let cool to 50 °C in a warm water bath. Add an appropriate antibiotic as a selection marker and gently shake to homogenize the solution, then pour plates, and let agar solidify at room temperature. Plates can be stored at 4 °C for up to 4 weeks.

## 3 Methods

### 3.1 Linker Design

The preparation of random CP libraries critically depends on the successful design of a peptide sequence to connect the native protein termini. The creation of linker sequences is simplified as many protein termini are within 20 Å of each other [22]. Our protocol focuses on this category of proteins. For termini distances of >20 Å, the design of extended linkers based on the described glycine-rich oligopeptides can be problematic as long, flexible sequences destabilize the target protein. In those cases, linker design might be aided by computational methods and by using sequences with defined secondary structure.

1. Measure the distance between the alpha-carbons of amino acids at the termini in the crystal structure of your target protein (*see* **Note 1**).
2. To bridge native termini distances of <20 Å, a glycine-rich oligopeptide sequence is designed (*see* **Note 2**). Calculate the number of amino acids in the oligopeptide based on an approximate length of 3.5 Å per amino acid. The corresponding DNA sequence should also encode a unique restriction endonuclease cleavage site.

### 3.2 Preparation of Random CP Libraries

1. Design and synthesize specific oligonucleotide primers for amplification of the gene sequence of the target protein (*see* **Note 3**).
2. Set up a 50 μl PCR containing 1× *Taq* DNA polymerase buffer, 1.25 mM $MgCl_2$, 0.5–1 μM of above primers, ~10 ng of template DNA, 5 units of *Taq* DNA polymerase, and 200 μM dNTP mix.
3. Amplify the template over 30 cycles of PCR (94 °C for 15 s, $T_A$ for 15 s, 72 °C for 1.2 min/kb) (*see* **Note 4**).
4. Check the quality of the PCR product by running 1–2 μl of the reaction mixture on an agarose gel next to a suitable size marker. Purify the remaining reaction with a Qiagen QIAquick PCR purification kit.
5. Spectrophotometrically determine the final DNA concentration of the sample (*see* **Note 5**).
6. Digest 10 μg of the PCR product in a 400 μl reaction with 1× restriction enzyme buffer and 20 units of restriction endonuclease at 37 °C for 6–8 h (*see* **Note 6**).
7. Purify the reaction mixture with a Qiagen QIAquick PCR purification kit.
8. Self-ligate the entire DNA sample in a 1 mL reaction with 1× T4 DNA ligase buffer and 8000 cohesive end units of T4 DNA ligase at 16 °C for 24 h (*see* **Notes 7** and **8**).
9. Concentrate DNA by ethanol precipitation: add 100 μl of 3 M sodium acetate and 2.5 mL of ice-cold 100 % ethanol. Mix well and incubate at −20 °C for 30 min. Centrifuge at 12,000 × *g* for 15 min. Carefully remove supernatant with pipette and wash the pellet with 1 mL of 70 % ethanol. Briefly centrifuge

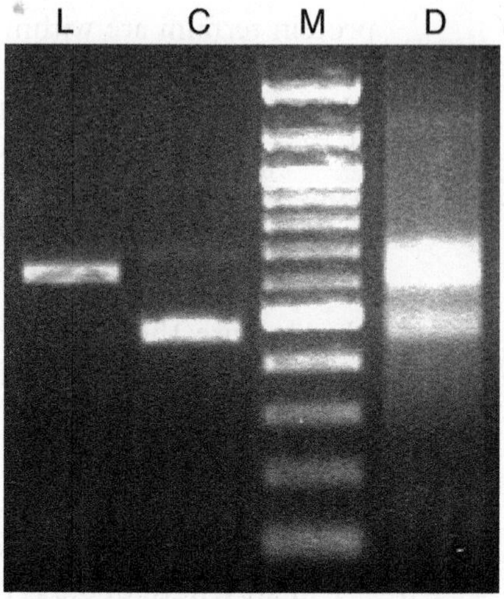


**Fig. 2** Agarose gel electrophoresis analysis of CP library preparation. *L* linear DNA sample, *C* intramolecularly circularized DNA sample, *M* 100-bp DNA ladder, *D* DNase I-treated sample. The DNA fragment in the DNase I-treated sample that corresponds in size to the linear DNA sample encodes the CP library. The library is recovered by excision of the gel block and extraction of the DNA

and then remove all traces of liquid. Air-dry the pellet for 15–20 min and resuspend in 50 μl of water.

10. Remove single-stranded DNA from the ligation mixture by incubating the DNA sample in a 200 μl reaction with 1× Exo III buffer and 400 units of Exo III at 37 °C for 30 min. Purify the reaction mixture with a Qiagen QIAquick PCR purification kit.
11. Check the quality of the PCR product by running 10–20 μl of the reaction mixture on an agarose gel next to a suitable size marker (*see* Fig. 2).
12. Spectrophotometrically determine the final DNA concentration of the sample (*see* **Note 5**).
13. Digest 2.5 μg of DNA in a 400 μl reaction of 1× DNase I buffer and 2 milli-units of DNase I (Mg-free) at 22 °C for 15 min (*see* **Notes 9** and **10**).
14. Quench the DNase digestion by adding 10 μl of 0.5 M EDTA. Purify the reaction mixture with a Qiagen QIAquick PCR purification kit.
15. Cure the DNA by incubating the sample in 1× T4 DNA ligase buffer with 0.5 units of T4 DNA polymerase, 200 μM dNTP, 800 cohesive end units of T4 DNA ligase at 22 °C for 1 h (*see* **Note 11**).
16. Purify linearized DNA by agarose gel electrophoresis: Pour a 1 % agarose gel with wells large enough to load the entire DNA

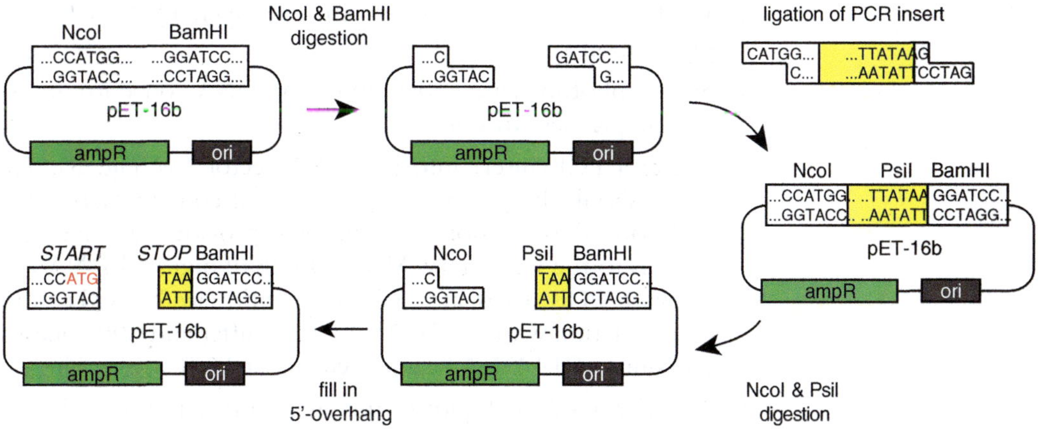


**Fig. 3** Preparation of a DNA vector suitable for CP library cloning. Genes encoding members of a random CP library lack a start and stop codon. For expression of the corresponding protein variants, library members must therefore be inserted in a plasmid with a specifically designed blunt-end cloning site which adds the missing ATG and TAA (or TGA, TAG), respectively. The preparation of such a vector is demonstrated on the example of pET-16b (Novagen). The vector is initially linearized by digestion with NcoI and BamHI, followed by ligation of a PCR insert which serves as adaptor and introduces a unique PsiI site. The PsiI restriction site is critically important as it provides a blunt-ended TAA terminus (highlighted in *yellow*) that will serve as the stop codon upon ligation of the CP library. Similarly, restriction digestion with NcoI generates a 5′-overhang that creates a blunt-end with an ATG start codon (marked in *red*) after DNA polymerase-catalyzed synthesis of the complementary strand

sample (*see* **Note 12**). Mix DNA 1:1 with glycerol, load aliquots of DNA, and proceed with gel electrophoresis. Visualize bands for linear, circularized, and concatemeric DNA to ensure sufficient separation of bands for subsequent excision of the gel block containing the linearized DNA (*see* **Note 13**). Recover DNA from the gel matrix using a Qiagen QIAquick gel purification kit.

17. Spectrophotometrically determine the final DNA concentration of the sample (*see* **Note 5**).

### *3.3 Preparation of the Recipient DNA Vector*

In parallel to preparing the random CP library, a suitable DNA vector must be assembled for blunt-end ligation of the DNA insert. More specifically, the vector must have start and stop codons directly preceding the cloning site (*see* Fig. 3). The preparation of DNA vectors for CP library expression in bacteria (*E. coli*) and yeast (*Pichia pastoris*) has been previously reported [19, 21]. To demonstrate the concept, the construction of a suitable recipient DNA vector based on pET-16b is described.

1. Digest 1 μg of pET-16b in a 50 μl reaction with 1× restriction enzyme buffer and 10 units each of NcoI and BamHI restriction endonucleases at 37 °C for 6–8 h (*see* **Note 14**).

2. Purify the reaction mixture with a Qiagen QIAquick PCR purification kit.
3. Spectrophotometrically determine the DNA concentration of the sample (*see* **Note 5**).
4. Ligate a PCR insert into the DNA vector (*see* Fig. 3). The insert should be prepared separately and contain NcoI, PsiI, and BamHI restriction sites that will provide start and stop codons to the CP library. Mix ~100 ng of digested DNA vector with PCR insert at a 3:1 molar ratio (*see* **Note 15**) in a 50 μl reaction with 1× T4 DNA ligase buffer and 800 cohesive end units of T4 DNA ligase. Incubate at 16 °C overnight.
5. Transform a 2 μl aliquot of the ligation reaction into electrocompetent *E. coli* cells (*see* **Note 16**). Thaw one frozen 50 μl aliquot of competent cells by storing the vial on ice. Add a 2 μl aliquot of the ligation reaction to the vial with competent cells and store the cells on ice for 1 min. Load the cuvette into the electroporator and apply an electrical pulse (for optimal settings, refer to the manufacturer's manual). Immediately after the electrical pulse, add 1 mL of SOC medium to the cuvette and incubate the cell suspension at 37 °C for 1 h. Transfer 10, 20, and 50 μl aliquots of the cell suspension onto LB agar plates containing the appropriate antibiotics and incubate the plates at 37 °C overnight (*see* **Note 17**).
6. Use sterile toothpicks to pick individual colonies and inoculate 10 mL of LB medium containing the appropriate antibiotics in a 50 mL Erlenmeyer flash. Incubate the liquid culture at 37 °C for 16 h.
7. Transfer the cell suspension into sterile centrifuge tubes and pellet cells at 10,000 × *g* at 4 °C for 15 min.
8. Recover the plasmid DNA from the cell pellet using a Qiagen QIAquick plasmid isolation kit.
9. Spectrophotometrically determine the DNA concentration of the sample (*see* **Note 5**).
10. Digest 1 μg of modified DNA vector in a 50 μl reaction with 1× restriction enzyme buffer and 10 units each of *Nco*I and *Psi*I restriction endonucleases at 37 °C for 6–8 h (*see* **Note 18**).
11. Purify the DNA from the reaction mixture with a Qiagen QIAquick PCR purification kit.
12. Dilute the DNA sample to approximately 0.1 μg/μl of DNA in 1× T4 DNA polymerase buffer and 100 μM dNTP. Incubate at 37 °C for 10 min.
13. Add 1 unit of T4 DNA polymerase and continue incubation at 37 °C for 5 min (*see* **Note 19**).

14. Quench the reaction by heat denaturation at 75 °C for 20 min. After equilibration at ambient temperature, the blunt-ended vector is ready for immediate use or storage at 4 °C (*see* **Note 20**).
15. Spectrophotometrically determine the final DNA concentration of the sample (*see* **Note 5**).

### 3.4 Library Ligation and Characterization

1. Dephosphorylate ~500 ng of blunt-ended vector (prepared in Subheading 3.3) in 1× shrimp alkaline phosphatase buffer with 3 units of shrimp alkaline phosphatase at 37 °C for 1 h (*see* **Note 21**).
2. Quench the reaction by heat denaturation at 65 °C for 15 min.
3. Mix ~60 ng of dephosphorylated blunt-ended DNA vector with random CP DNA library (prepared in Subheading 3.2) at a 10:1 molar ratio in a 15 μl reaction with 1× T4 DNA ligase buffer and 400 cohesive end units of T4 DNA ligase at 16 °C overnight (*see* **Note 22**).
4. Transform a 2 μl aliquot of the ligation reaction into electrocompetent *E. coli* cells as per Subheading 3.3, **step 5** (*see* **Note 16**).
5. Estimate the approximate library size by counting the number of colonies on the culture plates and considering the applied dilution.
6. To further characterize the library, use sterile toothpicks to pick 12–15 individual colonies and inoculate each in 1 mL of LB medium containing the appropriate antibiotics using sterile 15 mL Falcon tubes. Incubate the cultures at 37 °C overnight.
7. Resuspend a 1 μl aliquot of each culture in a 10 μl PCR mixture (*see* **Notes 23** and **24**). Use an initial 5 min cell lysis step at 95 °C, followed by the regular PCR cycling protocol, then mix the reaction with gel loading dye, and run an aliquot on an agarose gel. Determine the number of colony PCR products which correspond to the size of the gene fragment for the wild type protein.
8. Determine the precise position of the new termini in colony PCR products by DNA sequence analysis (*see* **Notes 25** and **26**).

## 4 Notes

1. Distance measurements can be performed by any structure analysis software. In addition to determining the distance between the actual native termini, one should review the structure for more proximal amino acid positions in a ten-residue region at the N- and C-termini [22]. Often, deletion of a few

amino acid residues at either terminus does not interfere with function but can significantly shorten the distance which simplifies linker design.

2. For example, a 15 Å distance between native termini can be bridged by a four- or five-amino-acid linker (15 Å/3.5 Å = 4.3). Assuming that the target gene does not have an internal SpeI restriction endonuclease cleavage site, a suitable hydrophilic and flexible linker sequence which also encodes the unique restriction endonuclease cleavage site could be -GlyThrSerGly-, corresponding to -GGC ACT AGT GGC-.
3. Oligonucleotide primers for the preparation of CP libraries carry a number of modifications. The 5′-end (forward) primer should start with the designed linker sequence (*see* **Note 2**), the start codon should be omitted, and the primer must have sufficient overlap with the target gene sequence to effectively anneal during the PCR. The 3′-end (reverse) primer should also encode the linker sequence, the stop codon should be removed, and the primer must have sufficient complementarity with the end region of the gene. An example for primers of an arbitrary gene with the linker sequence designed in **Note 2** (underlined) is shown below:

| | |
|---|---|
| *Gene* | 5′-ATG **GTA CAG CAT GGA AAT** ... **CTA TTA GAT CCG AGA GGA** TGA-3′<br>3′-TAC **CAT GTC GTA CCT TTA** ... **GAT AAT CTA GGC TCT CCT** ACT-5′ |
| *Forward primer* | 5′-GGC ACT AGT GGC **GTA CAG CAT GGA AAT** -3′ |
| *Reverse primer* | 5′-GCC ACT AGT GCC **TCC TCT CGG ATC TAA** -3′ |

4. The optimal annealing temperature ($T_A$) for primers needs to be determined prior to amplification. We recommend primers of between 20 and 25 nucleotides in length with $T_A$ of 55–60 °C.
5. We use the following rule for concentration measurements: 1 $A_{260}$ = 50 μg DNA.
6. The exact choice of restriction endonuclease is determined by the unique cleavage sequence in the linker sequence. In our previous example (*see* **Notes 2** and **3**), SpeI would be the chosen restriction enzyme.
7. To favor the intramolecular ligation over concatemerization of DNA, the concentration should be kept below 10 ng/μl. The volume of the ligation mixture can be adjusted accordingly.
8. In our experience, the self-ligation of digested PCR products can be inefficient. In those situations, the digested DNA

sample (*see* Subheading 3.2, **step 7**) can be ligated into a shuttle vector such as pSTBlue1 (Novagen) and transformed into a standard *E. coli* cloning strain. Following growth of the culture, larger quantities of pSTBlue-1 vector containing the CP library can be obtained by DNA plasmid midi- or maxiprep. The vector is then digested with the same restriction endonuclease used in Subheading 3.2, **step 6**, and the insert purified by preparative agarose gel electrophoresis. After recovery of the DNA from the gel matrix using a Qiagen QIAquick gel purification kit, the digested insert serves as DNA sample in the subsequent self-ligation step.

9. A magnesium-free preparation of DNase I must be used as the enzyme only introduces single-strand cuts in the presence of $Mg^{2+}$ but performs double-strand cuts with $Mn^{2+}$ counter ions [23]. The conditions for the DNase digestion require careful optimization of enzyme concentration and incubation time to maximize yields of linear, singly cut DNA. We typically run multiple digestions in parallel, varying the DNase I amount per reaction (0.5, 1, 2, and 5 milli-units). Further adjustments are possible by altering the incubation period (5, 10, 15, and 20 min). Aliquots of the reaction mixture are analyzed by agarose gel electrophoresis. We consider yields of ~40 % linearized DNA (with the remaining sample still circularized) suitable digestion conditions.
10. Linearization of the circular DNA by DNase I is a sequence-independent, double-strand cleavage event. Consequently, only one in three linearized gene sequences cut by the endonuclease will preserve the correct reading frame of the parental protein.
11. The curing step is critical to fill in termini and create blunt-ended DNA fragments, as well as to repair nicks in the DNA strands.
12. DNA samples can be concentrated by ethanol precipitation (*see* Subheading 3.2, **step 9**) prior to loading on the agarose gel.
13. We visualize the stained DNA bands in the agarose gel on a UV transilluminator. For simple and fast excision of the product band from the gel, we use a razor blade.
14. Typically, we directly use digested DNA vectors for subsequent ligation reactions. However, if vector self-ligation is a problem, the DNA sample can be treated with shrimp alkaline phosphatase after the digestion step.
15. The PCR insert is prepared separately by PCR amplification with oligonucleotide primers that contain NcoI as well as the PsiI/BamHI restriction sites as indicated in Fig. 3. The PCR product is digested with NcoI/BamHI prior to addition to the ligation mixture. The specific size and sequence composition of the PCR insert (besides the unique restriction sites

mentioned above) is not important, although fragments of 500–1,000 bp might be advantageous for handling the sample and monitoring the progress of vector digestion later in the protocol. The precise amount of PCR insert depends on the size of the DNA fragment.

16. We routinely use *E. coli* DH5α. Transformation by electroporation is generally more efficient than chemical transformation. However, either method can be applied.
17. pET-16b carries an ampicillin resistance marker.
18. The specific restriction endonuclease buffer is selected based on the enzyme manufacturer's recommendation. We chose NEB buffer #2 (New England Biolabs, Ipswich, MA) for the double digestion with NsiI/PsiI.
19. The choice of polymerase is important to avoid creation of A-overhangs. It is important not to exceed the indicated incubation time as consumption of dNTPs over time could lead to excessive truncation by the polymerase's exonuclease activity.
20. Storage of the digested vector for more than 2 days, as well as storage at temperatures below 4 °C is discouraged as we have noticed diminishing ligation efficiency under such conditions.
21. We have found that dephosphorylation of the DNA vector immediately before ligation proved more effective than usage of a stored sample. The dephosphorylation of the blunt-ended vector is not required but significantly reduces formation of vector-only side product.
22. Blunt-end ligation of the CP library results in library insertion into the DNA vector in the forward and reverse direction with equal probability. As a consequence, the evaluation of library quality should take into consideration that only roughly 50 % of transformants that contain a library member carry it in the correct (forward) orientation.
23. The colony PCR is a relatively quick and easy method for library analysis. However, the method sometimes fails to generate a DNA product despite of the presence of an appropriate template. A more reliable yet more laborious alternative method for library analysis is to miniprep the DNA plasmid of each liquid culture, digest the purified vector by the appropriate restriction enzymes, and determine DNA insert size by agarose gel electrophoresis.
24. PCR mixture: 1× *Taq* DNA polymerase buffer, 1.25 mM $MgCl_2$, 1 μM of universal primers (T7 promoter primer and T7 terminator primer for pET-16b), template DNA, 5 units of *Taq* DNA polymerase, and 200 μM dNTP mix.
25. DNA sequence analysis can be performed directly on the PCR products. Alternatively, the corresponding miniprep-purified DNA vectors can be submitted for analysis.

26. We typically estimate the size of a CP library and percent coverage of sequence space by the following calculations; the theoretical CP library size ($X$) of a target protein can be calculated by $X = (n - 1)$ where $n$ is the number of amino acids in the protein sequence. However, a complete experimental library ($\Upsilon$) increases in size due to practical limitations in library preparation involving the loss of correct reading frame upon DNase I treatment (*see* **Note 9**) and random insertion of the CP library during blunt-end ligation (*see* **Note 21**), multiplying the theoretical library size by threefold and twofold, respectively ($\Upsilon = 6X$). Finally, comprehensive coverage of sequence space necessitates ≥5-fold sampling of $\Upsilon$, hence requiring the analysis of a minimum of $5\Upsilon$ or $30\ (n - 1)$ CP library members.
27. Capitalizing on the recent advances in whole-gene synthesis, an alternative to the described random circular permutation protocol is the preparation of fully synthetic CP libraries [17, 24]. Briefly, a tandem gene sequence can be prepared by chemical DNA synthesis which can serve as template for the PCR amplification of specific CP variants. This strategy significantly reduces library size, eliminating out-of-frame and inversely cloned representatives in the sequence pool. Although initially more laborious, the synthetic approach further simplifies library analysis by eliminating the need for library oversampling.

## Acknowledgments

The authors like to thank the members of the Lutz Lab for many helpful comments and suggestions on the manuscript. This work was in part supported by funds from the US National Science Foundation (CBET 0730312 & 1159434).

### References

1. Bornscheuer UT et al (2012) Engineering the third wave of biocatalysis. Nature 485:185–194
2. Ha JH, Loh SN (2012) Protein conformational switches: from nature to design. Chemistry 18:7984–7999
3. Omenetto FG, Kaplan DL (2010) New opportunities for an ancient material. Science 329: 528–531
4. Boyle AL, Woolfson DN (2011) De novo designed peptides for biological applications. Chem Soc Rev 40:4295–4306
5. Lutz S, Bornscheuer UT (2009) Protein engineering handbook. Wiley-VCH, Weinheim
6. Arnold FH, Georgiou G (2003) Directed evolution library creation: methods and protocols. Humana Press, Totowa, NJ
7. Luger K et al (1989) Correct folding of circularly permuted variants of a beta alpha barrel enzyme in vivo. Science 243:206–210
8. Jaenicke R (1991) Protein folding—local structures, domains, subunits, and assemblies. Biochemistry 30:3147–3161
9. Pan T, Gutell RR, Uhlenbeck OC (1991) Folding of circularly permuted transfer-RNAs. Science 254:1361–1364
10. Ivarsson Y et al (2008) Folding and misfolding in a naturally occurring circularly permuted PDZ domain. J Biol Chem 283:8954–8960
11. Lindberg M, Tangrot J, Oliveberg M (2002) Complete change of the protein folding transition state upon circular permutation. Nat Struct Biol 9:818–822

12. Hahn M et al (1994) Native-like in vivo folding of a circularly permuted jellyroll protein shown by crystal structure analysis. Proc Natl Acad Sci U S A 91:10417–10421
13. Baird GS, Zacharias DA, Tsien RY (1999) Circular permutation and receptor insertion within green fluorescent proteins. Proc Natl Acad Sci U S A 96:11241–11246
14. Ostermeier M (2008) Engineering allosteric protein switches by domain insertion. Protein Eng Des Sel 18:359–364
15. Graf R, Schachman HK (1996) Random circular permutation of genes and expressed polypeptide chains: application of the method to the catalytic chains of aspartate transcarbamoylase. Proc Natl Acad Sci U S A 93:11591–11596
16. Hennecke J, Sebbel P, Glockshuber R (1999) Random circular permutation of DsbA reveals segments that are essential for protein folding and stability. J Mol Biol 286:1197–1215
17. Iwakura M et al (2000) Systematic circular permutation of an entire protein reveals essential folding elements. Nat Struct Biol 7: 580–585
18. Guntas G, Kanwar M, Ostermeier M (2012) Circular permutation in the Omega-loop of TEM-1 beta-lactamase results in improved activity and altered substrate specificity. PLoS One 7:e35998
19. Qian Z, Lutz S (2005) Improving the catalytic activity of Candida antarctica lipase B by circular permutation. J Am Chem Soc 127: 13466–13467
20. Qian Z et al (2009) Structural redesign of lipase B from Candida antarctica by circular permutation and incremental truncation. J Mol Biol 393:191–201
21. Reitinger S et al (2010) Circular permutation of Bacillus circulans xylanase: a kinetic and structural study. Biochemistry 49:2464–2474
22. Yu Y, Lutz S (2011) Circular permutation: a different way to engineer enzyme structure and function. Trends Biotechnol 29:18–25
23. Sambrook J, Fritsch EF, Maniatis T (1989) Molecular cloning: a laboratory manual, 2nd edn. CSH Laboratory Press, Cold Spring Harbor, NY
24. Daugherty AB, Govindarajan S, Lutz S (2013) Improved biocatalysts from a synthetic circular permutation library of the flavin-dependent oxidoreductase old yellow enzyme. J Am Chem Soc 135:14425–14432
25. Arnold FH (2006) Fancy footwork in the sequence space shuffle. Nat Biotechnol 24: 328–330

# Part III

## Structure-Guided Methods and Computational Tools for the Design and Analysis of Mutant Libraries

# Chapter 18

# Probabilistic Methods in Directed Evolution: Library Size, Mutation Rate, and Diversity

**Yuval Nov**


## Abstract

Directed evolution has emerged as an important tool for engineering proteins with improved or novel properties. Because of their inherent reliance on randomness, directed evolution protocols are amenable to probabilistic modeling and analysis. This chapter summarizes and reviews in a nonmathematical way some of the probabilistic works related to directed evolution, with particular focus on three of the most widely used methods: saturation mutagenesis, error-prone PCR, and in vitro recombination. The ultimate aim is to provide the reader with practical information to guide the planning and design of directed evolution studies. Importantly, the applications and locations of freely available computational resources to assist with this process are described in detail.


**Key words** Directed evolution, Probabilistic modeling, Library generation, Computational tools, Saturation mutagenesis, Error-prone PCR, DNA shuffling, Staggered extension process

## 1 Introduction

Over the past two decades, directed evolution has proven to be an invaluable tool in the hands of protein engineers. It has been successfully used in thousands of experiments to improve various desirable protein properties, including catalytic activity, thermostability, enantioselectivity, and binding affinity [1–4].

In a directed evolution experiment, researchers "breed" in vitro a population of protein variants via successive rounds of mutagenesis and screening. In each mutagenesis stage, the DNA sequences of one or several template genes are randomly altered, yielding a large number of slightly mutated genes. In the following screening stage, proteins expressed from a random sample of these genes are screened (or selected), and improved variants, if present, are identified; the set of variants sampled for screening is called *a library*. This process may continue by taking one or more of the best performing variants to serve as templates for the next round.

Elizabeth M.J. Gillam et al. (eds.), *Directed Evolution Library Creation: Methods and Protocols*, Methods in Molecular Biology, vol. 1179, DOI 10.1007/978-1-4939-1053-3_18, 

Because the very essence of directed evolution hinges on randomness, the process is amenable to probabilistic analysis. The goal of this chapter is to summarize and review in a nonmathematical way some of the works and tools related to the probabilistic analysis of directed evolution. Such an analysis allows the practitioner both to design better experiments and to interpret correctly the results of an experiment already undertaken. This chapter will focus on three common directed evolution methods: saturation mutagenesis, error-prone PCR, and in vitro recombination.

Regardless of the method used, screening multiple copies of the same protein variant is undesirable, as it expends laboratory resources without adding genetic diversity. The same is true for variants expressed from genes mutated to include a frameshift or a premature stop codon, which are essentially guaranteed to lack function. The total number of protein variants screened, here termed *the library size*, should therefore be distinguished from the library's *effective size*: the number of distinct, full-length protein variants in the library. The library's effective size is a central concept in this chapter. A related important concept is that of *sequence space*—the set of all protein sequences that *could possibly* be generated and included in the library; depending on experimental design, the sequence space may be hyper-astronomically large, well beyond that which could be plausibly generated or screened in a laboratory. Probability theory allows us to study how these three concepts are interrelated and to calculate, for example, what percentage of sequence space is expected to be represented in a library of given size, or by how much the effective library size is expected to increase if the library size is increased by, say, a thousand variants.

"All models are wrong, but some are useful," wrote the great statistician George E. P. Box. His words are apt. Like all mathematical models of real-world phenomena, the predictions of probabilistic models for directed evolution are only as valid as their assumptions. But these assumptions are never fully satisfied: different oligonucleotides may have different annealing probabilities; mutations, even at distant positions, do not occur completely independently; etc. Indeed, as Patrick and Firth [5] note, there is still a gap between the perfect library on paper and the perfect library in the lab. Fortunately, this gap is closing, and the analyses and results presented below may provide at least a good approximation for the experimental results.

## 2 Saturation Mutagenesis

In saturation mutagenesis, several protein positions are first identified as having a high potential to accommodate favorable mutations. The chosen positions are then randomized at the DNA level using degenerate oligonucleotides, so that each randomized position in each of the translated protein variants contains a random amino acid. The resulting library is then screened, in the hope of discovering among its members highly improved variants.

Mathematically, saturation mutagenesis is the most tractable of the methods discussed in this chapter. Let $M$ denote the number of randomized positions. When all 20 amino acids are allowed at each of the randomized positions, as is commonly the case, the size of the protein sequence space is $20^M$. This expression grows very rapidly with $M$, so that when randomizing more than 10 positions or so, even the most efficient screens or selections are unable to interrogate more than a minuscule fraction of the corresponding sequence space. In practice, however, saturation mutagenesis experiments usually involve $M \leq 4$ randomized positions, so that a manageable library may cover all or almost all of sequence space—hence the word "saturation" in the method's name.

### 2.1 Randomization Schemes

The simplest form of randomization in saturation mutagenesis is termed NNN: here, when synthesizing the degenerate oligonucleotides, each of the four bases is equally likely to appear at each of the three nucleotide positions of the randomized codon, so that in the resulting gene, all $4^3 = 64$ codons are equally likely to appear. In contrast, because of the redundancy of the genetic code, the resulting distribution across the 20 amino acids at the protein level is far from uniform, with probabilities ranging from 1/64 (for Met and Trp) to 6/64 (for Arg, Leu, and Ser); importantly, there is a 3/64 probability that NNN randomization will produce a premature stop codon at any randomized position.

Both these phenomena—uneven amino acid distribution and nonzero probability for a stop codon—can be shown mathematically to reduce the effective library size, and their negative effects amplify as more positions are randomized. Figure 1 shows how the probability of having at least one stop codon in the randomized gene increases as a function of the number of randomized positions.

To mitigate these problems, reduced codon sets are often used, i.e., only some of the four bases are allowed to appear (in equimolar proportions) at one or more nucleotide positions of the randomized codons. The most common such randomization schemes are NNK, NNS, and NNB; see Table 1 for the meaning of these abbreviations and the right panel of Fig. 1 for the resulting amino acid distributions. NNK randomization is in principle identical to NNS in terms of the translated sequences but is more efficient in practice when used in *E. coli* and *S. cerevisiae*, as it conforms better to the codon usage of these organisms [5].

An ideal randomization scheme generates a perfectly uniform distribution over the 20 amino acids while assigning zero probability for a stop codon. Such schemes, which we term UNIF, may be achieved either through the special MAX oligonucleotides [6] or through a proper mixture of several standard

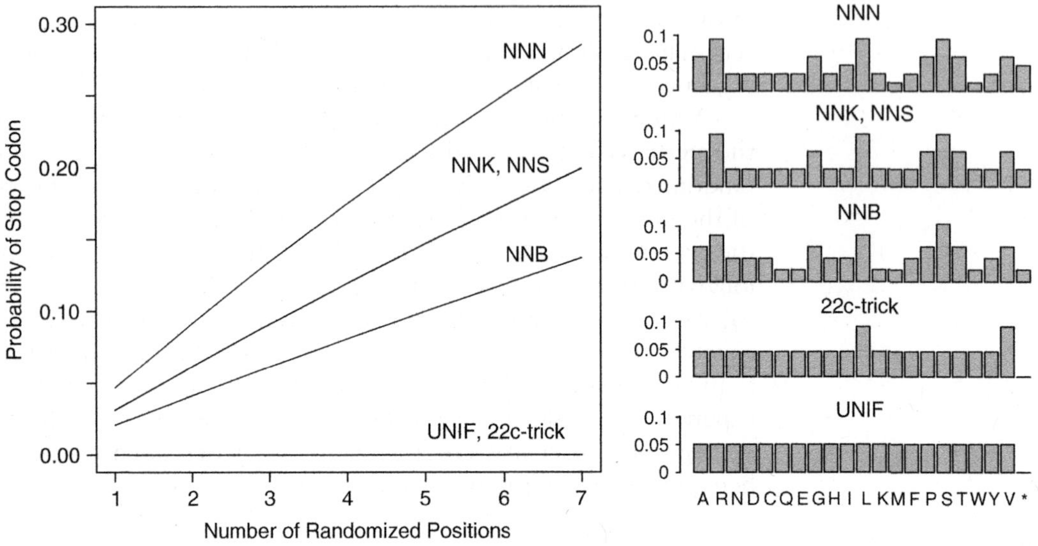


**Fig. 1** *Left*: The probability of introducing at least one premature stop codon in a gene, as a function of the number of randomized positions, for six saturation mutagenesis randomization schemes. *Right*: The probability distribution across the 20 amino acids and the stop codon (denoted by an *asterisk*) for six randomization schemes

**Table 1**
**Degenerate base abbreviations**

| N | V | H | D | B | M | R | W | S | Y | K |
|---|---|---|---|---|---|---|---|---|---|---|
| A/C/G/T | A/C/G | A/C/T | A/G/T | C/G/T | A/C | A/G | A/T | C/G | C/T | G/T |

degenerate oligonucleotides (e.g., a mixture of NDT, VMA, ATG, and TGG at 12:6:1:1 molar ratio [7]). Using the latter approach, one can reduce the number of required degenerate codons to three, at the expense of a slight deviation from a perfectly uniform distribution, by using the "22c-trick" of Kille et al. [8].

Yet another way to influence the amino acid distribution is via "doped" (also known as "spiked") oligonucleotides. Here, during oligonucleotide synthesis, non-equimolar proportions of the four bases are used at some (or all) of the three nucleotide positions of the randomized codon. This approach may be used in conjunction with the mixing approach mentioned above, to produce extra flexibility in shaping the resulting amino acid distribution. The probabilistic and algorithmic aspects of such randomization schemes have received attention in the literature [9–11], but perhaps because of the special equipment and expertise they require, they are seldom used [12].

Rather than allowing all 20 amino acids at each of the randomized positions, some authors advocate using a smaller, yet chemically balanced, subset of the amino acids. Sequence space can be reduced significantly in this way, thus making it easier to explore. See, for example, the NDT randomization approach of Reetz et al. [13].

### 2.2 Library Metrics

Once the randomized positions and the randomization scheme are determined, the main decision left to be made is of the library size. Clearly, the more variants screened, the higher the probability of discovering improved variants, but one needs to balance the benefits of this increased probability with the additional laboratory work and expense large libraries entail.

To choose judiciously the library size, the experimenter needs a metric to quantify how exhaustive is a library, a priori, given its size. Several such metrics have been suggested in the literature. The first is the probability of "full coverage," i.e., the probability that the library contains *all* variants in sequence space. The rationale behind this definition is that in saturation mutagenesis, the experimenter typically wishes to consider all possible variants, so the probability that all are indeed represented in the library appears initially a reasonable metric. The mathematical analysis of this metric is in essence the same as that of the "coupon collector's problem" from probability theory, which is also known as the "occupancy problem" [14]. The second metric is "expected coverage" (sometimes called "completeness" or "fractional completeness"), which is the fraction of sequence space that is expected to be represented in the library. The expected coverage is related to the expected effective library size via the simple formula:

$$\text{expected effective library size} = \text{expected coverage} \times \text{size of sequence space}.$$

To illustrate these concepts, consider a hypothetical randomization scheme for a single position, such that under a certain library size, there is a 0.8 probability for having all 20 amino acids in the library, a 0.15 probability for 19, and a 0.05 probability for only 18. Then, the probability of full coverage is 0.8, the expected coverage is $0.8 \cdot (20/20) + 0.15 \cdot (19/20) + 0.05 \cdot (18/20) = 0.9875$, and the expected effective library size is $0.9875 \cdot 20 = 19.75$. Exact and approximate formulae for computing the probability of full coverage and the expected coverage can be found in the pioneering works of Patrick et al. [15], Bosley and Ostermeier [16], and Firth and Patrick [17].

Recently, Nov [18] proposed a third metric, which is in fact a family of metrics: the probability that the library contains at least one of the best $k$ variants. When $k=1$, this metric is the probability that the library contains the single best variant in sequence space, which is clearly a desirable outcome of the experiment; this probability can be shown to always equal the expected coverage.

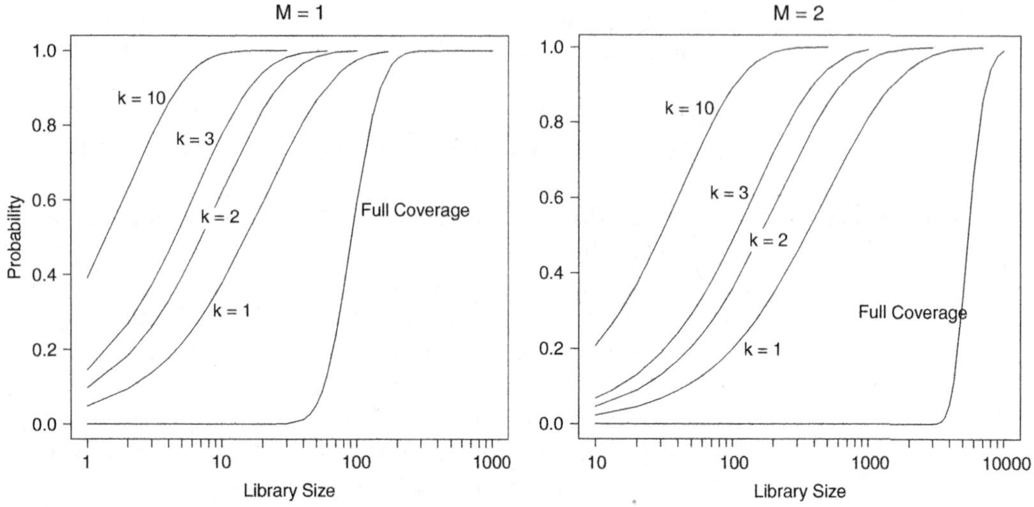


**Fig. 2** Saturation mutagenesis library metrics as a function of the library size, when randomizing a single NNK position (*left*) and two NNK positions (*right*). Note the logarithmic scale of the horizontal axis

The rationale behind this metric for $k \geq 2$ is that often, especially when sequence space is large (say, is at least in the hundreds), there are several variants that satisfy the design requirements of the protein at hand, so discovering any of them will be considered a success. By being willing to not restrict oneself to solely the best variant, the library size may be reduced significantly, and the experimenter may thus reduce labor and expenses.

Figure 2 shows the abovementioned metrics as a function of the library size, when randomizing one or two NNK positions. As the figure shows, the differences between the metrics at a given library size are substantial. For example, when screening 500 variants after randomizing two NNK positions, the probability of full coverage is essentially zero, the probability of discovering the single best variant is 0.61 (this is also the expected coverage), and the probability of discovering at least one of the top 3 variants is 0.94. More importantly, the differences are substantial also when reading the graph in inverted axis order: to ensure a 0.95 probability of full coverage, 8,128 variants are required; for a 0.95 probability of discovering the best variant, the number drops to 2,130; and for a 0.95 probability of discovering any of the top 3 variants, the number is 534.

Which metric should be used, then? The probability of full coverage is not recommended, as it leads to needlessly large libraries. The reason for this is that screening all—or almost all—of sequence space is not a goal in its own right, but rather just a means to an end: discovering the best variants. The search for the best variants, in turn, obeys the law of diminishing marginal utility—after, say, 90 % of sequence space has been explored, it

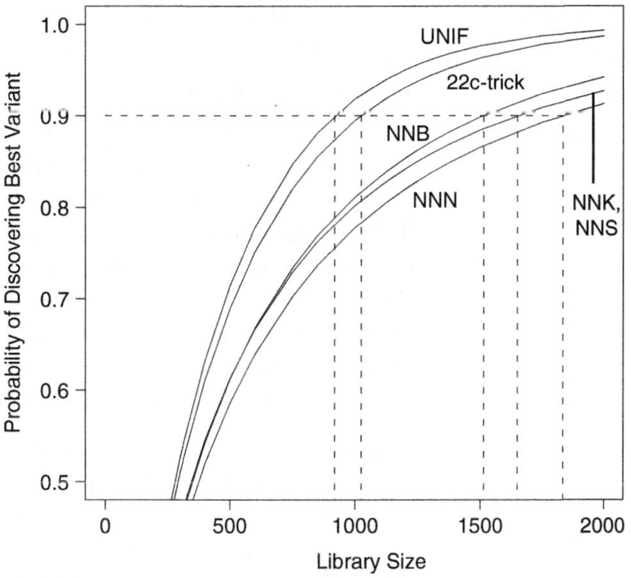


**Fig. 3** The probability of discovering the best variant as a function of the library size for six saturation mutagenesis randomization schemes

becomes increasingly rare for subsequent screened variants to turn out to be new ones, so more and more variants need to be screened to increase the coverage by one percent; at the same time, the probability that the remaining 10 % of sequence space will contain variants surpassing all those previously screened is greatly reduced.

In a high-stakes protein design process, the designer may not wish to settle for less than the best possible variant. The recommended strategy in this case is to choose a library size such that the probability of discovering the best variant (which is the same as the expected coverage) is very close to 1—say, equal to 0.99. As the design requirements become less stringent, the designer may lower this probability or change the metric to the probability of discovering any of the top *k* variants, for larger and larger *k*. In situations where there are likely to be many satisfactory solutions for the design problem—say, when trying to improve thermostability moderately—the designer may choose a rather large *k*.

Figure 3 is concerned with the role of the randomization scheme in a saturation mutagenesis experiment. The curves show the probability of discovering the best variant when randomizing two protein positions with various randomization schemes. It can be seen that to ensure a 0.9 probability of discovering the best variant, the most efficient method is UNIF, which requires a library of 920 variants; next best, by a small margin, is the "22c-trick" scheme (1,025 variants), followed by NNK or NNS (1,513), NNB (1,651), and NNN (1,834). The results are qualitatively similar for all other metrics and for any number of randomized positions.

### 2.3 Computational Resources

Several online computational resources concerning saturation mutagenesis are listed below, ordered by their publication date.

*GLUE* [15, 19] (http://guinevere.otago.ac.nz/cgi-bin/aef/glue.pl) is a web program through which users can compute three quantities: (1) the expected coverage corresponding to a given library size, (2) the library size corresponding to a given expected coverage (the inverse calculation), and (3) the library size corresponding to a given probability of full coverage. Importantly, all possible variants are assumed in GLUE to be equally probable, so the program applies, for example, to randomization at the DNA level and to MAX randomization at the protein level, but not to randomization at the protein level under nonuniform schemes such as NNN or NNK.

*CASTER* [20] (http://www.kofo.mpg.de/media/2/D1108347/0987095526/ISM_tools.zip) is an Excel worksheet that computes the library size required to ensure a specified expected coverage of DNA (not protein) sequence space, under arbitrary codon-based randomization.

*GLUE-IT* [17] (http://guinevere.otago.ac.nz/cgi-bin/aef/glue-IT.pl), where IT stands for "Including Translation," is an improved version of GLUE which computes library statistics at the protein level. On specifying a library size and any codon-based randomization scheme involving up to six codons, the program computes the expected effective library size, the expected coverage, the probability of full coverage, and the number of possible sequences both at the DNA and at the protein levels.

*Kong* [21] developed a web server (http://graphics.med.yale.edu/cgi-bin/lib_comp.pl) that allows users to compute the mean and variance of the effective library size. The reported results concern sequence diversity at the DNA (not protein) level. Of the servers reviewed in this chapter, this is the only one that computes also the variance and also the only one that allows arbitrary doping via direct specification of the molar ratio between the four bases. The number of randomized (DNA) positions is not restricted, but when this number is large and the doping varies across positions, the computation of the variance may take more than several minutes.

*TopLib* [18] (http://stat.haifa.ac.il/~yuval/toplib) is a web server that handles computations of the probability of discovering at least one of the top $k$ variants (recall that the case $k = 1$ gives the expected coverage), as well as the probability of full coverage. The server can compute either the probability corresponding to a given library size or the library size required to ensure a given probability (the inverse calculation). Any randomization scheme of up to 12 positions is allowed, either by specifying degenerate codons or by specifying directly the probabilities for the 20 amino acids at each position. The latter option is useful for MAX randomization, as well as for randomization via doped oligonucleotides (because each doping scheme induces a distribution over the 20 amino acids). Users can also specify the yield factor—the per-variant probability that the randomization succeeds.

*DC-Analyzer* [7] (http://cobi.uestc.edu.cn/resource/dc_analyzer/view) is a downloadable Java program that computes the optimal oligonucleotide mixtures (*see* Subheading 2.1) required to generate specified subsets of the 20 amino acids.

## 3 Error-Prone PCR

Error-prone PCR (epPCR) is a modification of the standard PCR protocol that deliberately introduces mutations during the elongation step. The mutation rate can be tuned by the experimenter via the choice of the polymerase used and by adjusting the reaction conditions. To avoid too many inactivated variants in the resulting library, the mutation rate is typically set to produce an average of no more than 5 mutations per variant at the protein level.

Unlike saturation mutagenesis, epPCR is therefore a whole-protein mutagenesis method, which targets the entire DNA sequence. Consequently, sequence space consists in principle of all $20^M$ protein variants (where $M$ now denotes the entire protein length); even for a short protein of length $M=100$, this number dwarfs the estimated number of atoms in the universe. But again unlike saturation mutagenesis, there are vast differences between the probabilities corresponding to the variants: only those that differ from the template protein in relatively few positions have a non-negligible probability to occur, whereas the vast majority of the theoretical sequence space is practically inaccessible.

### 3.1 Probabilistic Analysis

From the probabilistic point of view, epPCR is closely related to the well-studied theory of branching processes [22]. Unfortunately, certain characteristics of epPCR differentiate it from a pure branching process and thus complicate its analysis: parent sequences do not "die," but rather remain in the reaction tube; the mutation process is modulated on the amplification process, so that different mutations on the same sequence may have different ages; and the resulting DNA sequences need to be translated to protein sequences. In the words of Piau [23], epPCR, "although easy to describe, is not analytically tractable." Indeed, researchers invariably rely on various mathematical approximations and simplifying assumptions in its analysis.

Unlike in the case of saturation mutagenesis, the probabilistic analysis of epPCR depends crucially on parameters that need to be estimated from experimental data. These are the reaction's efficiency, which is the per cycle probability of successfully replicating a sequence, and the mutational spectrum, which is a set of parameters describing the mutation probabilities from each base to each of the other three during elongation, as well as the probabilities of insertions and deletions. The need to estimate these parameters, typically through a small, preliminary experiment, complicates the analysis further and weakens the prediction ability.

Early probabilistic works on epPCR analyzed the process only at the DNA level and tended to address questions having more of a theoretical flavor [24–27]. Examples of the mathematical results derived include the distribution of the number of DNA mutations in a randomly chosen sequence, the distribution of the distance (measured in the number of different bases) between two randomly chosen sequences, the distribution of the number of PCR replications separating two randomly chosen sequences, and the probability that a randomly chosen DNA molecule will follow a specified sequence. Piau used later mean field theory to provide a mathematically rigorous justification for some of the approximations used in the above studies [23].

Later works tended to focus on analysis that is more directly relevant to epPCR practitioners. Bosley and Ostermeier [16] appear to be the first to address the issue of DNA–protein translation in their analysis of epPCR. Following their earlier work on DNA diversity in epPCR libraries [15], Firth and Patrick [17] derived a method to estimate the expected effective library size at the protein level and to partition the expected library composition into sub-libraries according to the number of mutations: single-mutant variants, double mutants, etc. (Note that the term "effective library size" used here is called "number of unique, full-length variants" in their terminology, as they reserve the term "effective library size" to the total number of full-length variants, distinct or not.)

The left panel of Fig. 4 shows the expected effective library size, broken down into sub-libraries according to the number of mutations, as a function of the library size, as well as the expected number of distinct DNA sequences. The gene used in this example is *vs1* (GenBank accession number EU627691), an 891-bp-long gene encoding the tyrosinase VS1 of *Bacillus megaterium* [28]. The first important feature of this graph is that DNA diversity is significantly greater than protein diversity; since only the latter is relevant to practitioners, one is advised to not design epPCR experiments based on DNA diversity analysis. The second important feature is that the curves grow only slightly sublinearly, so that doubling the library size almost doubles the expected effective library size and also almost doubles the numbers corresponding to each of the sub-libraries. The deviations from perfect linearity are seen more clearly in the right panel of Fig. 4, which shows the expected fraction of each of the effective sub-libraries from the total expected effective library size. The number of distinct single mutants, for example, can never exceed $19 \cdot 297 = 5{,}643$ (19 substitutions for each of the 297 protein positions); as the library size increases, more and more single mutants are recurring variants

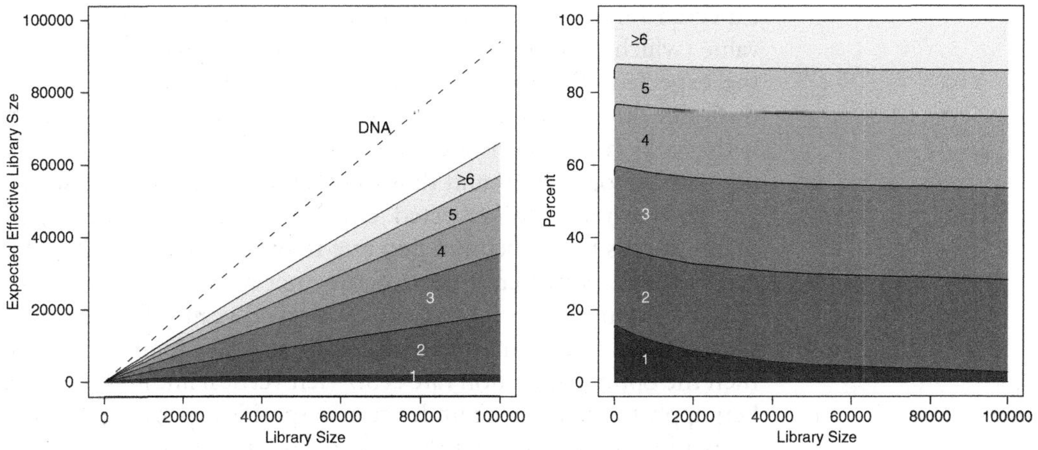


**Fig. 4** *Left*: Expected effective epPCR library size, broken into sub-libraries according to the number of mutations, as a function of the library size. *Right*: Expected fractions (in percent) of the effective sub-library sizes. The underlying gene is *vs1* [28], and the graphs were drawn based on the output of the PEDEL-AA program [17], using the PCR distribution and the server's default parameters

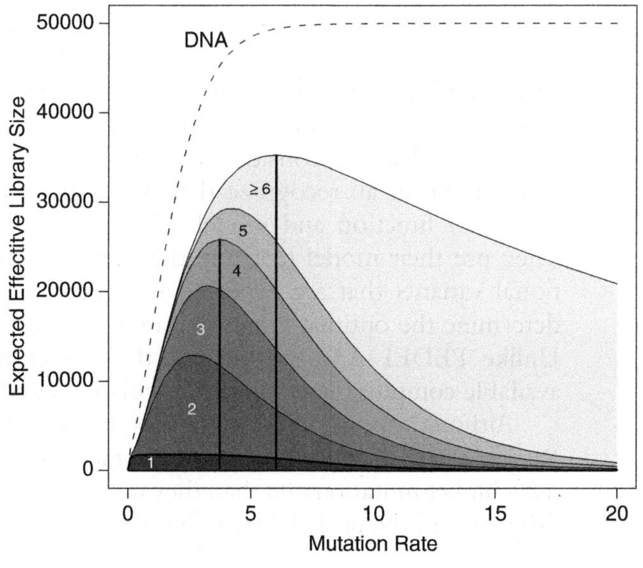


**Fig. 5** Expected effective epPCR library size for the *vs1* gene, broken into sub-libraries according to the number of mutations, as a function of the mutation rate. Library size is 50,000 DNA sequences, and other parameters are as in Fig. 4

already included in the library, and because this phenomenon happens more slowly to multiple-point mutants, the fraction of single-point mutants decreases as a function of the library size.

Figure 5 shows the influence of the mutation rate on epPCR library diversity, for a fixed-size library of 50,000 variants. DNA diversity is again much greater than protein diversity, and the expected number of distinct DNA sequences attains its maximal value (which is 50,000, the library size) rather quickly. In contrast, the expected effective protein library size not only never reaches this value but starts decreasing at a certain point. The reason for this is that a high mutation rate increases the fraction of variants with premature stop codons and thus lowers the effective library size.

To maximize the expected effective library size, the experimenter needs to set the mutation rate to 6.1 (indicated by the right vertical line). However, if the experimenter wishes to consider only variants having up to, say, four mutations (because further mutations are believed to inactivate even full-length, in-frame variants), then the optimal mutation rate is 3.7 (left vertical line). These numbers apply to the *vs1* gene under certain specific experimental conditions and will vary as either the gene or the conditions are changed.

The interrelations between mutation rate, library diversity, and retention of function are at the center of the work of Drummond et al. [29]. The authors model epPCR somewhat differently from Firth and Patrick [17]: instead of basing the analysis on the precise template DNA sequence and on the entire estimated mutational spectrum, they use one parameter for the probability that a mutation (regardless of its position) will result in a non-synonymous change and a second parameter for the probability that a mutation will result in a truncation (due to a premature stop codon, insertion, or deletion). Another important difference is that whereas Firth and Patrick consider all full-length, in-frame variants, Drummond et al. recognize that many of these variants may be devoid of function and model this event via a third parameter. They use their model to study the number of distinct and functional variants that are expected to appear in the library and to determine the optimal mutation rate that maximizes this number. Unlike PEDEL-AA (see below), there is currently no publicly available computer program for calculations under this model.

Although using different models, both Drummond et al. and Patrick et al. conclude that epPCR practitioners should generally use a higher mutation rate than they normally do, to increase library diversity and the probability of discovering improved variants.

### 3.2 Computational Resources

*PEDEL and PEDEL-AA.* PEDEL [15, 19] (http://guinevere.otago.ac.nz/cgi-bin/aef/pedel.pl) is a web server through which users can calculate the expected number of distinct DNA sequences in an epPCR library. PEDEL-AA [17] (http://guinevere.otago.ac.nz/cgi-bin/aef/pedel-AA.pl), where AA stands for "Amino Acid," is a much improved version of the program, which computes various statistics of epPCR libraries at the protein level. The most important of the many statistics reported are the effective library size and its breakdown into sub-libraries according to the number of mutations.

*Volles and Lansbury* [30] developed a C program (downloadable from http://lansbury.bwh.harvard.edu/Michael_Volles.htm) that computes various epPCR library statistics. The program can calculate some of these statistics analytically, but other statistics, among them the effective library size, are estimated only via simulation (the program simulates only a single realization of the process). The user needs to prepare two input files—one for the template DNA sequence and one for the mutational spectrum—and to answer, via a terminal interface, about 20 queries regarding the program's execution.

*MAP$^{2.0}$3D* [31] (http://map.jacobs-university.de/map3d.html), where MAP stands for "Mutagenesis Assistant Program," is a web server that compares the performance of 19 mutagenesis methods having different mutational spectra. Given a template DNA sequence, the server computes various protein-level statistics for each of the mutagenesis methods, e.g., the fraction of variants with a premature stop codon, the average number of amino acid substitution per residue, and a codon diversity coefficient. All statistics are calculated assuming a single nucleotide substitution in the mutagenized gene and thus do not incorporate parameters such as the library size and the reaction's efficiency. The server can also incorporate structural information about the target protein, when available, to aid in selecting the mutagenesis method. The reader is referred to chapter 19 by Verma et al. in this volume for further information.

## 4 In Vitro DNA Recombination

In in vitro DNA recombination, fragments from two or more homologous parent genes are assembled randomly into genes with novel DNA sequences. Common protocols for DNA recombination include DNA shuffling [32] and the staggered extension process (StEP) [33].

Variable positions that are nearby along the recombined sequence are likely to belong to the same fragment and hence to originate from the same parent gene. Consequently, the mutations in a recombined gene are in general not independent; the dependence between any two such mutations decreases as the distance between them increases and vanishes completely if this distance is greater than the maximal possible fragment length.

### 4.1 Probabilistic Analysis

DNA recombination experiments are more difficult to model and analyze than saturation mutagenesis and epPCR experiments, as they involve more parameters set by the experimenter (e.g., the number and composition of the parent genes and the experimental conditions that determine the distribution of the fragment lengths),

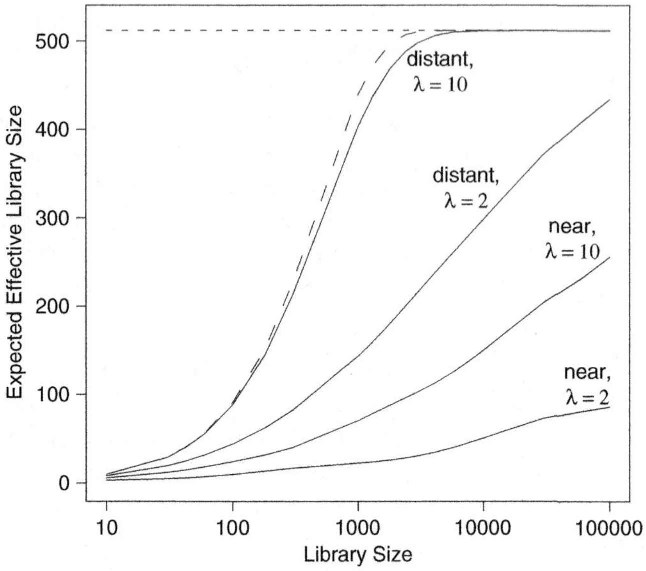


**Fig. 6** Expected effective library size as a function of the library size in a DNA recombination experiment with two genes, based on the model of Patrick et al. [15]. All genes are 600 bp long, and their variable positions are either at 100, 150, 200, 250, 300, 350, 400, 450, and 500 ("distant") or at 280, 285, 290, 295, 300, 305, 310, 315, and 320 ("near"). The mean number of crossovers per sequence is indicated by λ. The *dashed curve* corresponds to the diversity in a saturation mutagenesis library with nine randomized positions, each with two equiprobable possible amino acids

and these parameters influence the composition of the resulting genes in an intricate way. Further difficulties arise from the need to estimate some other parameters. As a result, the probabilistic modeling approaches in the literature vary greatly, and as in the case of epPCR, the analysis relies on various approximations and simplifying assumptions.

Early mathematical works on DNA recombination addressed questions such as the distribution of the number of mutations in a randomly chosen reassembled gene [34], the distribution of the fragment number and fragment length in the reassembled gene, the recombination probability between two mutations, and the probability of assembling a given target sequence [27, 35–37].

Patrick et al. [15] studied the expected effective library size resulting from a DNA recombination between two homologous genes, assuming a Poisson distribution for the number of crossovers between any two consecutive variable positions. Figure 6 is based on the model of Patrick et al., and shows the expected effective library size as a function of the library size. There are 9 variable positions in this example, so sequence space includes $2^9 = 512$ variants, a number indicated by the horizontal dotted line. The distance between adjacent variable positions is either 5 bp ("near") or

50 bp ("distant"), and the mean number of crossovers per sequence was either $\lambda = 2$ or $\lambda = 10$. It can be seen that diversity increases as the variable positions are further apart and as the crossovers are more frequent.

As Patrick et al. observe, in the extreme case of remote variable positions and highly frequent crossovers, the mutations are essentially independent of each other, and all possible variants become equally likely. This situation is similar to a saturation mutagenesis experiment in which there are two equiprobable possible amino acids at each randomized position. The dashed curve in Fig. 6 corresponds to the expected effective library size in such a saturation mutagenesis experiment; indeed, the curve corresponding to the high-crossover distant mutation follows it closely.

### 4.2 Computational Resources

*DRIVeR* [15, 19] (http://guinevere.otago.ac.nz/cgi-bin/aef/driver.pl), whose name stands for "Diversity Resulting from In Vitro Recombination," is to date the only publicly available computer program for calculations related to DNA recombination. This program uses the model and analysis of Patrick et al. [15] and thus applies only to experiments with two recombined genes. Given the library size, the gene length, the mean number of crossovers per gene, and the location of the variable positions, the program computes the expected effective library size. The program can also estimate the true mean number of crossovers from the observed mean number.

## 5 Concluding Remarks

In this chapter, we have reviewed the probabilistic models and analyses behind three widely used directed evolution methods—saturation mutagenesis, error-prone PCR, and DNA recombination. Of the many other existing methods, some fall into the same mathematical analysis category as one of the abovementioned methods; e.g., SISDC [38] is a shuffling method that achieves complete independence between the mutations at the variable positions and thus should be analyzed using the probabilistic methodology of saturation mutagenesis. Other methods require moderate modification of existing analyses, such as ISOR [39], which is mathematically similar—but not identical—to saturation mutagenesis. Still other methods, such as SeSaM [40], call for completely new modeling and analysis.

The existing probabilistic methods can certainly be improved. For example, the chemistry underlying saturation mutagenesis dictates that not all oligonucleotides designated to a given position are equally likely to anneal. The resulting bias can be mitigated by adjusting the reaction conditions [41] or by using a proper combination of polymerases [42] but should also be accounted for in the probabilistic analysis; Denault and Pelletier [43] suggest to use a chi-squared test to detect such a bias. Another possible improvement is *interval prediction*: rather than reporting only the expected value of the predicted quantity, it might be illuminating to account for sampling variability by reporting, e.g., "when screening 5,000 variants, there is a 95 % probability that the effective library size will be between 4,150 and 4,320." Kong [21], who analyzed saturation mutagenesis with doping, made an important step in this direction by deriving the variance of the expected effective library size but stopped short of deriving the prediction interval.

Probability theory can aid in protein design also beyond its role in directed evolution. Recent works demonstrate that given a moderate amount of sequence-activity data, probabilistic reasoning can help in deciding how to optimally combine existing mutations and point at few, highly targeted variants with a high predicted activity [44–47]. The mathematics underlying these methods is more involved than that behind the methods discussed in this chapter.

## References

1. Moore JC, Arnold FH (1996) Directed evolution of a para-nitrobenzyl esterase for aqueous-organic solvents. Nat Biotechnol 14(4):458
2. Giver L, Gershenson A, Freskgard PO, Arnold FH (1998) Directed evolution of a thermostable esterase. Proc Natl Acad Sci U S A 95(22):12809–12813
3. Reetz MT, Zonta A, Schimossek K, Jaeger K-E, Liebeton K (1997) Creation of enantioselective biocatalysts for organic chemistry by in vitro evolution. Angew Chem Int Ed 36(24):2830–2832
4. Boder ET, Midelfort KS, Wittrup KD (2000) Directed evolution of antibody fragments with monovalent femtomolar antigen-binding affinity. Proc Natl Acad Sci U S A 97(20): 10701–10705
5. Patrick WM, Firth AE (2005) Strategies and computational tools for improving randomized protein libraries. Biomol Eng 22(4):105–112
6. Hughes MD, Nagel DA, Santos AF, Sutherland AJ, Hine AV (2003) Removing the redundancy from randomised gene libraries. J Mol Biol 331(5):973–979
7. Tang L, Gao H, Zhu X, Wang X, Zhou M, Jiang R (2012) Construction of "small-intelligent" focused mutagenesis libraries using well-designed combinatorial degenerate primers. Biotechniques 52(3):149–158
8. Kille S, Acevedo-Rocha CG, Parra LP, Zhang Z-G, Opperman DJ, Reetz MT, Acevedo JP (2013) Reducing codon redundancy and screening effort of combinatorial protein libraries created by saturation mutagenesis. ACS Synth Biol 2(2):83–92
9. Tomandl D, Schober A, Schwienhorst A (1997) Optimizing doped libraries by using genetic algorithms. J Comput Aided Mol Des 11(1):29–38
10. Jensen LJ, Andersen KV, Svendsen A, Kretzschmar T (1998) Scoring functions for computational algorithms applicable to the design of spiked oligonucleotides. Nucleic Acids Res 26(3):697–702
11. Wolf E, Kim PS (1999) Combinatorial codons: a computer program to approximate amino acid probabilities with biased nucleotide usage. Protein Sci 8(3):680–688
12. Neylon C (2004) Chemical and biochemical strategies for the randomization of protein encoding DNA sequences: library construction methods for directed evolution. Nucleic Acids Res 32(4):1448–1459
13. Reetz MT, Kahakeaw D, Lohmer R (2008) Addressing the numbers problem in directed evolution. Chembiochem 9(11):1797–1804
14. Feller W (1971) An introduction to probability theory and its applications, vol 2. Wiley, New York
15. Patrick WM, Firth AE, Blackburn JM (2003) User-friendly algorithms for estimating completeness and diversity in randomized protein-encoding libraries. Protein Eng 16(6): 451–457

16. Bosley AD, Ostermeier M (2005) Mathematical expressions useful in the construction, description and evaluation of protein libraries. Biomol Eng 22(1–3):57–61
17. Firth AE, Patrick WM (2008) GLUE-IT and PEDEL-AA: new programmes for analyzing protein diversity in randomized libraries. Nucleic Acids Res 36(suppl 2):W281–W285
18. Nov Y (2012) When second best is good enough: another probabilistic look at saturation mutagenesis. Appl Environ Microbiol 78(1):258–262
19. Firth AE, Patrick WM (2005) Statistics of protein library construction. Bioinformatics 21(15):3314–3315
20. Reetz MT, Carballeira JD (2007) Iterative saturation mutagenesis (ISM) for rapid directed evolution of functional enzymes. Nat Protoc 2(4):891–903
21. Kong Y (2009) Calculating complexity of large randomized libraries. J Theor Biol 259(3): 641–645
22. Grimmett GR, Stirzaker DR (2001) Probability and random processes. Oxford University Press, Oxford
23. Piau D (2004) Mutation–replication statistics of polymerase chain reactions. J Comput Biol 9(6):831–847
24. Sun F (1995) The polymerase chain reaction and branching processes. J Comput Biol 2(1): 63–86
25. Wang D, Zhao C, Cheng R, Sun F (2000) Estimation of the mutation rate during error-prone polymerase chain reaction. J Comput Biol 7(1–2):143–158
26. Weiss G, von Haeseler A (1995) Modeling the polymerase chain reaction. J Comput Biol 2(1):49–61
27. Moore GL, Maranas CD (2000) Modeling DNA mutation and recombination for directed evolution experiments. J Theor Biol 205(3): 483–503
28. Shuster V, Fishman A (2009) Isolation, cloning and characterization of a tyrosinase with improved activity in organic solvents from *Bacillus megaterium*. J Mol Microbiol Biotechnol 17(4):188–200
29. Drummond DA, Iverson BL, Georgiou G, Arnold FH (2005) Why high-error-rate random mutagenesis libraries are enriched in functional and improved proteins. J Mol Biol 350(4):806–816
30. Volles MJ, Lansbury PT (2005) A computer program for the estimation of protein and nucleic acid sequence diversity in random point mutagenesis libraries. Nucleic Acids Res 33(11):3667–3677
31. Verma R, Schwaneberg U, Roccatano D (2012) MAP$^{2.0}$3D: a sequence/structure based server for protein engineering. ACS Synth Biol 1(4):139–150
32. Stemmer WP (1994) DNA shuffling by random fragmentation and reassembly: in vitro recombination for molecular evolution. Proc Natl Acad Sci U S A 91(22):10747–10751
33. Zhao H, Giver L, Shao Z, Affholter JA, Arnold FH (1998) Molecular evolution by staggered extension process (StEP) in vitro recombination. Nat Biotechnol 16(3):258–261
34. Moore JC, Jin H-M, Kuchner O, Arnold FH (1997) Strategies for the in vitro evolution of protein function: enzyme evolution by random recombination of improved sequences. J Mol Biol 272(3):336–347
35. Sun F (1999) Modeling DNA shuffling. J Comput Biol 6(1):77–90
36. Moore GL, Maranas CD, Gutshall KR, Brenchley JE (2000) Modeling and optimization of DNA recombination. Comput Chem Eng 24(2–7):693–699
37. Moore GL, Maranas CD, Lutz S, Benkovic SJ (2001) Predicting crossover generation in DNA shuffling. Proc Natl Acad Sci U S A 98(6):3226–3231
38. Hiraga K, Arnold FH (2003) General method for sequence-independent site-directed chimeragenesis. J Mol Biol 330(2):287–296
39. Herman A, Tawfik DS (2007) Incorporating Synthetic Oligonucleotides via Gene Reassembly (ISOR): a versatile tool for generating targeted libraries. Protein Eng Des Sel 20(5):219–226
40. Wong TS, Tee KL, Hauer B, Schwaneberg U (2004) Sequence saturation mutagenesis (SeSaM): a novel method for directed evolution. Nucleic Acids Res 32(3):e26
41. Airaksinen A, Hovi T (1998) Modified base compositions at degenerate positions of a mutagenic oligonucleotide enhance randomness in site-saturation mutagenesis. Nucleic Acids Res 26(2):576–581
42. Vanhercke T, Ampe C, Tirry L, Denolf P (2005) Reducing mutational bias in random protein libraries. Anal Biochem 339(1):9–14
43. Denault M, Pelletier JN (2007) Protein library design and screening: working out the probabilities. In: Arndt KM, Müller KM (eds) Protein engineering protocols, vol 352, Methods in molecular biology. Humana Press Inc., Totowa, NJ
44. Fox RJ, Davis SC, Mundorff EC, Newman LM, Gavrilovic V, Ma SK, Chung LM, Ching C, Tam S, Muley S, Grate J, Gruber J, Whitman JC, Sheldon RA, Huisman GW (2007) Improving

catalytic function by ProSAR-driven enzyme evolution. Nat Biotechnol 25(3):338–344

45. Nov Y, Wein LM (2005) Modeling and analysis of protein design under resource constraints. J Comput Biol 12(2):247–282
46. Barak Y, Nov Y, Ackerley DF, Matin A (2007) Enzyme improvement in the absence of structural knowledge: a novel statistical approach. ISME J 2(2):171–179
47. Brouk M, Nov Y, Fishman A (2010) Improving biocatalyst performance by integrating statistical methods into protein engineering. Appl Environ Microbiol 76(19): 6397–6403

# Chapter 19

# The Mutagenesis Assistant Program

**Rajni Verma, Tuck Seng Wong, Ulrich Schwaneberg, and Danilo Roccatano**


## Abstract

Mutagenesis Assistant Program (MAP) is a web-based statistical tool to develop directed evolution strategies by investigating the consequences at the amino acid level of the mutational biases of random mutagenesis methods on any given gene. The latest development of the program, the $MAP^{2.0}3D$ server, correlates the generated amino acid substitution patterns of a specific random mutagenesis method to the sequence and structural information of the target protein. The combined information can be used to select an experimental strategy that improves the chances of obtaining functionally efficient and/or stable enzyme variants. Hence, the $MAP^{2.0}3D$ server facilitates the "in silico" prescreening of the target gene by predicting the amino acid diversity generated in a random mutagenesis library. Here, we describe the features of $MAP^{2.0}3D$ server by analyzing, as an example, the cytochrome $P450_{BM3}$ monooxygenase (CYP102A1). The $MAP^{2.0}3D$ server is available publicly at http://map.jacobs-university.de/map3d.html.


**Key words** Directed evolution, Random mutagenesis methods, Mutational bias, Residue mutability indicator, Protein engineering

## 1 Introduction

The success of directed evolution methods [1, 2] to tailor properties of proteins via random mutagenesis methods depends upon the quality of the utilized mutant library [3, 4]. A non-biased random mutagenesis method would allow equal occurrence of all four transitions ($T_s$; purine to purine or pyrimidine to pyrimidine) and eight transversions ($T_v$; purine to pyrimidine or pyrimidine to purine), with a $T_s$ to $T_v$ ratio ($T_s/T_v$) of 0.5. However most methods reported to date are biased toward certain nucleotide exchanges (e.g., many error-prone PCR (epPCR) methods preferentially generate transition). Therefore, the organization of the genetic code and the mutational bias in random mutagenesis methods determine the generated amino acid diversity. Generally, a $T_s/T_v$ ratio is used to characterize the mutational propensity of any random mutagenesis method. However the $T_s/T_v$ ratio provides

Elizabeth M.J. Gillam et al. (eds.), *Directed Evolution Library Creation: Methods and Protocols*, Methods in Molecular Biology, vol. 1179, DOI 10.1007/978-1-4939-1053-3_19, 

information on the gene level, and hence it is insufficient to analyze the substitution patterns and to derive effective directed evolution strategies. The mapping at the protein level of the probability and the types of amino acid substitutions introduced by different mutagenesis methods on a given sequence allows the identification of functionally and structurally important regions of the protein that can be preferentially targeted by a specific method. We have developed a web-based tool, the Mutagenesis Assistant Program (MAP), that benchmarks different random mutagenesis methods at the level of protein sequence and structure [5]. MAP compares random mutagenesis methods by investigating the consequences of their mutational bias at the level of amino acid substitutions under the assumption of single nucleotide substitution in the codon. The current version of the server has a library of 19 random mutagenesis methods (*see* Wong et al. for their classification [6]) with the additional possibility of testing a user-defined method. Based on the subset of amino acid substitutions, MAP calculates, for each mutagenesis method, three indicators (a protein structure indicator, an amino acid diversity indicator with a codon diversity coefficient, and a chemical diversity indicator) to summarize the expected overall distribution of each amino acid substitution introduced into a given sequence by the different methods. In addition to these analyses, MAP provides an analysis of amino acid substitution patterns and substitution probabilities for each amino acid in a given sequence, and for each of the 19 random mutagenesis methods so far included in MAP$^{2.0}$3D.

The current version of the MAP$^{2.0}$3D server extends the protein sequence-based analysis with the possibility to map the sequence indicators on the tridimensional structure of the target protein for structure/function analysis [7]. In addition, MAP$^{2.0}$3D correlates the pattern of amino acid substitutions generated by the mutagenesis method with protein structural information like secondary structure elements, important local interactions, residual solvent accessibility, and residual motilities. The residue mutability indicator of MAP$^{2.0}$3D identifies hot spots generated due to the mutational preferences of the different mutagenesis methods. The implementation of a target-based query system in the server enables users to match the statistical analysis of amino acid substitution with the structural information of a specified region in the protein of interest. The diagram in Fig. 1 summarizes the types of analyses obtained from the MAP$^{2.0}$3D server upon input of a DNA sequence and atomic coordinates of the corresponding protein structure.

The capabilities of the server have been illustrated by comprehensive analysis and evaluation of random mutagenesis methods on the protein level, in silico screening of different enzymes and comparison with experimental results, and assistance in the design of novel random mutagenesis methods [7–9]. In the latter cases, the predicted results were in agreement with the experimental

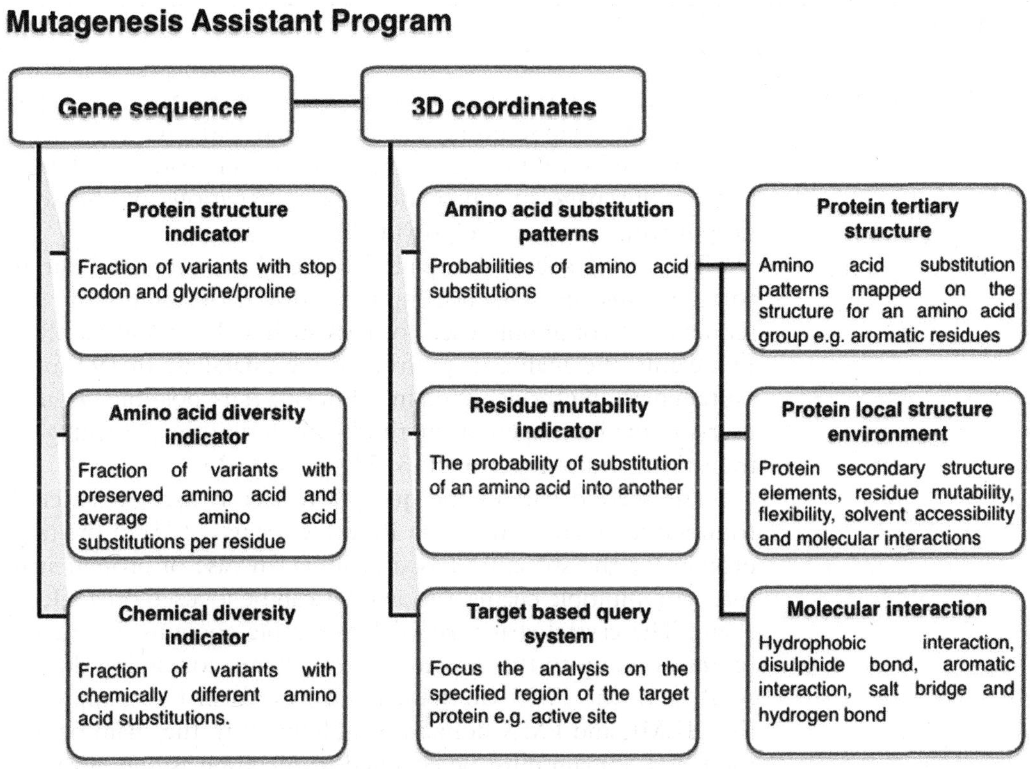


**Fig. 1** Schematic summary of the generated output information from sequence and structure data from the MAP$^{2.0}$3D server. This web-based statistical tool performs analyses of the consequences of mutational bias of random mutagenesis methods on protein sequence and structure

findings. In this chapter, as an example of a protocol to illustrate how to use the MAP$^{2.0}$3D server, the analysis of the cytochrome P450$_{BM3}$ is described. A detailed discussion of the MAP results is reported for P450$_{BM3}$ by Wong et al. [5, 10].

## 2 Materials

1. The MAP$^{2.0}$3D Server is available publicly at http://map.jacobs-university.de/map3d.html.
2. The gene sequence of the monooxygenase cytochrome P450$_{BM3}$ (EC 1.14.14.1) from *Bacillus megaterium* is available from GenBank (http://www.ncbi.nlm.nih.gov/genbank/) using accession number JO4832 [11]. The atomic coordinates used for the structural mapping are those of the non-stoichiometric complex of heme (HEME) and flavin mononucleotide (FMN) domains available from the Protein Data Bank (PDB; http://www.pdb.org/pdb/home/home.do) using identifier 1BVY [12].

## 3 Methods

### 3.1 Query Interface

The MAP$^{2.0}$3D server can be accessed at http://map.jacobs-university.de/map3d.html. The new MAP$^{2.0}$3D server welcome page has links to the query interface (Submission) and to an instructions page in which detailed explanations of the server output with examples are provided.

The server submission page is divided in two parts. The top part contains the sequence input window that accepts the gene sequence in commonly used formats such as Fasta and GenBank. By default, the analysis is performed on a database of 19 random mutagenesis methods, and, in addition, it is possible to add a user-specified mutagenesis method by entering the percentages of transitions and transversions [5, 7] (*see* **Note 1**).

The second part of the query interface is used to enter the tridimensional coordinates (in PDB file format [13]) of the target protein for the structure-based analysis. In case of protein structure with multiple chains, the analysis can be focused on a selected chain. The crystal structure of the example contains three chains (corresponding to the structure of two heme-containing domains A, B, and one FMN domain, F). To perform the mapping on the HEME and FMN domains simultaneously, the chain name of the FMN domain must be changed to the same as one of the two heme domains. This can be done with a text editor by selecting the column of the FMN chain identifier and replacing the letter F with A. The translated protein sequence from the gene sequence is automatically aligned with the protein sequence extracted from the PDB file of the target protein by the standard Smith–Waterman algorithm [14]. The alignment is provided in a separate downloadable log file. The MAP$^{2.0}$3D analysis is performed if the sequences have an identity greater than or equal to the minimum value of 70 %. The structural MAP analysis is performed for one selected/defined random mutagenesis method at a time. By default the analysis is performed on all the residues of the protein but a focused analysis of selected groups of amino acids (classified according to their chemical properties, solvent accessibilities, flexibility) or a specified set of residues (for example, those of active site of the HEME domain or of the substrate access channel of our tutorial protein P450$_{BM3}$) is also possible. The targeted analysis can be extended to neighboring residues within a given distance from the entered set of amino acids. Parameters for the calculation of molecular interactions can also be modified by selecting "Define molecular interaction parameters" at the end of the query page.

### 3.2 Output Page

The server generates a single scrollable interactive output page with the combined results of sequence- and structure-based analyses. Figure 1 shows the generated output information from the MAP$^{2.0}$3D analysis as sequence and structure data.

### 3.3 Sequence-Based Analysis

The statistical occurrence of an amino acid substitution in a sequence upon single nucleotide exchange is calculated using the published mutational spectrum (i.e., the probabilities of all 12 possible mutations expressed in percentages; *see* **Note 2**) of each random mutagenesis method (*see* ref. 7 for details). The obtained statistics are used to benchmark the performance of random mutagenesis methods using three types of indicator (*see* **Note 3**): protein structure indicators, amino acid diversity indicators, and chemical diversity indicators. These indicators are reported in the form of comparative histograms for all the random mutagenesis methods. In addition, the amino acid substitution pattern for each method is provided in the form of a color coded matrix. Figure 2 represents the output generated by the sequence-based analysis of $P450_{BM3}$.

#### 3.3.1 Protein Structure Indicators

*Protein structure indicators* reveal the likelihood of structure/function disrupting (stop codons, *see* Fig. 2a) and destabilizing (glycine and proline residues, *see* Fig. 2b) amino acid substitutions by indicating the fraction of single nucleotide substitutions that would result either in a stop codon, causing truncated or dysfunctional protein, or in a Gly/Pro substitution, destabilizing helical structure.

#### 3.3.2 Amino Acid Diversity Indicator

The *amino acid diversity indicator* shows the fraction of variants in which the amino acids encoded are not changed by a single nucleotide substitution (Fig. 2c) and the average amino acid substitutions per residue after single nucleotide exchange (Fig. 2d). This indicator is complemented by the *codon diversity coefficient* (*see* **Note 4**), which measures how random mutations are distributed among codons of a gene (reported in Fig. 2e).

#### 3.3.3 Chemical Diversity Indicator

*The chemical diversity indicator* reveals the chemical diversity generated by the random mutagenesis method in terms of the type of amino acid substitutions. The amino acids are grouped into four groups (aliphatic, aromatic, neutral, and charged; *see* **Note 5**) along with stop codons for this indicator. The result is presented as a histogram showing the percentage of amino acid group substitution generated by all 19 methods (*see* Fig. 2f) and, in tabular form, as the deviation of each method from the ideal chemical distribution (5/20 or 25 % aliphatic amino acids; 3/20 or 15 % aromatic amino acids; 7/20 or 35 % neutral amino acids; 5/20 or 25 % charged amino acids) after excluding stop codons. The detailed analysis of the amino acid substitution patterns for each of the 20 amino acids, along with the stop codon, provides an overview of conservative (i.e., substitution to chemically similar amino acids) and nonconservative amino acid substitutions created by a random mutagenesis method (*see* Fig. 3a) [10].

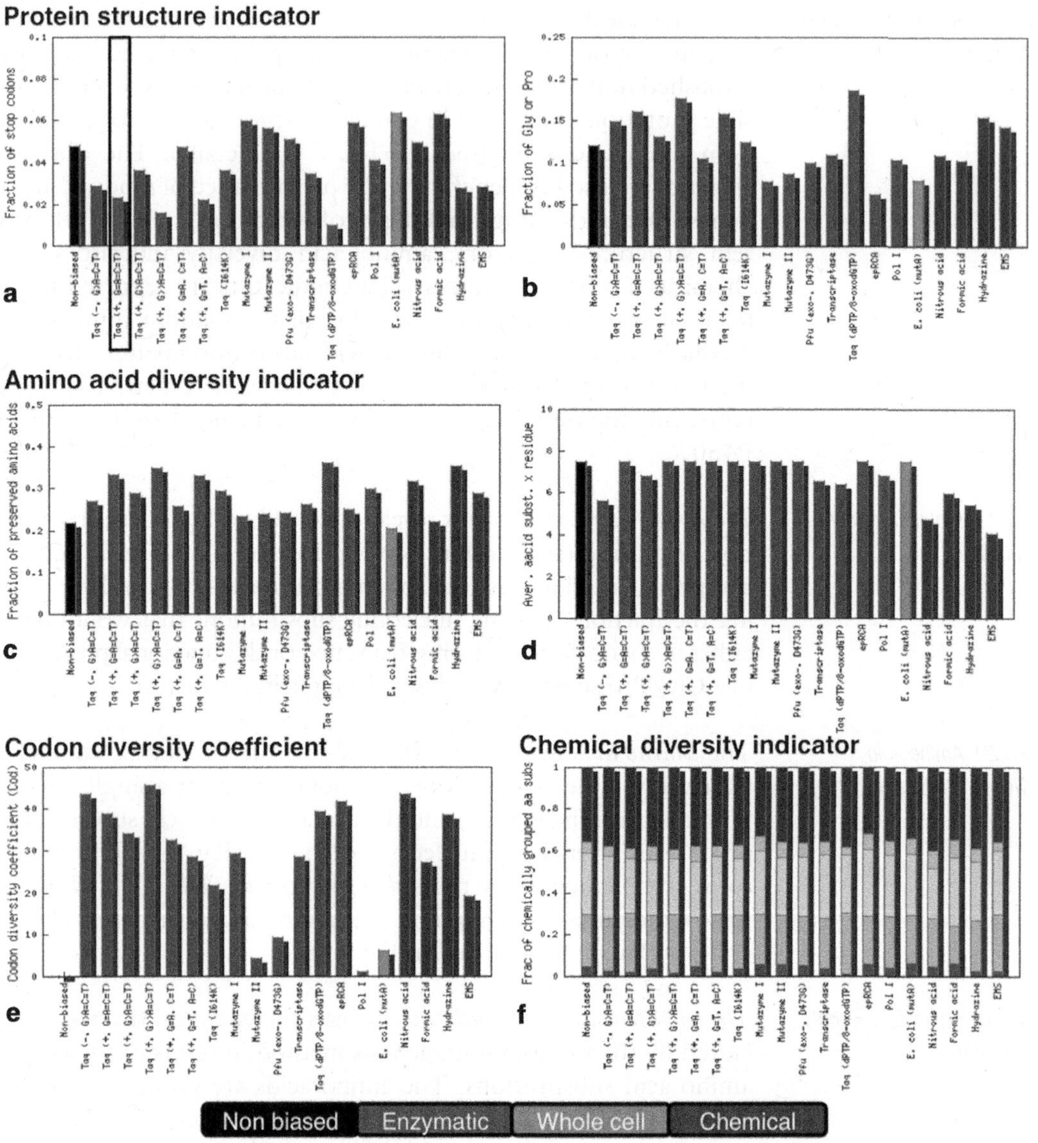


**Fig. 2** The statistical analysis of (**a**) stop codons frequencies, (**b**) Gly/Pro substitutions, (**c**) fraction of variants in which the originally encoded amino acid is unchanged upon single nucleotide exchange in a codon, (**d**) average amino acid substitutions per residue, (**e**) codon diversity coefficient (*see* **Note 4**), and (**f**) chemically grouped amino acid substitution for $P450_{BM3}$. In all the figures except (**f**), histograms are colored according to the method employed for nucleotide substitution: enzymatic methods are shown in *red*, whole cell methods in *green*, chemical methods in *blue*, and theoretically non-biased method in *black*. In the chemical diversity indicator, the histograms are colored according to the chemically different amino acid groups; charged groups are shown in *pink*, neutral in *yellow*, aromatic in *cyan*, aliphatic in *blue*, and stop codons in *red*. The random mutagenesis method (epPCR (Taq ($MnCl_2$, G = A = C = T))) enclosed in a *black rectangle* in figure (**a**) is used for structure-based analysis (Color figure online)

### 3.4 Structure-Based Analysis

Generally, the effect of amino acid substitution on the protein stability and function is strongly influenced by the perturbation of the local structural environment in which the amino acid is located. The secondary structure, residue flexibility, solvent accessibility, and molecular interactions are determinant factors that can be used to qualitatively characterize the local structural environment. The possibility of monitoring the effect of mutational bias of different mutagenesis methods on the structural properties of a target protein can assist in choosing the most efficient method to create a functionally enriched mutant library. In the $MAP^{2.0}3D$ server, the structural target-based query system is used to analyze the effect of the generated amino acid substitution patterns on protein structure and function. $MAP^{2.0}3D$ analysis is performed upon input of the gene sequence (protein coding region only) in conjunction with the 3D coordinates of the target protein to comprehensively analyze the mutational preferences of random mutagenesis methods. Figure 3 shows the results of the structure-based analysis of heme and FMN cofactor-binding domains of $P450_{BM3}$ using epPCR (employing *Taq* polymerase with a high error rate promoted by adding $Mn^{2+}$ ions, and using an equimolar ratio of each dNTP).

#### 3.4.1 Amino Acid Substitution Mapping

*Amino acid substitution mapping* uses amino acids classified according to their chemical properties as neutral, aromatic, aliphatic, and charged residues. The structure disrupting (stop codons) and destabilizing factors (Gly/Pro) are included as separate classes. The amino acid substitution pattern for each class is mapped on the 3D coordinates of the target protein using the Jmol applet [13]. A red-white-blue color scheme is used for protein color mapping where red indicates a higher probability of substitution and blue shows a lower probability. As an example, Fig. 3b shows the probabilities by which residues in the original protein sequence will be substituted with neutral amino acids (mapped onto the cartoon representation of $P450_{BM3}$ HEME-FMN complex). The latter is achieved by replacing the B-factor field in the corresponding PDB coordinate file with the calculated amino acid substitution probabilities. This altered PDB coordinate file is also available for downloading and customized visualization with different molecular visualization software.

#### 3.4.2 Amino Acid Substitution Patterns

*Amino acid substitution patterns* are provided in graphical format with local structural components (secondary structure elements (Fig. 3c, d), residue flexibility (Fig. 3c2), and solvent accessibility (Fig. 3c3)), molecular interactions (hydrophobic contacts, disulphide bonds, aromatic interactions, salt bridges, and hydrogen bonds), and residue mutability indicator (Fig. 3d). In this graphical format, the substitutions to different amino acid groups are represented by different colors, e.g., charged residues in pink,

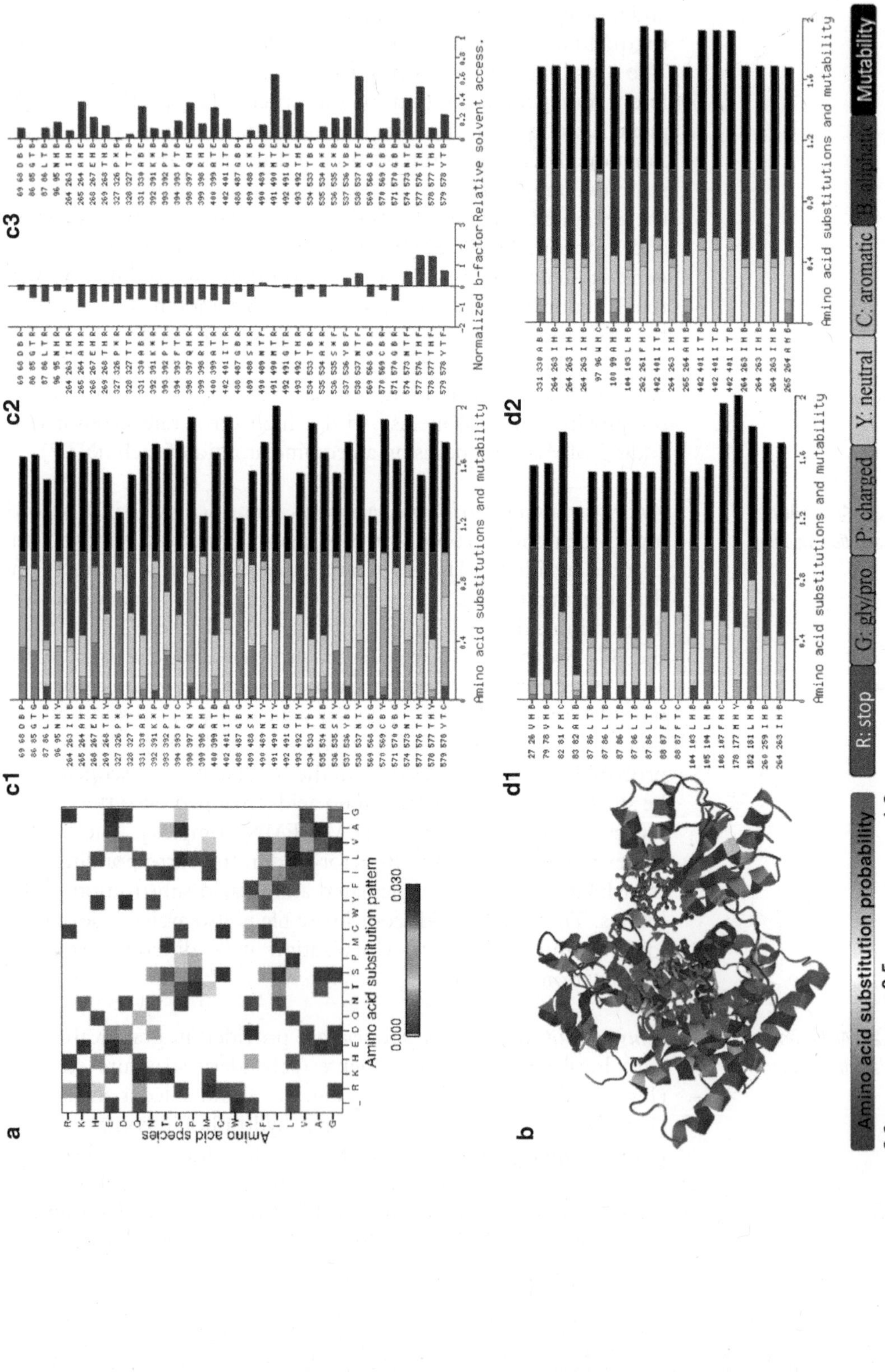
a
b
c1
c2
c3
d1
d2
Amino acid species
Amino acid substitution pattern
0.000
0.030
Amino acid substitution probability
0.0
0.5
1.0
Amino acid substitutions and mutability
Normalized b-factor
Relative solvent access.
R: stop
G: gly/pro
P: charged
Y: neutral
C: aromatic
B: aliphatic
Mutability

**Fig. 3** The MAP$^{2.0}$3D analysis for P450$_{BM3}$ using epPCR (Taq ($MnCl_2$, G = A = C = T)) method. All figures were reproduced from the output files generated by the program. (**a**) The generated amino acid diversity where the *Y*-axis shows the original amino acid species and the *X*-axis shows the amino acid substitution patterns indicated from *red* (lowest probability) to *blue* (highest probability). (**b**) The probability of substituting the original amino acid for a neutral amino acid projected on a cartoon representation of P450$_{BM3}$ with the heme- and FMN-binding residues shown in ball and stick format. The probability values increase from *blue* (lowest probability) to *red* (highest probability). (**c**) Amino acid substitution patterns (*pink* charged residues, *yellow* neutral residues, *cyan* aromatic residues, *blue* aliphatic residues, *red* stop codons, *green* Gly/Pro) for the heme- and FMN cofactor-binding residues in P450$_{BM3}$ further broken down in terms of (**c1**) residue mutability; (**c2**) residue flexibility; and (**c3**) relative solvent accessibility; of each of the original amino acids in the target protein. The *Y*-axis shows sequence id, PDB id, amino acid name, and the secondary structure elements in which the original residue is situated (*H* alpha helix, *B* beta bridge and extended strand, *T* hydrogen bonded turn and bend, *asterisk* loop or irregular structure). Further, in (**c1**) the original amino acid is classified according to the chemical property of the side chain (*P* charged, *Y* neutral, *C* aromatic, *B* aliphatic) with stop codon (R) and Gly/Pro (G) as separate classes. In (**c2**) the normalized Cα B-factor is listed to differentiate between flexible (F) and rigid (R) residues. In addition, in (**c3**) the relative solvent accessibility is indicated to identify exposed (E) or buried (B) residues. Finally in (**d**), the amino acid substitution pattern (*pink* charged residues, *yellow* neutral residues, *cyan* aromatic residues, *blue* aliphatic residues, *red* stop codons, *green* Gly/Pro) and the residue mutability (*black bar*) of heme-binding amino acids involved in hydrophobic interactions are shown in (**d1**) and those of their interacting partners in (**d2**) (e.g., the residue 87/86 interacts with residues 97/96, 100/99, 104/103, 262/261, and 402/401) (Color figure online)

neutral residues in yellow, aromatic residues in cyan, aliphatic residues in blue, stop codons in red, and Gly/Pro in green. The probability of one amino acid being exchanged to another is represented by the residue mutability indicator shown as a black bar after the amino acid substitution pattern. Figure 3c, d represents the amino acid substitution patterns and residue mutability indicator for the residues in the heme and FMN-binding sites in $P450_{BM3}$. Figure 3c2 and c3 correlates these indicators with the flexibility and relative solvent accessibility of the original residue in the target protein, respectively. Residue flexibility is estimated qualitatively using the normalized crystallographic B-factors (*see* ref. 7 for details). This indicator helps to identify the flexible and rigid regions in the target protein (*see* Fig. 3c2). The relative solvent accessibility shows which of the original residues are in surface exposed and buried regions in the target protein (*see* Fig. 3c3; *see* **Note 6**). The numerical results of these analyses are also available as downloadable text file.

### 3.4.3 Amino Acid Substitution Patterns with Molecular Interactions

*Amino acid substitution patterns with molecular interactions* provide information about the consequences of mutational bias on the internal environment of the target protein. Here, we examined those heme-binding residues that are involved in hydrophobic interactions in the original protein structure. We analyzed the amino acid substitution patterns of each interacting pair (i.e., selected residues (Fig. 3d1) and their partners (Fig. 3d2)) and correlated these patterns with their secondary structure elements and residue mutability indicators.

## 4 Notes

1. To specify a random mutagenesis method, the user is required to input the percentages of four transitions (A→G, T→C, G→A, C→T) and eight transversions (A→T, T→A, A→C, T→G, G→C, C→G, G→T, C→A). This data is either obtained from the published mutational spectrum of a random mutagenesis method or from DNA sequencing results of a random mutagenesis library. For a theoretically non-biased method, as an example, the percentage of each nucleotide substitution is 8.33 % (i.e., one in 12). Detailed information is provided in our previous papers [5, 7].
2. The MAP library contains 19 random mutagenesis methods. The mutational spectra of these methods were modified in two ways compared to the reported data. Firstly, insertion and

deletion mutations were neglected and the occurrence of the remaining nucleotide substitutions was scaled proportionally to 100 %. Secondly, mutations in the sense and antisense DNA strand resulting in equivalent substitutions (e.g., A → C (sense strand), T → G (antisense strand)) were considered to occur with equal frequency.

3. These indicators are defined as the probability with which (i.e., the fraction of protein variants in which) a specific amino acid substitution (e.g., S → G) occurs upon a single nucleotide substitution in a codon. These indicators are used to compare the performance of random mutagenesis methods.

4. The codon diversity coefficient is defined as the coefficient of variation calculated on the probabilities of generating single nucleotide substitutions in each codon. For each mutagenesis method, it is proportional to the ratio between the standard deviation of the probability of mutation of each codon along the protein sequence, divided by the corresponding average probability. In a method with a non-biased mutational spectrum, the codon diversity coefficient has a value of 0, i.e., each nucleotide and each codon substitution is considered to occur with equal frequency. For a full mathematical definition of the codon diversity coefficient and the other MAP indicators, *see* [7] and the manual page of the $MAP^{2.0}$3D server.

5. In MAP analysis, amino acids are grouped according to the chemical properties of their side chains: aliphatic (G, A, V, L, I; 5 out of 20), aromatic (F, Y, W; 3 out of 20), neutral (C, M, P, S, T, N, Q; 7 out of 20), and charged (D, E, H, K, R; 5 out of 20).

6. The relative solvent accessibility (RSA) is calculated by dividing the water accessible surface area of each residue in the protein (calculated using the DSSP program [15]) by the total surface area of the isolated amino acid [16]. A threshold value of 0.16 is used to differentiate between exposed ($RSA \geq 0.16$) or buried ($RSA < 0.16$) residues.

## Acknowledgements

The authors thank the European Union seventh framework program (project "OXYGREEN", Ref. 212281), Royal Society UK (RG110465), ChELSI and EPSRC (EP/E036252/1) for the financial support.

## References


1. Bornscheuer UT, Pohl M (2001) Improved biocatalysts by directed evolution and rational protein design. Curr Opin Chem Biol 5(2): 137–143
2. Brakmann S (2001) Discovery of superior enzymes by directed molecular evolution. Chembiochem 2(12):865–871. doi:10.1002/1439-7633(20011203)2:12<865::AID-CBIC865>3.0.CO;2-6
3. Wong TS, Roccatano D, Schwaneberg U (2007) Steering directed protein evolution: strategies to manage combinatorial complexity of mutant libraries. Environ Microbiol 9(11): 2645–2659. doi:10.1111/j.1462-2920.2007.01411.x
4. Shivange AV, Marienhagen J, Mundhada H, Schenk A, Schwaneberg U (2009) Advances in generating functional diversity for directed protein evolution. Curr Opin Chem Biol 13(1):19–25. doi:10.1016/j.cbpa.2009.01.019
5. Wong TS, Roccatano D, Zacharias M, Schwaneberg U (2006) A statistical analysis of random mutagenesis methods used for directed protein evolution. J Mol Biol 355(4):858–871. doi:10.1016/j.jmb.2005.10.082
6. Wong TS, Zhurina D, Schwaneberg U (2006) The diversity challenge in directed protein evolution. Comb Chem High Throughput Screen 9(4):271–288
7. Verma R, Schwaneberg U, Roccatano D (2012) MAP2.03D: A Sequence/Structure Based Server for Protein Engineering. ACS Synth Biol 1(4):139–150. doi:10.1021/sb200019x
8. Rasila TS, Pajunen MI, Savilahti H (2009) Critical evaluation of random mutagenesis by error-prone polymerase chain reaction protocols, Escherichia coli mutator strain, and hydroxylamine treatment. Anal Biochem 388(1):71–80. doi:10.1016/j.ab.2009.02.008
9. Ruff AJ, Marienhagen J, Verma R, Roccatano D, Genieser H-G, Niemann P, Shivange AV, Schwaneberg U (2012) dRTP and dPTP a complementary nucleotide couple for the Sequence Saturation Mutagenesis (SeSaM) method. J Mol Catal B Enzym 84:40–47
10. Wong TS, Roccatano D, Schwaneberg U (2007) Are transversion mutations better? A Mutagenesis Assistant Program analysis on P450 BM-3 heme domain. Biotechnol J 2(1):133–142. doi:10.1002/biot.200600201
11. Ruettinger RT, Wen LP, Fulco AJ (1989) Coding nucleotide, 5′ regulatory, and deduced amino acid sequences of P-450BM-3, a single peptide cytochrome P-450:NADPH-P-450 reductase from Bacillus megaterium. J Biol Chem 264(19):10987–10995
12. Sevrioukova IF, Li HY, Zhang H, Peterson JA, Poulos TL (1999) Structure of a cytochrome P450-redox partner electron-transfer complex. Proc Natl Acad Sci U S A 96(5):1863–1868
13. Berman HM, Westbrook J, Feng Z, Gilliland G, Bhat TN, Weissig H, Shindyalov IN, Bourne PE (2000) The Protein Data Bank. Nucleic Acids Res 28(1):235–242
14. Smith TF, Waterman MS (1981) Identification of common molecular subsequences. J Mol Biol 147(1):195–197
15. Kabsch W, Sander C (1983) Dictionary of protein secondary structure: pattern recognition of hydrogen-bonded and geometrical features. Biopolymers 22(12):2577–2637. doi:10.1002/bip.360221211
16. Chothia C (1976) The nature of the accessible and buried surfaces in proteins. J Mol Biol 105(1):1–12

# Chapter 20

# Computational Tools for Designing Smart Libraries

**Eva Sebestova, Jaroslav Bendl, Jan Brezovsky, and Jiri Damborsky**

## Abstract

Traditional directed evolution experiments are often time-, labor- and cost-intensive because they involve repeated rounds of random mutagenesis and the selection or screening of large mutant libraries. The efficiency of directed evolution experiments can be significantly improved by targeting mutagenesis to a limited number of hot-spot positions and/or selecting a limited set of substitutions. The design of such "smart" libraries can be greatly facilitated by in silico analyses and predictions. Here we provide an overview of computational tools applicable for (a) the identification of hot-spots for engineering enzyme properties, and (b) the evaluation of predicted hot-spots and selection of suitable amino acids for substitutions. The selected tools do not require any specific expertise and can easily be implemented by the wider scientific community.

**Key words** Computational tool, Rational design, Smart library, Focused library, Directed evolution, Mutation, Protein engineering, Software, Web tool

## 1 Introduction

Directed evolution is a widely used approach for modification of protein structure and function. Traditionally, directed evolution involves the iterative introduction of random mutations in a target gene, followed by selection or high-throughput screening of the resulting mutant libraries for protein variants with improved properties. The potential disadvantage of this entirely random approach is that very large libraries usually need to be screened to obtain desired variants, which is time-consuming, laborious, and costly [1–3].

The efficiency of directed evolution can be increased by creating smaller, higher quality libraries that are more likely to yield positive results. Such "smart" or "focused" libraries can be generated by restricting the amino acid changes to only a limited number of target positions, which are likely to affect the property under study, and/or selecting a limited set of substitutions [1, 3–6]. These promising target sites, often referred to as "hot-spots," are usually identified in a rational way, utilizing information on protein

Elizabeth M.J. Gillam et al. (eds.), *Directed Evolution Library Creation: Methods and Protocols*, Methods in Molecular Biology, vol. 1179, DOI 10.1007/978-1-4939-1053-3_20, © Springer Science+Business Media New York 2014

sequence-structure-function relationships or by using computational methods [2, 3, 5]. Compared to conventional random mutagenesis libraries, the focus on specific amino acid positions will significantly reduce the size of generated libraries [1, 3–6] and the screening efficiency can be further improved by the use of appropriate degenerate codons. Degenerate codons can be used to restrict mutations at each target position to a rationally selected set of amino acids, or a reduced set of diverse amino acids [4, 6, 7].

Here, we provide an overview of easy-to-use computational tools that can be utilized to (a) identify typical hot-spots for engineering protein stability and enzyme catalytic properties, (b) evaluate selected hot-spots, and (c) design suitable sets of amino acids to be introduced into these positions (Table 1). We note that although

**Table 1**
**Selected computational tools applicable for the design of smart libraries**

| Computational tool | Aim[a] | URL | References |
|---|---|---|---|
| 3DM | E | http://3dmcsis.systemsbiology.nl/ | [74] |
| 3D-SURFER | C | http://dragon.bio.purdue.edu/3d-surfer/ | [37] |
| AA-Calculator | D | http://guinevere.otago.ac.nz/cgi-bin/aef/AA-Calculator.pl | [87] |
| B-FITTER | S | http://www.kofo.mpg.de/en/research/organic-synthesis/ | [52] |
| CASTER | D | http://www.kofo.mpg.de/en/research/organic-synthesis/ | [52] |
| CASTp | C | http://cast.engr.uic.edu/ | [31] |
| CAVER | C, S | http://www.caver.cz/ | [42] |
| ConSurf | E | http://consurf.tau.ac.il/ | [43] |
| ConSurf-DB | E | http://consurfdb.tau.ac.il/ | [72] |
| CUPSAT | E | http://cupsat.tu-bs.de/ | [77] |
| Dmutant | E | http://sparks.informatics.iupui.edu/hzhou/mutation.html | [76] |
| Evolutionary Trace | E | http://mammoth.bcm.tmc.edu/ETserver.html | [65] |
| Evolutionary Trace Database | E | http://mammoth.bcm.tmc.edu/cgi-bin/report_maker_ls/traceServerResults.pl?identifier=reports | [66] |
| FoldX | S, E | http://foldx.crg.es/ | [60] |
| Fpocket | C | http://fpocket.sourceforge.net/ | [36] |
| GLUE | D | http://guinevere.otago.ac.nz/cgi-bin/aef/glue.pl | [88] |
| GLUE-IT | D | http://guinevere.otago.ac.nz/cgi-bin/aef/glue-IT.pl | [87] |
| HotSpot Wizard | C, S, E | http://loschmidt.chemi.muni.cz/hotspotwizard/ | [39] |

(continued)

**Table 1**
**(continued)**

| Computational tool | Aim[a] | URL | References |
|---|---|---|---|
| HSSP | E | http://swift.cmbi.ru.nl/gv/hssp/ | [73] |
| I-Mutant | E | http://folding.uib.es/i-mutant/i-mutant2.0.html | [78] |
| LigPlot+ | C | http://www.ebi.ac.uk/thornton-srv/software/LigPlus/ | [19] |
| LPC | C | http://bip.weizmann.ac.il/oca-bin/lpccsu/ | [21] |
| MBLOSUM | E | http://apps.cbu.uib.no/mblosum/ | [67] |
| metaPocket | C | http://projects.biotec.tu-dresden.de/metapocket/ | [32] |
| MolAxis | C, S | http://bioinfo3d.cs.tau.ac.il/MolAxis/ | [48] |
| MOLE | C, S | http://mole.upol.cz/online/ | [47] |
| PANTHER | E | http://www.pantherdb.org/tools/csnpScoreForm.jsp | [84] |
| PDBePISA | C | http://www.ebi.ac.uk/msd-srv/prot_int/pistart.html | [22] |
| Pocket-Finder | C | http://www.modelling.leeds.ac.uk/pocketfinder/ | [33, 34] |
| PoPMuSiC | S, E | http://babylone.ulb.ac.be/popmusic/ | [58] |
| PoseView | C | http://poseview.zbh.uni-hamburg.de/ | [20] |
| PROVEAN | E | http://provean.jcvi.org/ | [83] |
| Q-SiteFinder | C | http://www.modelling.leeds.ac.uk/qsitefinder/ | [34] |
| Scorecons | E | http://www.ebi.ac.uk/thornton-srv/databases/cgi-bin/valdar/scorecons_server.pl | [63] |
| SIFT | E | http://sift.jcvi.org/ | [82] |
| SITEHOUND | C | http://scbx.mssm.edu/sitehound/sitehound-web/Input.html | [35] |
| TopLib | D | http://stat.haifa.ac.il/~yuval/toplib/ | [86] |
| WebLogo | E | http://weblogo.berkeley.edu/ | [68] |

[a]C—identification of hot-spots for engineering catalytic properties; S—identification of hot-spots for engineering thermostability; E—evaluation of hot-spots and design of amino acid sets; D—design of library diversity

any method used in rational protein design is, in principle, applicable to the design of smart libraries, we have highlighted only the methods that are implemented and provided to the community as web servers or programs equipped with appropriate graphical user interfaces, and requiring no specific expertise to setup calculations or interpret results. The presented tools can be used by research teams that do not have prior knowledge of structural and bioinformatic analyses, as well as by experts. For other methods and strategies potentially useful in designing smart libraries, readers are referred to previously published reviews [2–5, 8–14].

## 2 Identification of Hot-Spots for Engineering Catalytic Properties

The amino acid residues that mediate substrate binding, transition-state stabilization, or product release are frequently selected as hot-spots for modification of enzyme specificity, stereoselectivity, or activity [6, 15–18]. The easiest way to determine some of these hot-spots is to analyze the interactions between enzyme and bound ligands, or identify residues located in binding pockets or access tunnels. However, it is very important to keep in mind that the hot-spots identified using most methods in this section are also likely to include catalytic or other essential residues, which generally should not be mutated.

### 2.1 Analysis of Protein–Ligand Interactions

The amino acid residues in direct contact with the ligand can be identified from the structure of enzyme–ligand complex. Computational tools facilitating the investigation of protein–ligand interactions make this kind of analysis accessible even to researchers inexperienced in structural analyses. These tools include LigPlot+ [19] or PoseView [20] for generating easy-to-interpret 2D diagrams of protein–ligand interactions, as well as LPC [21] or PDBePISA [22], which provide information about interatomic contacts between the protein and bound ligands. These methods require reliable 3D coordinates of the enzyme–ligand complex (*see* **Note 1**). However the experimental structure of the target complex is often not available. In such cases, one can predict the probable binding orientation of the ligand in the active site by molecular docking [23–25].

*LigPlot+* (http://www.ebi.ac.uk/thornton-srv/software/LigPlus) is a program that generates schematic diagrams of the interactions between the ligand and individual protein residues from the 3D coordinates of the protein–ligand complex [19]. All potential hydrogen bonds and nonbonded contacts between the protein and the bound ligand are calculated by the HBPLUS program [26]. The ligand and interacting protein residues are then flattened out into a 2D diagram, while minimizing atom clashes and overlaps. The program is provided with a graphical user interface. Users specify the structure of a protein–ligand complex as the input, and choose the ligands for which interactions should be depicted. The program enables the user to plot and superimpose multiple sets of protein–ligand interactions, and thus highlights differences and similarities in the binding of various ligands to the same protein target, or in the binding of a single ligand to a set of related proteins. The interactions can also be inspected in 3D by launching the results in PyMOL (http://www.pymol.org) or RasMol (http://rasmol.org/) visualization software, automatically showing the superimposed ligands, interacting residues, hydrogen bonds, and nonbonded contacts [19].

*LPC* web server (http://ligin.weizmann.ac.il/cgi-bin/lpccsu/LpcCsu.cgi) analyzes protein–ligand contacts using the surface complementarity approach [21, 27]. Users specify the structure of the protein–ligand complex and select the ligand for which the analysis should be conducted. The server outputs a list of residues and atoms in contact with the ligand and indicates the type of contacts and the distance and contact surface area between the ligand and individual residues. The listed residues and ligand can be interactively visualized in the web browser [21].

*PDBePISA* web server (http://www.ebi.ac.uk/msd-srv/prot_int/pistart.html) enables the analysis of macromolecular interfaces, surfaces, and assemblies predicted from protein structures. Users can either search the PISA database for pre-calculated results, or request a new calculation by uploading their own structure of a macromolecular complex. After launching the interface analysis, all interfaces found in the query structure are listed, including interfaces between the ligand and macromolecule. Selected interfaces can be visualized in the web browser. For each interface, detailed information is provided, including overall characteristics of the interface (number of atoms and residues, solvent accessible area, solvation energy), a list of all interface atoms and residues, and their potential interactions (i.e., hydrogen/disulfide bonds, salt bridges, or covalent links) [22].

### 2.2 Analysis of Binding Pockets

The binding and active sites of enzymes are often associated with structural pockets and cavities [28], both hereafter referred to as pockets. Many amino acid residues located in these pockets may come into contact with the ligands during the catalytic cycle. Thus, even without precise knowledge of ligand orientation in the active site, one can accurately predict which residues may interact with the ligand just by identifying the active site pocket. Many tools for the identification and analysis of pockets have been developed [28–30]. Some of these tools, e.g., CASTp [31], MetaPocket 2.0 [32], Pocket-Finder [33, 34], Q-SiteFinder [34], SITEHOUND [35], fpocket [36], or 3D-SURFER [37], are available as easy-to-use web servers with minimal user demands and interactive visualization of results. Importantly, all these servers delineate residues forming each identified pocket. Users should be able to recognize the active site pocket as long as they know the approximate location of the active site (e.g., 1–2 catalytic residues), and thus easily obtain a list of suitable targets for saturation mutagenesis (*see* **Note 2**).

*CASTp* web server (http://cast.engr.uic.edu) provides a comprehensive and detailed characterization of all pockets in the protein structure [31, 38]. The identification of pockets is based on computational geometry methods and the volume and area of each pocket is measured analytically [38]. It is possible to access either the pre-calculated data from a database, or request a new calculation. Users only have to specify the target structure in the input and can optionally adjust the radius of the solvent probe used for the delineation of internal pockets. Calculated pockets are interactively visualized in the web browser or can be loaded in PyMOL software. Additional outputs include information about the volume and area of individual pockets, a list of their lining atoms and residues, and the annotations of biologically important residues derived from various databases (if they are available) [31]. Based on the data provided, users can easily select residues for saturation mutagenesis—they can either focus on all residues forming the target pocket or select only the residues that contribute several side-chain atoms to the pocket.

*HotSpot Wizard* web server (http://loschmidt.chemi.muni.cz/hotspotwizard) uses CASTp as one of the computational tools for automatic identification of hot-spots for engineering enzyme catalytic properties. The web server integrates the structural and evolutionary information obtained from several bioinformatic databases and tools. Users need to specify the protein structure for the input, and optionally adjust other parameters [39]. HotSpot Wizard then uses the Catalytic Site Atlas [40] and UniProtKB [41] databases to determine the residues indispensable for enzyme function. In the next step, the CASTp [31] and CAVER [42] tools identify the residues located in the active site pockets and tunnels. The ConSurf method [43], with a slightly modified protocol for sequence dataset selection, is utilized to estimate the evolutionary conservation of individual amino acid positions, which is then used to assign mutability to individual residues of the query enzyme [39]. In the output, HotSpot Wizard reports predicted hot-spots, as well as other functional residues. All residues are accompanied with information about their estimated mutability, potential structural and functional importance, available mutagenesis data, and amino acids occurring at a given position in related proteins. Results are mapped on the enzyme structure and can be visualized either directly in the web browser or in PyMOL software using the provided script. Altogether, HotSpot Wizard enables the user to perform several structural and evolutionary analyses at once, with minimal demands [39]. For a detailed description of HotSpot Wizard usage and interpretation of results, see the protocol at the end of this chapter.

*MetaPocket 2.0* web server (http://projects.biotec.tu-dresden.de/metapocket/) predicts ligand binding sites and binding residues from protein 3D structures. It is a consensus method combining results from eight pocket detection programs [32, 44]. Users specify a protein structure as the input and, optionally, the number of pockets to be found (*see* **Note 3**). The server outputs the predicted meta-pocket sites, as well as pocket sites predicted by each individual method. The pockets can be visualized directly in the web browser, or loaded in PyMOL software using the provided scripts. For each meta-pocket, the potential ligand binding residues are listed [32]. While the consensus prediction improves the success rate compared to individual prediction methods, the potential disadvantage of this server is that it only utilizes the top three pocket sites from each method. Therefore, if the active site pocket does not rank within the top three pockets in any of the integrated methods, it will not be reported by MetaPocket 2.0.

### 2.3 Analysis of Access Tunnels

In many enzymes, the active site is buried inside the protein core. The exchange of ligands between the buried active site and bulk solvent (or between two buried active sites in multifunctional enzymes) is mediated by access tunnels. Several studies have shown that substitutions modulating the shape, physicochemical properties, or dynamics of the access tunnels can lead to significant changes in enzyme activity, substrate specificity, and stereoselectivity, reflecting the importance of ligand entrance and release for enzymatic catalysis [45]. The residues located in access tunnels therefore represent hot-spots for the modulation of catalytic properties in enzymes with buried active sites.

The tunnel-forming residues can easily be obtained by analyzing protein structure using the available computational tools for the identification of access tunnels [46]. The tools suitable for the analysis of access tunnels include CAVER 3.0 [42], MOLE 2.0 [47], and MolAxis [48]. All these tools are based on computational geometry methods and search the target protein structure for possible pathways connecting the starting point inside the protein with the bulk solvent. The starting point is typically defined by several atoms or residues located within the active site and should be placed into a free space in the active site pocket (*see* **Note 4**). Prior to calculation, all unnecessary ligands and water molecules should be removed. As the main output, the tools provide geometry of all tunnels with the radius wider than the radius specified by the user (*see* **Note 5**). All identified tunnels are accompanied with information regarding tunnel characteristics, tunnel profiles, a list of tunnel-lining residues, and residues forming the tunnel bottleneck, i.e., the narrowest part of the access tunnel [46] (*see* **Note 6** and 7). The bottleneck residues represent particularly promising hot-spots for the modification of tunnel geometry and their substitutions tend to have a high impact on the function of the tunnel [45].

Identification of tunnels by geometry-based tools is very fast, but the disadvantage of this approach is that usually only a single structure is analyzed and thus enzyme dynamics are neglected. Only the tunnels that are open in the analyzed enzyme structure are identified, while temporarily closed tunnels will be missed by this analysis. Moreover, it is sometimes difficult to distinguish biologically relevant tunnels based on the analysis of a single static structure [42]. These problems can be overcome by analyzing the tunnels in a large ensemble of protein conformations, which can be obtained using molecular dynamics simulations [49, 50].

However, the setup, computing, and analysis of molecular dynamics simulations require a lot of time and expertise (this technique is beyond the scope of this chapter).

*CAVER 3.0* (http://www.caver.cz/) is a widely used tool for the identification and characterization of access pathways in static and dynamic macromolecular structures. It is available as a PyMOL plug-in for the analysis of single static structures and as a command-line application for the analysis of both static and dynamic structures. The plug-in provides a graphical interface for setting up the calculation and interactive visualization of results. The starting point position can be specified by x, y, and z coordinates, or derived from any PyMOL selection. The calculation can be performed for any structure loaded into PyMOL and users can thus easily analyze and compare tunnels from multiple superimposed structures. The residues and atoms forming individual tunnels and bottleneck residues are provided in text files and can be visualized using the provided script [42]. When selecting suitable hot-spots for saturation mutagenesis, users can choose to focus on all residues forming relevant tunnels, select residues that contribute by several side-chain atoms to the tunnel, or focus on bottleneck residues. The calculations can be conveniently run and the results visualized using the user-friendly graphical interface CAVER Analyst. CAVER is also integrated into the *HotSpot Wizard* web server (described in Subheading 2.2).

*MOLE 2.0* web server (http://mole.upol.cz/online/) enables the identification and interactive analysis of access pathways in a single macromolecular structure. For the input, users have to specify the target structure and can optionally adjust calculation parameters. The calculation can be run automatically from multiple internal pockets detected by the MOLE 2.0 algorithm. While this can provide a useful overview of possible pathways, users are advised to check carefully that the calculation was also performed from their pocket of interest. Alternatively, the starting point can be specified by x, y, and z coordinates or by a set of residues [47]. Users can choose the active site residues which are automatically derived from the Catalytic Site Atlas database when working with known and annotated enzyme structures; otherwise users must specify or interactively select residues from the protein sequence. Identified tunnels are visualized in the web browser. Alternatively, results can be downloaded and visualized with PyMOL software. Tunnel-lining residues are depicted along the tunnel centerline with bottlenecks and local minima indicated, and can be highlighted in the structure [47].

## 3 Identification of Hot-Spots for Engineering Thermostability

Thermostability is a very important property that often determines the applicability of proteins in technologies and products [51]. Protein thermostability can be improved by introducing rigidifying mutations at highly flexible positions. Such amino acid positions can be predicted based on crystallographic B-factors, as these reflect the degree of thermal motion, and thus the flexibility of individual residues [52, 53]. A quick calculation of residue B-factors can be done using *B-FITTER* (http://www.kofo.mpg.de/en/research/organic-synthesis). This program calculates the average B-factor of each amino acid residue in the target protein by averaging the B-factors of all atoms of this residue, excluding hydrogen. For the input, users provide the crystal structure of the protein of interest—preferably a high-resolution crystal structure with a minimum number of non-resolved atoms and without any external molecules bound. For the output, residues are ranked according to their average B-factors—the residues with the highest B-factors are good targets for saturation mutagenesis [52, 53].

Residues forming protein access tunnels represent another type of potential hot-spots to enhance protein thermostability—computational predictions indicate that saturation mutagenesis targeting the tunnel residues has approximately two times better chance to produce protein variants with significantly improved stability than mutagenesis in other protein regions [54]. The computational tools aiding in the identification of tunnels and tunnel-lining residues—CAVER, MOLE, MolAxis, and Hotspot Wizard—have already been described with respect to their usage in the identification of hot-spots for engineering of catalytic properties.

A large number of computational tools for the prediction of effect of amino acid substitutions on protein stability have been developed [55–57]. Two of these tools—PoPMuSiC [58, 59] and FoldX [60]—can be easily utilized for systematic in silico saturation mutagenesis, which can help reveal amino acid positions with a potential for stabilization. Predictions are made by replacing each residue in the protein structure by all other possible amino acids and estimating the change in protein stability ($\Delta\Delta G$) upon mutation. In both FoldX and PoPMuSiC, a negative $\Delta\Delta G$ value corresponds to predicted stabilizing mutations. Considering the calculation errors of computational methods, we recommend dividing the mutations into at least three categories—stabilizing (predicted $\Delta\Delta G < -1$ kcal/mol), neutral ($-1$ kcal/mol $\leq$ predicted $\Delta\Delta G < 1$ kcal/mol), and destabilizing (predicted $\Delta\Delta G \geq 1$ kcal/mol)—to minimize the number of destabilizing mutations classified as stabilizing and vice versa. The positions with a high proportion of stabilizing mutations and/or low proportion of destabilizing mutations are good candidates for randomization by saturation mutagenesis. It is important to keep in mind that some hot-spots for protein stabilization may at the same time be indispensable for protein function, e.g., catalytic or binding residues, and thus should not be mutated unless the risk of destroying protein function is not relevant for the study under consideration.

*PoPMuSiC 2.1* web server (http://babylone.ulb.ac.be/popmusic/) enables very fast predictions of the stability changes caused by single-point mutations in globular proteins. The estimation of the folding free energy change upon mutation is calculated using a linear combination of 13 statistical potentials, two volume-dependent terms, and one independent term, the coefficients of which are optimized via a neural network and depend on solvent accessibility of the mutated residue [58, 59]. The systematic scan of all possible single-point mutations can be launched by simply specifying the target structure. The outputs from the systematic scan include a file with predicted ΔΔGs for all mutations. The potential hot-spots for stabilization can be selected based on the so-called sequence optimality score (i.e., the sum of all negative ΔΔGs at a given position) provided in a text file and as a bar chart. A low sequence optimality score indicates that the corresponding position is poorly optimized and thus represents a promising target to improve stability of a given protein [58].

*FoldX 3.0* (http://foldx.crg.es/) is a widely used program designed to estimate the effect of mutations on the stability of proteins, nucleic acids, and macromolecular complexes. The free energy is calculated using linear combinations of different energy terms weighted to fit experimental data [60, 61]. While the less demanding FoldX analyses can be run via a graphical user interface provided by a FoldX plug-in for YASARA software (www.yasara.org) [62], the systematic mutagenesis scan can only be efficiently performed by using the stand-alone command line application. For the input, users have to provide a high-resolution structure of the target protein, list the mutations to be evaluated by FoldX, and set up the calculation. Prior to analysis, the input structure should be repaired using the FoldX RepairPDB module. The BuildModel module can be then used to predict the free energy change resulting from each specified mutation (*see* **Note 8**). The output provides the difference in the free energy between each mutant and the wild-type structure decomposed into the FoldX energy terms [61]. The hot-spots for saturation mutagenesis can be selected, for example, as sites with the best ratio between the proportion of highly stabilizing and destabilizing mutations.

## 4 Evaluation of Hot-Spots and Selection of Substitutions

### 4.1 Analysis of Evolutionary Conservation

The hot-spots identified by the computational tools described in previous sections can be further evaluated to prevent the replacement of indispensable amino acid residues and to prioritize the hot-spots, i.e., order the hot-spots based on their suitability for mutagenesis. Analysis of the evolutionary conservation of individual amino acid positions within a group of related proteins can be used to predict highly important residues, as well as estimate mutability, i.e., tolerance to substitutions, of individual amino acid positions. Amino acid residues that are essential for the maintenance of the structural or functional properties of a protein tend to be conserved during evolution [63, 64]. If a researcher wants to generate a library with a high proportion of correctly folded and active variants, conserved residues are generally not recommended as suitable targets for mutagenesis because their replacement often leads to the loss of protein function (*see* **Note 9**). On the other hand, it has been shown that mutagenesis targeting highly mutable positions provides a significantly higher proportion of viable variants than random mutagenesis [39]. Targeting moderately or highly variable positions, which are expected to be tolerant to a wide range of substitutions, thus represents a good approach for producing efficient smart libraries [18, 39]. Alternatively, the conservation level can be used only to discard highly conserved residues from the design, e.g., catalytic residues, and the remaining hot-spots can be prioritized based on other criteria.

Many different methods for scoring residue conservation, variability, or mutability have been proposed [63, 64]. The available web servers include ConSurf [43], HotSpot Wizard, Evolutionary Trace [65, 66], Scorecons [63], MBLOSUM [67], or WebLogo [68]. Residue conservation is calculated from a multiple alignment of a set of related protein sequences (*see* **Note 10**). The ConSurf, Evolutionary Trace, and HotSpot Wizard tools can automatically perform database searches for sequence homologs of the user-provided protein, select a suitable sequence dataset, and construct the multiple sequence alignment. Nonetheless, advanced users can still upload their own multiple sequence alignments. In addition to conservation score, these tools also provide a list of residues occurring at individual positions in the analyzed protein sequences. This information can be used to limit substitutions at each amino acid site to those existing in known natural proteins, likely increasing the proportion of viable variants in the resulting library, while decreasing its size [1, 69].

*ConSurf* web server (http://consurf.tau.ac.il/) calculates the level of evolutionary conservation of individual amino acid positions in proteins and nucleic acids [43]. The conservation scores are derived from site-specific evolutionary rates, which are computed for each amino acid position by Rate4Site program using empirical Bayesian or maximum-likelihood algorithms [70, 71]. For the input, users should ideally provide the structure of a target protein; however the calculation can also be conducted by submitting only the protein sequence. Optionally, users can also upload their own multiple sequence alignment and phylogenetic tree, or adjust parameters for their calculation—they can choose the database and algorithm for sequence searching, criteria for inclusion of sequences to the dataset, alignment method, substitution model, or algorithm for the estimation of evolutionary rates.

The normalized conservation scores are mapped onto the query structure and visualized in the web browser. Other essential outputs include scripts for the visualization of results in different structure visualization programs, the sequence alignment colored by ConSurf conservation scores, and files providing the normalized conservation scores and frequencies of individual residues at each position of the multiple sequence alignment [43].

*HotSpot Wizard* web server (also discussed in the Subheading 2.2) implements a slightly modified ConSurf method for calculation of conservation grades, which are in turn converted into mutability grades. To prevent mutagenesis of residues indispensable for protein function, catalytic residues are assigned as immutable. All functional residues, i.e., residues located in the active site pocket or access tunnels, with high mutability are assigned as hot-spots. The hot-spots are prioritized based on their mutability, starting with the most mutable. If users want to select or prioritize hot-spots based on other criteria than mutability, they are advised to use the list of annotated functional residues [39]. Based on the frequencies of individual residues at each individual position, users can estimate which substitutions are expected to be well tolerated.

Pre-calculated conservation scores for proteins of publically known structures can be obtained from specialized databases, e.g., ConSurf-DB [72], Evolutionary Trace [66], HSSP [73], or 3DM [74].

*3DM* database (http://3dmcsis.systemsbiology.nl/) is a commercial expert system which enables effective exploration of protein sequence-structure-function relationships. It provides platforms for the collection, combination, and integration of a wide variety of protein-related data from different sources. A structure-based multiple sequence alignment is available for each protein superfamily within the database, and represents one of the possible access points to the underlying data. The integrated data include sequences, 3D structures, active site positions, mutation data extracted from the literature, protein–protein and protein–ligand contacts, conservation patterns, correlated mutations, etc. [74]. Users can easily obtain information about highly conserved positions, correlated mutations, or amino acid distributions at individual positions to utilize for the design of smart libraries. 3DM also offers the possibility to predict the effect of individual mutations at a selected position based on superfamily alignment statistics [75]. Protein structures and homology models from the superfamily together with selected 3DM data can be visualized in YASARA software [74].

### 4.2 Analysis of the Effects of Mutations on Protein Stability

In addition to the analysis of conservation data, identified hot-spots can be subjected to in silico site-saturation mutagenesis to check if mutations at a given site are likely to be tolerated. This task can be performed using computational tools for the prediction of the effects of amino acid substitutions on protein stability or protein function.

The computational tools for the calculation of stability changes upon mutation have already been discussed with respect to their usage in the identification of hot-spots. Both PoPMuSiC and FoldX can also be employed for the evaluation and prioritization of hot-spots, provided that the hot-spots were predicted by another approach, not by these tools. Besides these two methods, computational tools enabling in silico site-saturation mutagenesis, i.e., mutation of a single selected residue to all possible amino acids, can also be used. These include several web servers, e.g., Dmutant [76], CUPSAT [77], or I-Mutant2.0 [78]. The input data for these calculations consist of the structure of the target protein and the position to be saturated. Outputs include the predicted free energy changes for all possible substitutions at a given position (*see* **Note 11**). Many highly destabilizing mutations predicted for a certain position indicate that this site is not a very good target for randomization by saturation mutagenesis. On the other hand, if only a few highly destabilizing mutations are predicted, such sites can still represent promising hot-spots—the amino acids with potentially destabilizing effects can be discarded from the library by the appropriate selection of degenerate codons. Alternatively, stability predictions can be used to prioritize hot-spots, e.g., by favoring the positions where no, or a minimum number of, highly destabilizing mutations were predicted.

### 4.3 Analysis of the Effects of Mutations on Protein Function

Computational tools for the prediction of the effect of mutations on protein function can be used in a similar manner as those calculating stability. Each identified hot-spot is mutated to all possible amino acids and the effect of each substitution estimated. Positions with a high proportion of deleterious substitutions should be omitted from the design, or the results can be used as an indicator of "allowed" or "forbidden" amino acid replacements at a given position, or for prioritization of hot-spots.

A large number of computational tools for the prediction of mutation effects on protein function have been developed [79–81]. It is important to note that the applicability of some of these tools for the evaluation of protein engineering hot-spots can be limited, since they were primarily developed to reveal disease-associated mutations. Web servers that are potentially useful for the prediction of mutation effects on protein function include SIFT [82], PROVEAN [83], and PANTHER [84], all of which enable in silico site-saturation mutagenesis.

*SIFT* web server (http://sift.jcvi.org/) predicts the effects of all possible single-point mutations on protein function based on a multiple sequence alignment of related proteins. For the input, users are only required to provide the target protein sequence. SIFT automatically performs a database search to collect the sequence dataset and constructs the multiple sequence alignment. Optionally, users can upload their own sequence dataset or multiple sequence alignment, or specify criteria for inclusion of sequences into the dataset. For each possible substitution, SIFT calculates a normalized probability that this substitution will be tolerated. The server outputs an overview of tolerated and deleterious substitutions depicted along the protein sequence, and the matrix of SIFT scores containing predictions for all possible substitutions in a given protein. SIFT also provides a measure of confidence of the prediction to indicate potentially problematic classifications [82, 85].

*PROVEAN* web server (http://provean.jcvi.org/) enables users to predict the functional effects of single amino acid substitutions as well as multiple amino acid substitutions, in-frame insertions and deletions. For the input, users provide the target protein sequence and specify amino acid variants of interest. PROVEAN performs a database search and sequence clustering to identify sequence homologs of the query. The change in pairwise sequence similarity between the query protein and each homolog before and after the introduction of mutation is calculated. The obtained scores are averaged to generate a final PROVEAN score, which is used to distinguish deleterious and neutral mutations. For each mutation, a PROVEAN score and prediction is provided in the output [83].

## 5 Design of Library Diversity

After the evaluation and prioritization of identified hot-spots, a final library can be designed. One has to decide how many and which positions will be targeted, whether positions will be randomized simultaneously or separately, as well as whether to introduce all, or only a reduced set of amino acids at individual positions. These decisions dramatically affect the size of the resulting library and should therefore be made rationally and with great care. The design of appropriate degenerate codons and calculation of the number of variants that will have to be screened for covering a significant fraction of the generated library are facilitated by tools such as TopLib [86], CASTER [52], AA-Calculator [87], GLUE [88], or GLUE-IT [87] and the reader is referred to the review of these methods in Chapter 18 of this book.

## 6 Protocol for Identification of Hot-Spots by HotSpot Wizard

### 6.1 Calculation Setup

1. Go to the HotSpot Wizard calculation page at http://loschmidt.chemi.muni.cz/hotspotwizard.
2. In the "Query submission" form, specify the PDB-ID code of the target structure or upload the PDB file from your computer.

In this example, we used the structure of haloalkane dehalogenase LinB from *Sphingobium japonicum* (formerly *Sphingomonas paucimobilis*) UT26, complexed with the reaction product 2 bromo 2 propene 1 ol (PDB ID code 1K63).

3. Optionally, specify chains for which the calculation should be performed (*see* **Note 12**). This is not relevant for the example LinB structure, which contains only one chain.
4. Specify your e-mail address for receiving the notification about finished calculation.
5. Check "Keep results private" if you want to prevent other users from seeing your results.
6. Optionally, go to the "Advanced settings" form to manually set catalytic residues, specify parameters for conservation analysis, upload pre-calculated multiple sequence alignment or phylogenetic tree, or set up calculation of access tunnels. In this example, we increased the maximum number of homologs used for the calculation of conservation from the default value of 50 to 100 (*see* **Note 13**).
7. Submit the calculation.

### 6.2 Interpretation of Results

1. The "Job" panel appears once the calculation is finished. If you receive a warning message about a failure in the identification of catalytic residues, we recommend you resubmit the calculation and manually specify some of the functional residues in "Advanced settings" (*see* **Note 14**). If you do not know any functionally important residues, you may still proceed to analysis of the results (*see* **Note 15**).
2. Open the "Results browser" by clicking on the "View hot spot sites" link in the "Job" panel (*see* **Note 16**).
3. Explore the annotated protein structure with highlighted hot-spots and other functional residues (Fig. 1a) (*see* **Note 17**). Hot-spots are colored with red, orange, dark yellow, and yellow colors (corresponding to mutability grades 9, 8, 7, and 6, respectively). Expand the "Color coding" panel on the left to get the full mutability color scale. Basic operations for manipulation of the structure are accessible through the "Controls" panel on the left.
4. Expand the "Tunnels" panel on the left to check and visualize identified tunnels (*see* **Note 18**). One tunnel with a bottleneck radius of 1.5 Å was identified in our exemplary LinB structure (*see* **Note 19**).
5. Go to the "Mutagenesis Hot Spots" panel to explore identified hot-spots (*see* **Note 20**). The following information is provided for each residue: information about its mutability, potential function (catalytic, located in the active site pocket or in the access tunnel), experimental mutagenesis data, existing amino

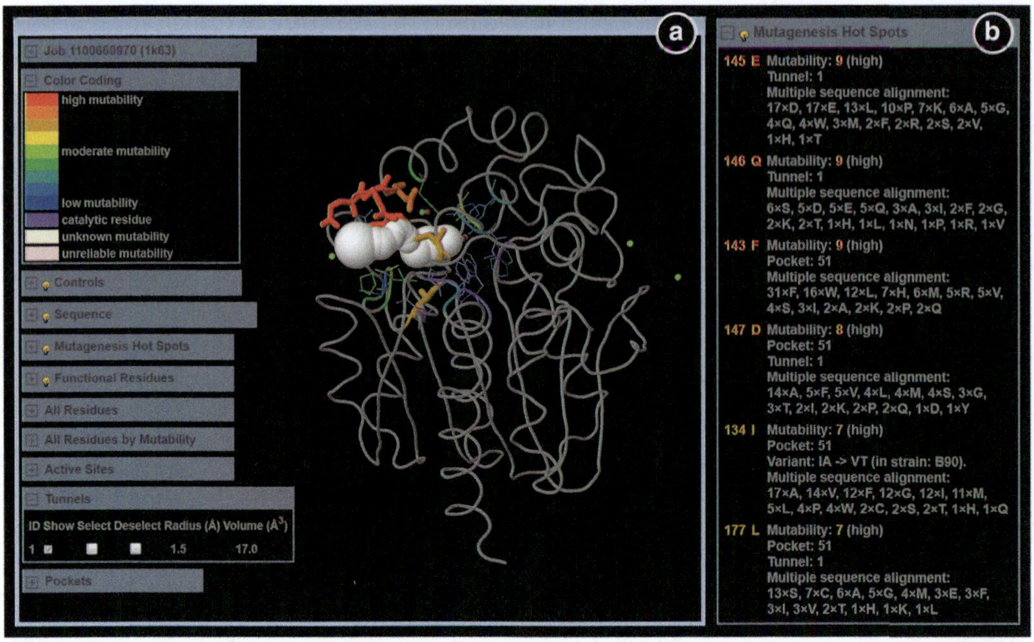


**Fig. 1** Graphical user interface of HotSpot Wizard results. (**a**) The "Results Browser" includes the embedded Jmol applet visualizing the annotated input structure. The hot-spot residues identified in the structure of halo alkane dehalogenase LinB (PDB-ID 1K63) are shown as sticks, other functional residues are shown as lines, and the tunnel is visualized by spheres. Residues are colored according to their estimated mutability. The "Results Browser" also provides the "Job" panel facilitating navigation through the results, "Color coding" panel defining the mutability color scale, "Controls" panel providing basic operations for manipulation of the structure, annotated interactive sequence and summary panels of mutagenesis hot-spots, functional residues, all residues, active sites, pockets, and tunnels. (**b**) All identified hot-spots are summarized in the "Mutagenesis Hot Spots" panel. The provided annotations include information about residues' mutability, structural location, functional role, existing sequence variants, experimental mutagenesis data, and amino acids occurring at the corresponding position in the sequence alignment

acid variants in related characterized proteins, and amino acids occurring at the corresponding position of the sequence alignment. In our example, six hot-spots were identified (Fig. 1b): E145, Q146, F143 (all estimated as highly mutable with a grade of 9), D147 (mutability grade 8), I134 and L177 (mutability grade 7). Based on the provided annotations, residues D147 and L177 were identified as part of both the tunnel and the active site pocket, E145 and Q146 as part of the tunnel, and F143 and I134 as part of the active site pocket. We also learn that residue I134 is replaced by valine in the experimentally characterized LinB variant from *S. japonicum* B90.

6. Explore the hot-spots in the structure and check whether their side chains are oriented towards the pocket or the tunnel. The left-side panels are interactively connected with the structure. Click on the residue in the "Mutagenesis Hot-Spots"

panel (or any other panel or interactive sequence) to highlight it on the structure (*see* **Note 21**). You can also display the residue annotation by positioning the mouse cursor over the residue in the Jmol applet.

All six hot-spots from our example are oriented towards the pocket or the access tunnel. However, residues E145 and Q146 are located at the very end of the tunnel and from this point of view may be regarded as less promising targets than the four other residues. On the other hand, the mutagenesis of the residue F143 located in the active site pocket close to the ligand (2-bromo-2-propene-1-ol) and the residue L177 located in the bottleneck of the tunnel has the potential to significantly influence the catalytic properties of LinB (*see* **Note 22**). Based on structural analysis, the selected hot-spots can be prioritized as follows: F143, L177 > I134, D147 > E145, Q146.

7. Analyze the frequencies of individual amino acids at selected positions to identify the most frequent and thus the most likely safe mutations. For example, the most frequent residue at the position corresponding to I134 was alanine, followed by valine, phenylalanine, glycine, isoleucine (wild-type residue), and methionine. To decrease the library size and increase the proportion of viable mutants in the library, you could use this data to design an appropriate degenerate codon covering these five residues.
8. Go to the "Functional residues" panel if you want to select hot-spots based on criteria other than mutability (*see* **Note 23**). Repeat **steps 5–7**, only this time focus on the residues with moderate mutability. Five such residues were identified in LinB: A247, V173 (mutability grade 5), P245, L248, and F151 (mutability grade 4). Residues F151 and L248, which are in contact with the ligand, and residue A247, located in the tunnel bottleneck, are the most promising mutagenesis targets. Analysis of amino acid frequencies reveals that leucine, isoleucine, or tryptophan residues are relatively frequent at the position corresponding to the residue F151. As is obvious from the annotation of F151, the F151L, F151W, and F151Y mutations were previously experimentally constructed and led to an increased activity of LinB [89].

## 7 Notes

1. We recommend that you check if the ligand orientation in the crystal structure is sufficiently supported by electron density data. Electron density maps for many publically available structures can be downloaded from the Electron Density Server (http://eds.bmc.uu.se/eds/) [90] and visualized in several

molecular viewers, e.g., PyMOL or Chimera (http://www.cgl.ucsf.edu/chimera/).

2. If users have no prior knowledge regarding the location of the active site and this information cannot be found in databases such as UniProtKB [41] or Catalytic Site Atlas [40], they should carefully analyze all identified pockets, and preferably compare results obtained from different methods [32] and/or perform additional analyses to identify possible functional sites [91, 92].
3. Users should increase the default value if their target pocket is not found. If the target pocket is still not found even after setting this parameter to a high value, users will have to use a different tool.
4. Incorrect positioning of the starting point can significantly influence the results of the calculation. The tools automatically optimize the starting point position to avoid its collision with protein atoms; however if an incorrect initial starting point is provided, the starting point optimization procedure does not help and consequently none, or irrelevant pathways are found.
5. Users should always adjust the radius for their target system—setting the radius to a value which is too high can lead to relevant tunnels being overlooked. Conversely, setting the radius to a value which is too small can lead to the identification of irrelevant pathways.
6. Since the calculation starting point is usually placed close to the active site, a portion of the residues forming the active site pocket is also detected as tunnel-lining residues.
7. Under certain circumstances, the tunnel bottleneck may be detected at the beginning of the tunnel, i.e., close to the starting point. This is usually an artifact of the calculation when the starting point is placed too close to a protein atom. This problem can be solved by placing the starting point in free space.
8. Due to the stochastic nature of the calculation, individual mutations should preferably be evaluated multiple times to check whether the algorithm achieved convergence.
9. A highly conserved residue does not necessarily mean that the residue is conserved for stability or functional reasons and hence immutable. Residues can be highly conserved, e.g., to maintain a particular substrate specificity or other protein properties [18]. If these properties are not important for the use of the enzyme under consideration, the respective residues can be mutated safely. However, the reasons for high conservation often remain elusive.
10. The overview of alignment methods together with practical advice for their usage is provided in the following reviews [93–95].

Additionally, alignments pre-calculated for different sets of proteins can be obtained from databases such as Pfam [96], 3DM [74], HSSP [73], or several "engineering" databases, e.g., the Laccase Engineering Database [97], the Lipase Engineering Database [98], or the Cytochrome P450 Engineering Database [99].

11. In Dmutant, a negative ΔΔG value corresponds to the prediction of a stabilizing mutation, while in CUPSAT and I-MUTANT 2.0, a negative ΔΔG value corresponds to the mutation predicted as destabilizing.
12. Provide a comma-separated list of chains, e.g., "A, B, C" or "1, 2, 3".
13. Using a higher number of sequences should slightly increase the accuracy of the evolutionary rate estimation and provide more informative statistics of amino acid distributions at individual positions of the multiple sequence alignment. However, increasing the number of sequences will lead to a longer computational time.
14. Provide a comma-separated list of residue numbers as provided in the PDB file, e.g., "127, 159, 203". If you want to specify multiple active sites, separate them by a semicolon, e.g., "127, 159, 203; 16, 250".
15. In instances where no catalytic or functionally important residues were found, the results should be interpreted with caution. Catalytic residues are used as a starting point for the calculation of tunnels and to assign the active site pocket. Consequently, if no such residue is available, the calculation of tunnels will not be performed and the largest pocket will automatically be assigned as the active site pocket.
16. The results can also be downloaded. The "Job" panel provides further links to external databases and servers providing additional information about the query protein.
17. The structure is visualized using the Jmol Java applet. If you have problems with visualization of the structure, make sure that Java is properly installed on your system and allowed in your web browser.
18. If some of the relevant tunnels are missing, resubmit the calculation and decrease the value of "Minimal tunnel radius" in "Advanced settings". Values of 0.9–1.1 Å should solve this problem. Keep in mind that setting the radius to small values can lead to identification of irrelevant pathways. In cases where irrelevant tunnels are reported by HotSpot Wizard the value of "Minimal tunnel radius" should be increased.
19. Additional LinB tunnels could be identified by decreasing the "Minimal tunnel radius" to 0.95 Å. Such auxiliary tunnels also

represent good targets for mutagenesis; however we decided to only focus on the active site pocket and main tunnel in this example.

20. All residues located in the active site pocket or access tunnel that have high mutability (grades between 6 and 9) and are *not* annotated as catalytic are assigned as hot-spots.
21. By clicking on residues in the left hand panels or interactive sequence, the residue is added to the selection. The visualization style or color of selections can easily be changed in the "Controls" panel.
22. L177 was experimentally mutated into all 19 possible amino acids, yielding 15 active LinB variants with distinct catalytic activities and substrate specificities [100].
23. All residues located in the active site pocket or access tunnel are annotated as functional residues. This panel thus also includes residues from the "Mutagenesis Hot Spots" panel. Selecting targets for mutagenesis from the "Functional residues" panel increases the risk of producing inactive mutants, especially when positions with low mutability are targeted.

## Acknowledgements

The research work of the authors is supported by the Grant Agency of the Czech Republic (P207/12/0775 and P503/12/0572), the Czech Ministry of Education (LO1214, LH14027, CZ.1.07/2.3.00/30.0037), and the European Regional Development Fund (CZ.1.05/2.1.00/01.0001). MetaCentrum is acknowledged for providing access to computing facilities, supported by the Czech Ministry of Education of the Czech Republic (LM2010005).

### References

1. Bornscheuer UT, Huisman GW, Kazlauskas RJ et al (2012) Engineering the third wave of biocatalysis. Nature 485:185–194
2. Chica RA, Doucet N, Pelletier JN (2005) Semi-rational approaches to engineering enzyme activity: combining the benefits of directed evolution and rational design. Curr Opin Biotechnol 16:378–384
3. Lutz S (2010) Beyond directed evolution–semi-rational protein engineering and design. Curr Opin Biotechnol 21:734–743
4. Bommarius AS, Blum JK, Abrahamson MJ (2011) Status of protein engineering for biocatalysts: how to design an industrially useful biocatalyst. Curr Opin Chem Biol 15: 194–200
5. Goldsmith M, Tawfik DS (2012) Directed enzyme evolution: beyond the low-hanging fruit. Curr Opin Struct Biol 22:406–412
6. Reetz MT, Kahakeaw D, Lohmer R (2008) Addressing the numbers problem in directed evolution. Chembiochem 9:1797–1804
7. Reetz MT, Wu S (2008) Greatly reduced amino acid alphabets in directed evolution: making the right choice for saturation mutagenesis at homologous enzyme positions. Chem Commun 43:5499–5501. http://www.ncbi.nlm.nih.gov/pubmed/18997931
8. Barrozo A, Borstnar R, Marloie G et al (2012) Computational protein engineering: bridging the gap between rational design and laboratory evolution. Int J Mol Sci 13:12428–12460

9. Damborsky J, Brezovsky J (2014) Computational tools for designing and engineering enzymes. Curr Opin Chem Biol 19C:8–16
10. Chaparro-Riggers JF, Polizzi KM, Bommarius AS (2007) Better library design: data-driven protein engineering. Biotechnol J 2:180–191
11. Moore GL, Maranas CD (2004) Computational challenges in combinatorial library design for protein engineering. AIChE J 50:262–272
12. Wong TS, Roccatano D, Schwaneberg U (2007) Steering directed protein evolution: strategies to manage combinatorial complexity of mutant libraries. Environ Microbiol 9: 2645–2659
13. Fox RJ, Huisman GW (2008) Enzyme optimization: moving from blind evolution to statistical exploration of sequence-function space. Trends Biotechnol 26:132–138
14. Verma R, Schwaneberg U, Roccatano D (2012) Computer-aided protein directed evolution: a review of web servers, databases and other computational tools for protein engineering. Comput Struct Biotechnol J 2:e201209008
15. Dalby PA (2003) Optimising enzyme function by directed evolution. Curr Opin Struct Biol 13:500–505
16. Hibbert EG, Dalby PA (2005) Directed evolution strategies for improved enzymatic performance. Microb Cell Fact 4:29
17. Morley KL, Kazlauskas RJ (2005) Improving enzyme properties: when are closer mutations better? Trends Biotechnol 23:231–237
18. Paramesvaran J, Hibbert EG, Russell AJ et al (2009) Distributions of enzyme residues yielding mutants with improved substrate specificities from two different directed evolution strategies. Protein Eng Des Sel 22:401–411
19. Laskowski RA, Swindells MB (2011) LigPlot+: multiple ligand-protein interaction diagrams for drug discovery. J Chem Inf Model 51:2778–2786
20. Stierand K, Rarey M (2010) Drawing the PDB: protein-ligand complexes in two dimensions. ACS Med Chem Lett 1:540–545
21. Sobolev V, Sorokine A, Prilusky J et al (1999) Automated analysis of interatomic contacts in proteins. Bioinformatics 15:327–332
22. Krissinel E, Henrick K (2007) Inference of macromolecular assemblies from crystalline state. J Mol Biol 372:774–797
23. Biesiada J, Porollo A, Velayutham P et al (2011) Survey of public domain software for docking simulations and virtual screening. Hum Genomics 5:497–505
24. Li X, Li Y, Cheng T et al (2010) Evaluation of the performance of four molecular docking programs on a diverse set of protein-ligand complexes. J Comput Chem 31:2109–2125
25. Morris GM, Lim-Wilby M (2008) Molecular docking. In: Kukol A (ed) Molecular modeling of proteins, Methods in molecular biology. Humana Press, Totowa, NJ, pp 3–23
26. McDonald IK, Thornton JM (1994) Satisfying hydrogen bonding potential in proteins. J Mol Biol 238:777–793
27. Sobolev V, Wade RC, Vriend G et al (1996) Molecular docking using surface complementarity. Proteins 25:120–129
28. Henrich S, Salo-Ahen OMH, Huang B et al (2010) Computational approaches to identifying and characterizing protein binding sites for ligand design. J Mol Recognit 23:209–219
29. Laurie ATR, Jackson RM (2006) Methods for the prediction of protein-ligand binding sites for structure-based drug design and virtual ligand screening. Curr Protein Pept Sci 7: 395–406
30. Pérot S, Sperandio O, Miteva MA et al (2010) Druggable pockets and binding site centric chemical space: a paradigm shift in drug discovery. Drug Discov Today 15:656–667
31. Dundas J, Ouyang Z, Tseng J et al (2006) CASTp: computed atlas of surface topography of proteins with structural and topographical mapping of functionally annotated residues. Nucleic Acids Res 34:W116–W118
32. Zhang Z, Li Y, Lin B et al (2011) Identification of cavities on protein surface using multiple computational approaches for drug binding site prediction. Bioinformatics 27:2083–2088
33. Hendlich M, Rippmann F, Barnickel G (1997) LIGSITE: automatic and efficient detection of potential small molecule-binding sites in proteins. J Mol Graph Model 15: 359–363
34. Laurie ATR, Jackson RM (2005) Q-SiteFinder: an energy-based method for the prediction of protein-ligand binding sites. Bioinformatics 21:1908–1916
35. Hernandez M, Ghersi D, Sanchez R (2009) SITEHOUND-web: a server for ligand binding site identification in protein structures. Nucleic Acids Res 37:W413–W416
36. Schmidtke P, Le Guilloux V, Maupetit J et al (2010) fpocket: online tools for protein ensemble pocket detection and tracking. Nucleic Acids Res 38:W582–W589
37. La D, Esquivel-Rodríguez J, Venkatraman V et al (2009) 3D-SURFER: software for high-throughput protein surface comparison and analysis. Bioinformatics 25:2843–2844

38. Binkowski TA, Naghibzadeh S, Liang J (2003) CASTp: Computed Atlas of Surface Topography of proteins. Nucleic Acids Res 31:3352–3355
39. Pavelka A, Chovancova E, Damborsky J (2009) HotSpot Wizard: a web server for identification of hot spots in protein engineering. Nucleic Acids Res 37:W376–W383
40. Porter CT, Bartlett GJ, Thornton JM (2004) The Catalytic Site Atlas: a resource of catalytic sites and residues identified in enzymes using structural data. Nucleic Acids Res 32: D129–D133
41. Magrane M, Uniprot Consortium (2011) UniProt Knowledgebase: a hub of integrated protein data. Database (Oxford) 2011:bar009
42. Chovancova E, Pavelka A, Benes P et al (2012) CAVER 3.0: a tool for the analysis of transport pathways in dynamic protein structures. PLoS Comput Biol 8:e1002708
43. Ashkenazy H, Erez E, Martz E et al (2010) ConSurf 2010: calculating evolutionary conservation in sequence and structure of proteins and nucleic acids. Nucleic Acids Res 38: W529–W533
44. Huang B (2009) MetaPocket: a meta approach to improve protein ligand binding site prediction. OMICS 13:325–330
45. Prokop Z, Gora A, Brezovsky J et al (2012) Engineering of protein tunnels: keyhole-lock-key model for catalysis by the enzymes with buried active sites. In: Lutz S, Bornscheuer UT (eds) Protein engineering handbook, vol 3. Wiley-VCH, Weinheim, pp 421–464
46. Brezovsky J, Chovancova E, Gora A et al (2013) Software tools for identification, visualization and analysis of protein tunnels and channels. Biotechnol Adv. 31:38–49
47. Berka K, Hanák O, Sehnal D et al (2012) MOLEonline 2.0: interactive web-based analysis of biomacromolecular channels. Nucleic Acids Res 40:W222–W227
48. Yaffe E, Fishelovitch D, Wolfson HJ et al (2008) MolAxis: a server for identification of channels in macromolecules. Nucleic Acids Res 36:W210–W215
49. Becker OM, Karplus M (2005) Guide to biomolecular simulations. Springer, New York
50. Lindahl ER (2008) Molecular dynamics simulations. In: Kukol A (ed) Molecular modeling of proteins, Methods in molecular biology. Humana Press, Totowa, NJ, pp 3–23
51. Polizzi KM, Bommarius AS, Broering JM et al (2007) Stability of biocatalysts. Curr Opin Chem Biol 11:220–225
52. Reetz MT, Carballeira JD (2007) Iterative saturation mutagenesis (ISM) for rapid directed evolution of functional enzymes. Nat Protoc 2:891–903
53. Reetz MT, Carballeira JD, Vogel A (2006) Iterative saturation mutagenesis on the basis of B factors as a strategy for increasing protein thermostability. Angew Chem Int Ed Engl 45:7745–7751
54. Koudelakova T, Chaloupkova R, Brezovsky J et al (2013) Engineering enzyme stability and resistance to an organic cosolvent by modification of residues in the access tunnel. Angew Chem Int Ed Engl 52(7):1959–1963. doi:10.1002/anie.201206708
55. Potapov V, Cohen M, Schreiber G (2009) Assessing computational methods for predicting protein stability upon mutation: good on average but not in the details. Protein Eng Des Sel 22:553–560
56. Khan S, Vihinen M (2010) Performance of protein stability predictors. Hum Mutat 31: 675–684
57. Thiltgen G, Goldstein RA (2012) Assessing predictors of changes in protein stability upon mutation using self-consistency. PLoS One 7:e46084
58. Dehouck Y, Kwasigroch JM, Gilis D et al (2011) PoPMuSiC 2.1: a web server for the estimation of protein stability changes upon mutation and sequence optimality. BMC Bioinformatics 12:151
59. Dehouck Y, Grosfils A, Folch B et al (2009) Fast and accurate predictions of protein stability changes upon mutations using statistical potentials and neural networks: PoPMuSiC-2.0. Bioinformatics 25:2537–2543
60. Guerois R, Nielsen JE, Serrano L (2002) Predicting changes in the stability of proteins and protein complexes: a study of more than 1000 mutations. J Mol Biol 320:369–387
61. Schymkowitz J, Borg J, Stricher F et al (2005) The FoldX web server: an online force field. Nucleic Acids Res 33:W382–W388
62. Van Durme J, Delgado J, Stricher F et al (2011) A graphical interface for the FoldX forcefield. Bioinformatics 27:1711–1712
63. Valdar WSJ (2002) Scoring residue conservation. Proteins 48:227–241
64. Johansson F, Toh H (2010) A comparative study of conservation and variation scores. BMC Bioinformatics 11:388
65. Mihalek I, Res I, Lichtarge O (2004) A family of evolution-entropy hybrid methods for ranking protein residues by importance. J Mol Biol 336:1265–1282

66. Morgan DH, Kristensen DM, Mittelman D et al (2006) ET viewer: an application for predicting and visualizing functional sites in protein structures. Bioinformatics 22:2049–2050
67. Ma B-G, Berezovsky IN (2010) The MBLOSUM: a server for deriving mutation targets and position-specific substitution rates. J Biomol Struct Dyn 28:415–419
68. Crooks GE, Hon G, Chandonia JM et al (2004) WebLogo: a sequence logo generator. Genome Res 14:1188–1190
69. Jochens H, Bornscheuer UT (2010) Natural diversity to guide focused directed evolution. Chembiochem 11:1861–1866
70. Mayrose I, Graur D, Ben-Tal N et al (2004) Comparison of site-specific rate-inference methods for protein sequences: empirical Bayesian methods are superior. Mol Biol Evol 21:1781–1791
71. Pupko T, Bell RE, Mayrose I et al (2002) Rate4Site: an algorithmic tool for the identification of functional regions in proteins by surface mapping of evolutionary determinants within their homologues. Bioinformatics 18: S71–S77
72. Goldenberg O, Erez E, Nimrod G et al (2009) The ConSurf-DB: pre-calculated evolutionary conservation profiles of protein structures. Nucleic Acids Res 37:D323–D327
73. Joosten RP, Te Beek TAH, Krieger E et al (2011) A series of PDB related databases for everyday needs. Nucleic Acids Res 39: D411–D419
74. Kuipers RK, Joosten H-J, Van Berkel WJH et al (2010) 3DM: systematic analysis of heterogeneous superfamily data to discover protein functionalities. Proteins 78:2101–2113
75. Kuipers R, Van den Bergh T, Joosten H-J et al (2010) Novel tools for extraction and validation of disease-related mutations applied to Fabry disease. Hum Mutat 31:1026–1032
76. Zhou H, Zhou Y (2002) Distance-scaled, finite ideal-gas reference state improves structure-derived potentials of mean force for structure selection and stability prediction. Protein Sci 11:2714–2726
77. Parthiban V, Gromiha MM, Schomburg D (2006) CUPSAT: prediction of protein stability upon point mutations. Nucleic Acids Res 34:W239–W242
78. Capriotti E, Fariselli P, Casadio R (2005) I-Mutant2.0: predicting stability changes upon mutation from the protein sequence or structure. Nucleic Acids Res 33:W306–W310
79. Ng PC, Henikoff S (2006) Predicting the effects of amino acid substitutions on protein function. Annu Rev Genomics Hum Genet 7:61–80
80. Thusberg J, Olatubosun A, Vihinen M (2011) Performance of mutation pathogenicity prediction methods on missense variants. Hum Mutat 32:358–368
81. Thusberg J, Vihinen M (2009) Pathogenic or not? And if so, then how? Studying the effects of missense mutations using bioinformatics methods. Hum Mutat 30:703–714
82. Ng PC, Henikoff S (2003) SIFT: predicting amino acid changes that affect protein function. Nucleic Acids Res 31:3812–3814
83. Choi Y, Sims GE, Murphy S et al (2012) Predicting the functional effect of amino acid substitutions and indels. PLoS One 7:e46688
84. Thomas PD, Campbell MJ, Kejariwal A et al (2003) PANTHER: a library of protein families and subfamilies indexed by function. Genome Res 13:2129–2141
85. Kumar P, Henikoff S, Ng PC (2009) Predicting the effects of coding non-synonymous variants on protein function using the SIFT algorithm. Nat Protoc 4: 1073–1081
86. Nov Y (2012) When second best is good enough: another probabilistic look at saturation mutagenesis. Appl Environ Microbiol 78:258–262
87. Firth AE, Patrick WM (2008) GLUE-IT and PEDEL-AA: new programmes for analyzing protein diversity in randomized libraries. Nucleic Acids Res 36:W281–W285
88. Firth AE, Patrick WM (2005) Statistics of protein library construction. Bioinformatics 21:3314–3315
89. Bohác M, Nagata Y, Prokop Z et al (2002) Halide-stabilizing residues of haloalkane dehalogenases studied by quantum mechanic calculations and site-directed mutagenesis. Biochemistry 41:14272–14280
90. Kleywegt GJ, Harris MR, Zou JY et al (2004) The Uppsala Electron-Density Server. Acta Crystallogr D Biol Crystallogr 60:2240–2249
91. Leis S, Schneider S, Zacharias M (2010) In silico prediction of binding sites on proteins. Curr Med Chem 17:1550–1562
92. Xin F, Radivojac P (2011) Computational methods for identification of functional residues in protein structures. Curr Protein Pept Sci 12:456–469

93. Do CB, Katoh K (2008) Protein multiple sequence alignment. In: Thompson JD, Ueffing M, Schaeffer-Reiss C (eds) Functional proteomics, Methods in molecular biology. Humana Press, Totowa, NJ, pp 379–413
94. Pirovano W, Heringa J (2008) Multiple sequence alignment. In: Keith JM (ed) Bioinformatics, Methods in molecular biology. Humana Press, Totowa, NJ, pp 143–161
95. Pei J (2008) Multiple protein sequence alignment. Curr Opin Struct Biol 18:382–386
96. Finn RD, Tate J, Mistry J et al (2008) The Pfam protein families database. Nucleic Acids Res 36:D281–D288
97. Sirim D, Wagner F, Wang L et al (2011) The Laccase Engineering Database: a classification and analysis system for laccases and related multicopper oxidases. Database (Oxford) 2011:bar006
98. Fischer M, Pleiss J (2003) The Lipase Engineering Database: a navigation and analysis tool for protein families. Nucleic Acids Res 31:319–321
99. Fischer M, Knoll M, Sirim D et al (2007) The Cytochrome P450 Engineering Database: a navigation and prediction tool for the cytochrome P450 protein family. Bioinformatics 23:2015–2017
100. Chaloupková R, Sýkorová J, Prokop Z et al (2003) Modification of activity and specificity of haloalkane dehalogenase from *Sphingomonas paucimobilis* UT26 by engineering of its entrance tunnel. J Biol Chem 278: 52622–52628

# Chapter 21

# Computational Tools for Directed Evolution: A Comparison of Prospective and Retrospective Strategies

**Julian Zaugg, Yosephine Gumulya, Elizabeth M.J. Gillam, and Mikael Bodén**

## Abstract

Directed evolution methods have proved to be highly effective in the design of novel proteins and in the generation of large libraries of diverse sequences. However, searching through the vast number of mutants produced during such experiments in order to find the best represents a daunting and difficult task. In recent years, a number of computational tools have been developed to provide guidance during this exploratory process. It can, however, be unclear as to which tool or tools best complement the chosen library design strategy. In this review, we describe and critically evaluate some of the more notable tools in this area, discussing the rationale behind each, the requirements for their implementation, and potential issues faced when using them. Some examples of their application in an experimental setting are also provided. The tools have been classified based on contrasting strategies as to how they function: prospective tools SCHEMA and OPTCOMB use extant sequence and structural data to predict optimal locations for crossover sites, whereas retrospective tools ProSAR and ASRA use property data from the mutant library to predict beneficial mutations and features. From our evaluation, we suggest that each tool can play a role in the design process; however this is largely dictated by the data available and the desired experimental strategy for the project.

**Key words** Directed evolution, SCHEMA, OPTCOMB, ProSAR, ASRA, Computational tools, Library generation, Mutant, Protein, Scoring

## 1 Introduction

Protein engineering has typically been performed by trial-and-error methods, such as rational design, based on scant information about how protein structure determines function. Such methods limit sequence solutions to only a minute portion of the explorable "sequence space" and risk limiting the functional capabilities of potential sequence variants. To address such inefficiencies, methods to computationally capture and model nature's selectivity have been developed and have been incorporated in a variety of tools. Such tools help navigate the vast space of artificially evolved sequences before an investment is made in their synthesis and

Elizabeth M.J. Gillam et al. (eds.), *Directed Evolution Library Creation: Methods and Protocols*, Methods in Molecular Biology, vol. 1179, DOI 10.1007/978-1-4939-1053-3_21, 

subsequent screening. The protein designs produced during this exploratory process can be extremely desirable to biochemists and protein engineers, representing viable starting points for further optimization through techniques such as directed evolution [1].

Directed evolution mimics nature's optimizing strategy by applying iterative rounds of random sequence changes in one or more extant genes to produce a library of mutants. These libraries can then be screened to discover those with desired properties. The insertion of multiple mutations can be effectively achieved through a variety of methods, the most popular of which include random mutagenesis (epPCR), DNA shuffling, and saturation mutagenesis. Some methods are more appropriate than others depending on the data available and library requirements. Thus, epPCR is usually performed when there is no structural guidance available, i.e., no crystal structure. By contrast, saturation mutagenesis is frequently selected when structural information is available; however this tends to produce higher numbers of inactive mutants. Recombination/gene shuffling is used when a number of homologues are available to recombine. This approach tends to produce higher proportions of active mutants but depends heavily on the degree of sequence identity in the homologues.

## 2 Computational Tools

Directed evolution methods have enormous potential to leverage and amplify combinations of existing traits but, unless carefully guided, are fraught by the sheer numbers of possible mutants that need to be screened to find the best variants. Designing proteins becomes an exercise in choosing the best starting materials and recombination or substitution sites and trading on the diversity found within their combinations. Guidance can come from computational models. Tools such as SCHEMA [2], OPTCOMB [3, 4], ASRA [5], and ProSAR [6] try—each in their own way—to capture and integrate the rich information that is inherently available in sequences and structures of existing proteins and libraries of mutants.

These tools are divided by contrasting strategies in how they function—either *prospectively* (SCHEMA and OPTCOMB), by using information within extant sequence and structural data, or *retrospectively* (ProSAR and ASRA), by using property data observed within the mutant library (*see* Fig. 1). Given the limited number of published reports using these methods on which to base an extensive critical evaluation, within this review we attempt to provide a simple, yet thorough, analysis of each of these tools, detailing how their functional differences make them suitable for a variety of design strategies in directed evolution experiments.

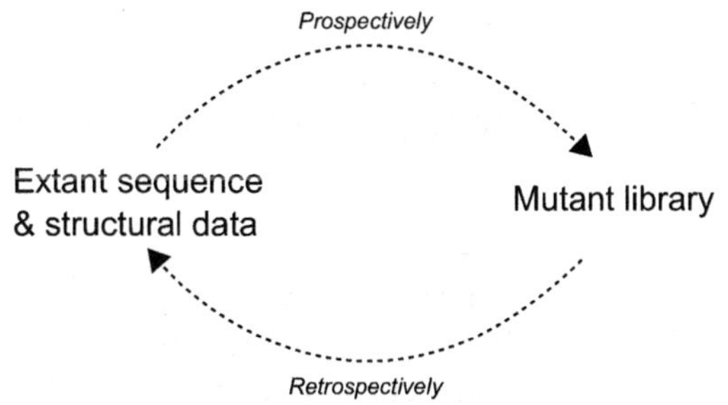


**Fig. 1** Generalized model of how computational tools function in directed evolution experiments. Tools can apply a "prospective" strategy by using existing sequence or structural data to design mutant variants/libraries; alternatively, a "retrospective" strategy can be used by applying properties observed from a selection of mutants to select the best features from the extant data

### 2.1 SCHEMA

The SCHEMA energy function can be used to design libraries of chimera/mutant protein sequences, with the aim to minimize structural disruption. The disruption within a mutant is reflected by its "*E*-score," a parameter computed using a contact map of a representative structure of one of the parents and an alignment of parental protein sequences. The *E*-score indicates the level of disruption within a mutant, i.e., the number of residue–residue contacts that are displaced within the mutant due to novel combinations of parental sequences vis-á-vis structural combinations. Low levels of such displacement are found for mutants that recombine parents at sites with low structural connectivity, i.e., boundaries of structural "building blocks."

Sequence regions within parental sequences, representing likely boundaries of structural building blocks, can be predicted with SCHEMA through the generation of a disruption profile. Unlike the *E*-score, the disruption profile is generated across the whole sequence through a "sliding-window" function, with a disruption "*S*-score" calculated for each residue. Each residue's *S*-score corresponds to the level of predicted structural disruption that a crossover would cause at that location. Residues with low *S*-scores are likely boundary sites, more tolerable to crossovers [7].

SCHEMA's *E*-score works well in retaining the integrity of densely interconnected blocks and, complementary to this, the SCHEMA profile function *S* can provide accurate predictions of likely block boundary regions. There are many examples of where SCHEMA has identified advantageous settings for site-directed recombination of, for instance, beta lactamases and cytochromes P450 [8, 9].

#### 2.1.1 Technical Background

Two inputs are required for SCHEMA—a multiple sequence alignment (MSA) of parental sequences and a representative structure—this is under the assumption that parental sequences will be structurally similar. Given a mutant sequence constructed with fragments $\alpha$ and $\beta$ inherited from parent sequences A and B respectively, the disruption of the mutant, $E_{\alpha,\beta}$, can be calculated by

$$E_{\alpha,\beta} = \sum_{i\in\alpha}\sum_{j\in\beta} c_{ij}\Delta_{ij} \quad (1)$$

where $c_{ij}=1$ when residues $i$ and $j$ are within a specified distance of each other in the representative structure, else $c_{ij}=0$. A frequently used cutoff distance is 4.5 Å. The SCHEMA $\Delta_{ij}$ function is calculated solely from the parental sequence alignment and accounts for the fact that there is no disruption when the amino acids found at positions $i$ and $j$ in the mutant sequence are the same as those at equivalent positions in any parent sequence ($\Delta_{ij}=0$) [7, 9]. In the case where a novel amino acid pair is observed in the mutant sequence, the contact is said to be disrupted ($\Delta_{ij}=1$).

The large number of possible sequence solutions or sequence space sampled by directed evolution makes exhaustive disruption calculations for every mutant sequence not feasible. To avoid such a computationally expensive problem, SCHEMA's $S$-score can be used to identify viable crossover locations and in turn aid in reducing the sequence space that needs to be explored to discover functional variants [10] (*see* Fig. 2). The function achieves this by analyzing the protein structure with a sliding window ($w$ = width in residues), considering only residue–residue interactions of the local sequence region. It should be noted that distant residue–residue interactions, outside the defined window region, are not considered.

Given a representative protein structure and an alignment of parental sequences, the SCHEMA disruption profile $S$ for each residue $i$ is

$$S_i = \sum_{j=i-w+1}^{i} \sum_{k=j}^{j+w-2} \sum_{l=k+1}^{j+w-1} c_{kl}\Delta_{kl} \quad (2)$$

Similar to the $E$-score calculation, $c_{kl}=1$ when the residue pair $kl$ are within the defined interacting distance, else $c_{kl}=0$. Additionally, $\Delta_{kl}$ is calculated in the same fashion as $\Delta_{ij}$ for the $E$-score function.

The Recombination as Shortest-Path Problem (RASPP) algorithm, given constraints on the number and size of sequence fragments, can be used in combination with SCHEMA to identify crossover locations that minimize the average disruption $E$ [11]. These optimal crossover locations define the sequence blocks which, when shuffled among the parental homologues, should define the most functionally enriched recombination libraries with the lowest average $E$ across a range of mutation levels.

### 2.1.2 Applications

SCHEMA has been successfully applied to a variety of protein design problems, encompassing not only improvements to protein activity levels [9, 12, 13] but also thermostability [9, 12, 14–17] and substrate specificity [18]. For example, Otey et al. implemented SCHEMA to calculate the potential residue–residue disruption

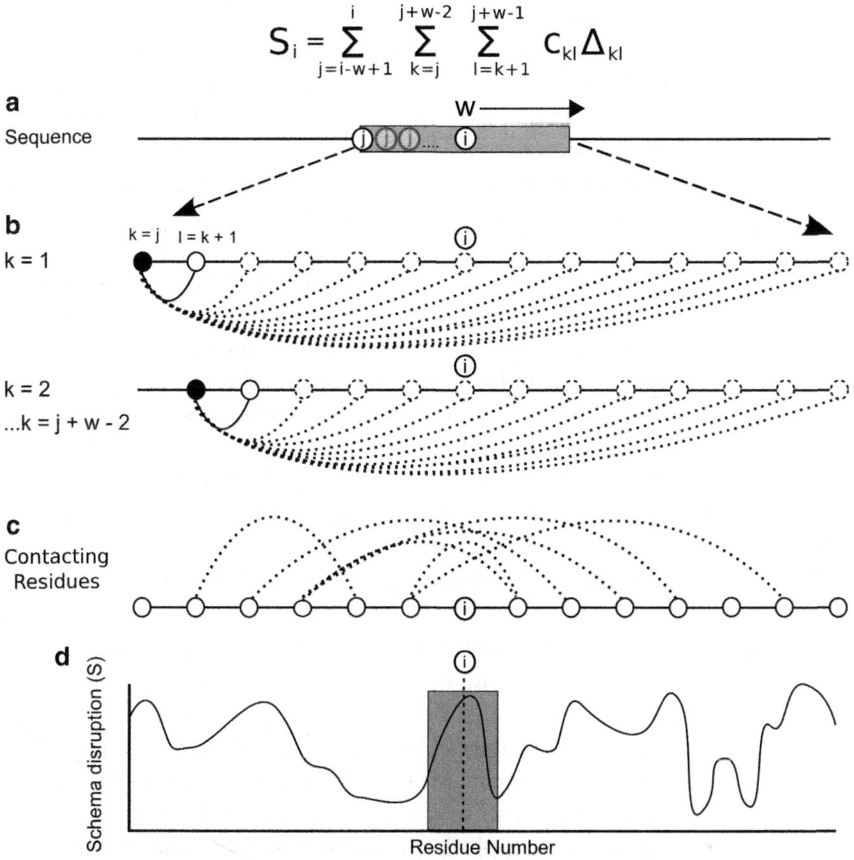


**Fig. 2** Steps for the generation of a SCHEMA disruption profile, given a representative protein structure and an alignment of parental sequences. (**a**) The SCHEMA disruption *S* for each residue *i* is calculated by counting the number of residue–residue interactions within a sliding window of *w* residues. (**b**) $S_i$ is determined by iteratively summing and scoring over each potential residue pair *kl* within the window. The score for each residue pair is based on their interaction ($c_{kl}$), where $c_{kl}=1$ when residues *k* and *l* are within a specified distance of each other in the representative structure, else $c_{kl}=0$. Also influencing the score, $\Delta_{kl}$ represents the presence, or lack thereof, of a novel pair of amino acid types for residues *kl*. $\Delta_{kl}=1$ when the amino acid combination for residues *kl* are not observed to occur in any of the parental sequences at equivalent positions, otherwise $\Delta_{kl}=0$. (**c**) Not all pairs of residues evaluated in a given window will contribute to the $S_i$ score of a given residue, specifically when $c_{kl}=0$ or $\Delta_{kl}=0$. In this example, *dotted lines* represent the subset of evaluated residue pairs where a $S_i$ contributing interaction was observed. (**d**) The disruption profile, generated for each residue in the sequence of the representative structure, represents the number of residue–residue interactions potentially disrupted if a crossover was made. In this example, residue *i* is involved in a compact structural region, representing a location SCHEMA would predict to be unsuitable for a crossover site

scores for the structure of the heme-binding domain of CYP102A1 and applied these scores, in combination with the sequences for the CYP102A1-A3 homologues, to guide the placement of crossover locations [9, 13]. These optimized crossover locations allowed the researchers to successfully produce a functionally enriched, and mutationally diverse, recombinatorial library of 6,561 mutants, of which ~3,000 were observed to properly fold.

Experimental application of SCHEMA has demonstrated its use in producing functionally enriched libraries; however methodological development is still needed for identifying the most beneficial sequence elements in relation to desired protein properties. Some recent work has attempted to address this by focusing on identifying the contribution that specific sequence features or blocks provide to protein properties [10, 12]. For example, Romero et al. focused on identifying what sequence features contribute to arginase stability [10]. A SCHEMA-guided recombination approach was used to design a library of chimeric proteins derived from arginase I and II. An active learning algorithm was then applied to select a sample of functional and highly informative sequences. Following this, the researchers were able to use a simple linear regression model, in combination with experimental data from the sampled mutants, to predict how specific sequence elements contributed to arginase properties. The researchers were able to use these proposed element–property relationships to further hypothesize potential solutions to arginase stability issues.

#### 2.1.3 Access/License/Requirements

Software for performing SCHEMA calculations and accompanying usage documentation is provided online by the Arnold group as a license-free downloadable toolbox (http://cheme.che.caltech.edu/groups/fha/index.html) [2]. Documentation included with the toolbox provides step-by-step instructions for use of the software along with example input data. Additionally, a detailed walkthrough example is provided online at the group's website and elsewhere in this volume.

The SCHEMA software itself comes as a collection of Python scripts (http://www.python.org) that can be easily run through the command line. For this reason, it is advisable that users of the software have some basic familiarity with running Python scripts from the command line. To simplify the use of the SCHEMA tools, it is advisable to run the scripts on a UNIX-based computer.

#### 2.1.4 Data Preparation

SCHEMA recombination has been applied to sequences sharing as little as 34 % sequence identity [8]. However, to reduce the presence of deleterious (to protein function), novel, residue contacts, selected parental sequences to be recombined should generally be closely related. The parental sequences are provided as input into SCHEMA as a MSA and should be in a typical ALN format, such as that produced by CLUSTALW [19].

To accurately determine viable crossover locations, the representative protein structure should be high resolution (<2 Å), and to avoid hindering the library design, should correspond to one of the parental sequences. If no parental structure is available, the structure of a closely related homologue can be used. Such structures can be modeled using tools such as MODELLER [20], SWISS-model [21–23], or PHYRE [24].

#### 2.1.5 Tool Performance

Unlike OPTCOMB, SCHEMA does not consider the subtler residue–residue physicochemical interactions that, additively, may influence protein properties, i.e., SCHEMA weighs all interactions the same. For example, SCHEMA would score a close approach between methyl groups the same as a hydrogen bond or electrostatic interaction, yet the latter are stronger, more specific interactions that serve to hold proteins together in a correctly folded form.

SCHEMA relies on a single representative structure in order to generate the contact map that underlies disruption scores. This imposes a rather inflexible limitation on potential residue–residue interactions. A single structure may not fully encapsulate the motion and conformational flexibility shown by other protein homologues, which may play a pivotal role in their overall functionality. By using a structural alignment to produce the "best" structure in terms of similarity to each parent, it is possible to accommodate major differences between parents. However, this still risks introducing a degree of "noise" into the results.

The windowed scanning approach implemented by SCHEMA does not take into account residue–residue interactions that occur outside of the profiling window. Disruption of such interactions can prove detrimental to crossover site prediction, i.e., when many crossovers can occur, the profile minima of SCHEMA can become a poor predictive guide [7]. RASPP may compensate for this by averaging out the disruption $E$ for the recombinatorial libraries.

### 2.2 OPTCOMB

Optimal Pattern of Tiling for COMBinatorial library design (OPTCOMB) also aims to predict the "fittest" mutants by inspecting properties of the parents, but in contrast to SCHEMA, the scoring model used in OPTCOMB captures the level of disruption present based on a sequence alignment of other proteins in the same family [4]. The $S1$-score in OPTCOMB finds what amino acids are acceptable in a mutant by inspecting the frequency profile of each position ("column") in the amino acid alignment independently. The $S2$-score extends $S1$ by also considering statistical dependencies between pairs of columns. In both cases, mutants with few "clashes" against the family profile are considered more likely to be functional (and thus preferred in a library). $S2$ discriminates very well between functional and nonfunctional products, even outperforming SCHEMA on a SCHEMA-guided library [4].

#### 2.2.1 Technical Background

The $S1$-scoring system relies solely on a screened MSA containing well-aligned sequences (with no duplicates) for the protein family of interest. The frequency of occurrence of each amino acid in each column of the MSA is calculated while columns where gaps are present are ignored. These amino acid frequencies are in turn used to calculate the frequency distributions of each property $k$ (charge, hydrophobicity, and volume) at each position $i$ for each bin $b$—represented as $F_{i,k,b}$. Bins here are based on the concept of "property bins" described in earlier work by Saraf et al. [25] where amino acids were binned together based on similarity between their individual property values, e.g., alanine and tryptophan would be placed in the same bin for charge ($b=0$) but placed in different volume bins ($b=20$ and $b=140$ respectively).

Binned amino acid frequencies can be used to calculate the likelihood a mutant sequence will share similar functionality to the parent protein family. Each mutant is encoded as a binary sequence, where each position is represented by $Y_{i,k,b}=1$ when the amino acid at $i$ position falls within $b$ bin for property $k$, i.e., when the mutant residue matches the property value for the equivalent parental residue.

The protein family information, $F_{i,k,b}$, and mutant sequence information, $Y_{i,k,b}$, form the basis of the $S$-scores. $S1$ is calculated as

$$S1 = \frac{\sum_{i}\sum_{k}\sum_{b}\log_{10}\left(F_{i,k,b}\right)Y_{i,k,b}}{L} \quad (3)$$

where $L$ is the number of scored column positions in the mutant sequence. A high $S1$-score indicates that positions in the mutant sequence have amino acids within mostly high frequency bins, suggesting more parent-like functionality is likely to have been retained.

Similar to SCHEMA [2], $S2$, by incorporating the relationship between two positions, takes account of the potential disruption caused by the interchange of contacting amino acids during recombination. $S2$ borrows from SCHEMA by evaluating residues based on how similar their paired property values are to those observed in the parental sequences, with penalties assigned in proportion to the degree of dissimilarity [4].

A representative structure is required to calculate the $S2$-score. Summing over each residue pair in contact ($C_{i1,i2}=1$), each property ($k$) and each property bin ($b1$ and $b2$), and by incorporating the frequency distribution for property $k$ ($\mathrm{FF}_{i1,i2,k,b1,b2}$), $S2$ is calculated as:

$$S2 = \frac{\sum_{i1}\sum_{\substack{i2=i1+1\\ C_{i1i2}=1}}\sum_{k}\sum_{b1}\sum_{b2} Y_{i,k,b1}Y_{i2,k,b2}\log_{10}\left(\mathrm{FF}_{i1,i2,k,b1,b2}\right)}{T} \quad (4)$$

where $T$ is the total number of residue pairs in contact in the mutant sequence. Similar to $S1$, a high $S2$-score indicates the mutant sequence includes more pairs of high frequency bins from the protein family.

The $S$-scoring systems are directed towards the scoring of individual mutant sequences; however it is often preferable to design a whole, optimized, library that incorporates a constraint on its size, i.e., the number of mutants included. OPTCOMB uses two optimization models, M1 and M2, based on alternative library

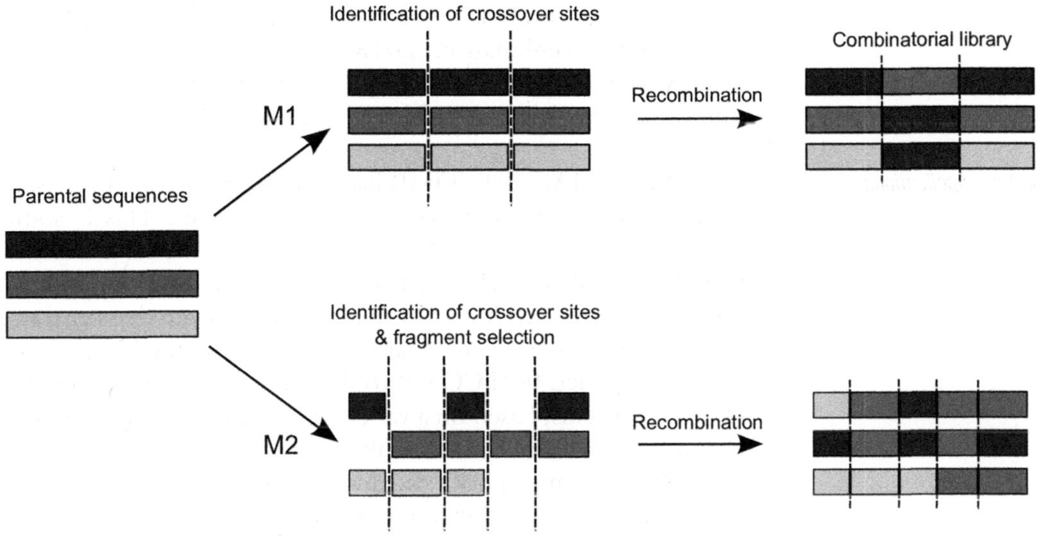


**Fig. 3** The two optimization models of OPTCOMB, M1 and M2. Both models involve the selection of crossover locations; however M2 allows the omission of parental sequence fragments that would reduce the score of the mutant sequence. Figure adapted from [4]

construction strategies (*see* Fig. 3). Both models employ the *S*2-scoring system to identify optimal crossover locations, which maximize the overall score of the library.

M1 works on the premise that there is no restriction on the fragment contributed to a sequence, in that the sequence fragment can originate from any of the parental sequences. However, the model does stipulate that all parental sequences must be able to contribute a sequence fragment for each crossover location [3]. Increasing the number of crossover locations, and hence fragments, will be likely to reduce the *S*2-score for, and increase the size of, the recombinatorial library. This score reduction is caused by an increase in the number of novel residue–residue contacts formed between different parental fragments. An M1 approach may be suitable when employing a method such as SISDC, where crossover locations can be specified largely independent of sequence identity [26].

In contrast to M1, M2 is designed to accommodate construction methods that allow for the exclusion of fragments at each crossover location. In this model, parental sequence fragments can be omitted if their inclusion would result in a lower *S*2-score for the library. In respect to a variety of factors, library size will not increase significantly in response to a change in the number of crossovers. The ability to omit parental sequence fragments that would lower the *S*2-score of the library provides a degree of flexibility in the design of the library and can lead to a lower retention of clash contributing residues [3, 4]. M2 is suited for construction methods where there are restrictions on the oligomers that can be contributed by each parent for recombination, e.g., synthetic shuffling or degenerate homoduplex recombination (DHR) [27, 28].

Further information on the specific details of the M1 and M2 optimization models is available in [3] and [4].

*2.2.2 Applications*

Unlike SCHEMA, OPTCOMB has seen little experimental use in the design of optimized recombinatorial libraries. This is somewhat surprising given that it was only recently demonstrated that SCHEMA could be used to generate an optimally sized library [10]. Nonetheless, the usefulness of OPTCOMB has been demonstrated during its development by Maranas and colleagues [3] and [4]. Saraf et al. applied OPTCOMB to design recombinatorial libraries of DHFR proteins, based on a variety of bacterial sequence variants. They were able to computationally verify the optimal size for the library that maximized both the diversity and quality of the library.

Pantazes et al. benchmarked OPTCOMB against the results of Otey et al. [9] where the SCHEMA algorithm was used to identify optimized crossover locations of a recombinatorial P450 library (*see* Subheading 2.1.2). They were able to demonstrate that the *S*2-scoring system of OPTCOMB was able to effectively distinguish functional and nonfunctional sequences from each other. Additionally, they observed that by using both M1 and M2 in conjunction with each other, a significantly higher library score (~60 % more than M1 alone) could be achieved with an increased number of fragments.

*2.2.3 Access/License/Requirements*

To the best of our knowledge, there is no software/webtool for OPTCOMB available online; it is up to the prospective user to implement OPTCOMB locally. The original research papers detailing the gradual development of OPTCOMB [3, 4, 29] provide the only information for implementation of OPTCOMB for potential users.

*2.2.4 Data Preparation*

Parental sequences included in the MSA should be those that contribute, either through recombination or mutations, to the mutants being scored. The scoring system itself is not particularly sensitive to the number of sequences in the parental MSA; however there should still be enough sequences included to provide a good statistical basis for the chosen protein family.

In order to cluster amino acids into the appropriate property bins and ultimately detect the presence of clashes between mutant and parental sequences, property values encoding charge, volume, and hydrophobicity need to be specified. Values used in OPTCOMB and FamClash are detailed in the respective papers [4, 25] and the premise behind their use is discussed in depth by Saraf et al. [29]. These property values, or alternatives to them, can also be obtained from alternative sources, such as the AAindex database [30].

#### 2.2.5 Tool Performance

Despite the fact that the *S*1- and *S*2-scoring systems in OPTCOMB appear to work well together, they do not consider existing structural data and do not directly measure or optimize mutant structural or functional viability—they only indicate how much a mutant differs from known members in the protein family. Both SCHEMA and OPTCOMB determine the viability of a mutant indirectly via comparison with its parents. Simple additive scoring models (employed by both methods) are not suited to properties that are confined to very specific sites or to the promotion of traits not applicable in the natural environment of the enzymes in question. Essentially, scores are based on the "summation" of properties across the whole sequence, with critical properties being mixed with conserved properties irrespective of the desired applications.

### 2.3 ProSAR

Ultimately directed evolution involves many rounds of breeding and selection (the screening and subsequent assessment of structural or functional properties). Traditional "hit-shuffling" [31, 32] uses a selection of the best mutants (chosen according to a particular assay and subject to experimental noise) as a template for subsequent rounds, thereby progressively refining the library design. Though not required by this simple strategy, sequencing of mutants provides explicit information of their genetic composition, offering an opportunity to establish a more detailed understanding of what enables (or disables) the activity in question. ProSAR (Protein Sequence Activity Relationship) mines the information present in the library as a whole, to "learn" a function from sequence to activity [6].

Briefly, ProSAR detects whether mutation of a specific residue is beneficial, neutral, or deleterious. It uses data collected progressively over an engineering experiment, to train a statistical regression model. In general, ProSAR is an example of a statistical analysis that attempts to view the problem prospectively from tendencies that are available in the library at hand: what is the fate of a mutant given particular changes to a sequence? To answer this, it is informative to look at the relationship between sequence composition and fitness, over all mutants.

#### 2.3.1 Technical Background

Given a set of mutants and data on their sampled activities, ProSAR attempts to build a statistical model that approximates the *local* fitness landscape and that can be used to extrapolate across the local sequence space, i.e., the sequence space defined by the set of residues being varied. Typically, large sections of mutant sequences will not display any differences to their parental sequences. Such sequence regions are not considered by ProSAR in the development of the local landscape model since they are invariant [33].

A pool of mutations, generated from a variety of diversification methods, can define the library for the initial mutant sample. For example, Fox and colleagues constructed between one and four libraries per round from a pool of 10 to 50 mutations [33].

Mutants are then sampled from these libraries and their activities characterized. The size of the mutation pools (or source of diversity) directly affects the efficiency of the ProSAR process—small pools do not provide the diversity required for substantial evolution whereas large pools present a more exhaustive space through which to search.

ProSAR finds the coefficients $c_{i,a}$ to fit sampled activities $y$ of mutants by regression, using the function

$$y = \sum_{i=1}^{N} \sum_{a \in A} c_{i,a} x_i(a) \tag{5}$$

where $A$ is the amino acid alphabet, and $N$ is the number of positions that form a mutant sequence $X$. $x_i(a) = 1$ if the $i$th amino acid in $X$ is $a$, $x_i(a) = 0$ otherwise. Note that sequences need to be standardized with respect to length, through a MSA. As there is a large number of coefficients and usually a relatively small number of samples, to avoid over-fitting issues ProSAR uses a partial least squares (PLS) approach which finds a linear regression model by projecting $y$ and $X$ onto a lower-dimensional space. Additionally, a user may wish to nominate a smaller subset of residues to be explored in this way.

Once the coefficients have been found they can be used to *predict* activities of novel sequences. ProSAR primarily uses the coefficients in Eq. 5 to determine which residues in a mutant should receive particular attention when designing the next round of mutagenesis (*see* Fig. 4). To maximize fitness, deleterious mutations are avoided, and mutations associated with increased activity receive preferential treatment. Neutral mutations can be either discarded [6] or retained to find potentially beneficial epistatic interactions [34].

#### 2.3.2 Applications

ProSAR has been used successfully for a number of applications [6, 35–38]. For instance, Fox and colleagues were able to evolve a halohydrin dehalogenase (HHDH) to the point where the volumetric productivity of a biocatalytic cyanation process was improved by ~4,000-fold [6]. This HHDH variant was improved sufficiently that it became economically viable for use in an industrial process [39].

#### 2.3.3 Access/License/Requirements

To the best of our knowledge, there is no implementation of ProSAR freely available online; rather it is used as a proprietary method in biocatalyst development by Codexis Inc. (Redwood City, CA). For information on the development of ProSAR *see* the original research papers [6, 33, 34].

#### 2.3.4 Data Preparation

ProSAR requires sequence data, in MSA format for ease of analysis, for a small sample of mutants from a library. Sampled mutants should also be assayed for their associated activity levels.

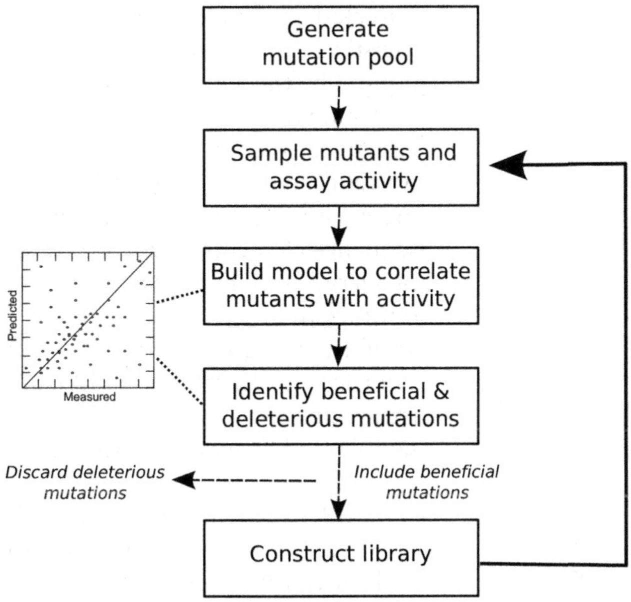


**Fig. 4** A diagram of the ProSAR work flow. A mutation pool, generated through a variety of mutation-inducing methods, provides the mutations for the initial library. Mutants are sampled from this library, sequenced, and their activities screened. Using the sequence-activity data, a statistical model is constructed to associate each mutation with a regression coefficient. Residues with the largest regression coefficients (most beneficial) are fixed for inclusion in the next library, with less beneficial mutations remaining variable, i.e., potentially beneficial mutations are re-assessed. Neutral mutations may also be retained while deleterious mutations are discarded

#### 2.3.5 Tool Performance

The coverage of the statistical analysis of ProSAR is not free from limitations. ProSAR treats all changes in isolation (a simple additive score)—it scores a mutant over all the chosen mutation positions independently. Fox and colleagues note, however, that combinations of positions can be included in the regression function [33], but give no guidance as to how to determine which combinations should be included.

The rationale for discarding neutral mutations is not discussed in any detail in recent publications on ProSAR [6], while in earlier work it has been suggested that such mutations can be retained to discover further, potentially beneficial, epistatic, interactions [34]. Excluding neutral mutations may be acceptable if the goal is to quickly discover and produce the variant with the best properties in the defined sequence space. However, later work by others has discussed (and demonstrated) that inclusion of neutral mutations can improve overall genetic diversity of the library and the evolvability of enzymes (thus avoiding possible evolutionary dead ends) and promote the potential for latent adaptations to a range of promiscuous activities [1, 40–42].

## 2.4 ASRA

Directed evolution experiments are usually limited by a major logistical issue—the number of mutants that can be analyzed is limited by the availability of high-throughput screening and selection systems. This means it is often difficult to efficiently gauge the effectiveness of protein design strategies, so that ultimately further, wasteful (both in time and cost) sampling and synthesis experiments are required to determine the most functionally beneficial sequence elements. Various other tools often rely upon the analysis of a large amount of sequence-function data in order to predict an optimized library, making them unsuitable for highly efficient small-sampling operations. In contrast, the Adaptive Substituent Reordering Algorithm (ASRA) attempts to address limits on sampling efficiency by identifying potential mutants from minimal sampling of focused libraries [5, 43–45]. In this respect, it is similar to ProSAR.

ASRA works on the premise that by synthesizing a small sample of a library with $N$ substitution positions (based on a scaffold) and measuring the property value for each sampled mutant, it is possible to derive the optimal substituent ordering for each substitution position, i.e., determine the amino acids with the most beneficial properties at that position. Substituents at each position are "reordered" to achieve the most regular (i.e., the smoothest) property landscape across all substitution positions. This reordering reveals the structure of the property landscape, allowing a search across it to efficiently discover the mutants with most desired properties. A simple visual observation of the landscape demonstrates this idea well, with an ordered landscape much easier to traverse than a highly disordered one.

### 2.4.1 Technical Background

The operation of ASRA occurs in three iterative steps.

Given a set of $N$ substitution positions and a scaffold sequence, at each substitution position ($i$; $i = 1,2,\ldots,N$) each unique amino acid (S) is given a random integer assignment ($X$), i.e., $X_i \in [1\ldots S_j]$. Typically with proteins, $S_j = 20$. This integer assignment is initially randomized—it neither requires nor assumes the existence of prior data. A subset of the possible library $\left(S_j^N\right)$ should be synthesized with simultaneous random mutations at each substitution position. Each sampled mutant should then have their property values measured (*see* Eq. 6).

In the work by Feng et al. [5], property values were a measurement of enantioselectivity for each mutant, obtained through HPLC. These were calculated as "$E$-values" using the following relationship [46]:

$$E = \frac{\ln\left[(1-\mathrm{ee}_s)/(1+\mathrm{ee}_s/\mathrm{ee}_p)\right]}{\ln\left[(1+\mathrm{ee}_s)/(1+\mathrm{ee}_s/\mathrm{ee}_p)\right]} = \frac{(k_{\mathrm{cat}}/k_M)\mathrm{fast}}{(k_{\mathrm{cat}}/k_M)\mathrm{slow}} \quad (6)$$

By the end of this step, each mutant should be associated with a $E$-value and an integer vector $[X_1,\ldots,X_i,\ldots,X_N]$. The sampled mutants, with their associated $E$-values and assignment vectors, form an $N$-dimensional space. Visual inspection of the property landscape for this space between two substitution positions ($X_{iA}$ and $X_{iB}$) would reveal it to be highly irregular (typically), due to the initial random integer assignment. This landscape will provide little to no predictive power. ASRA attempts to identify the optimal integer assignment for each amino acid at each substitution position that provides the most regular landscape. As suggested by Feng et al. [5], a measure of landscape regularity is calculated for each $m$-th amino acid ($Q_m$) with the following relationship:

$$Q_m = \sum_{m' \neq m} \sum_{n=1}^{20} \left(a_{mn} a_{m'n}\right) \frac{1}{1+\omega\sigma_{mn}} \frac{1}{1+\omega\sigma_{m'n}} \quad (7)$$

where $a_{mn}$ and $a_{m'n}$ are the property values observed for two variants with position 2 indexed by $n$ and position 1 indexed by $m$ and $m'$ respectively. The terms $\sigma_{mn}$ and $\sigma_{m'n}$ are the standard deviations of $a_{mn}$ and $a_{m'n}$ respectively. The coefficient $\omega$ is simply a weighting value. The $Q_m$ measurement indicates the influence each $m$th amino acid has on the property value relative to each other amino acid at position 1, averaged over all the amino acids on position 2 [5]. Minimization of $Q$ tends to group substituents with similar property values together and, with ordering of all the $Q$ measurements, results in the most regular landscape [45].

Based on the features of the optimized landscape/s, it is possible to predict which of the mutants are the "best" and interpolate the properties of those mutants that remained unsampled. Optimal mutants can be located either through a simple visual inspection of the property landscape/s or through alternative interpolation methods [43, 44]. Sampled mutants can either be accepted as satisfactory final products or can be used as parents for the next round of directed evolution.

#### 2.4.2 Applications

Although ASRA has been in development for some time [43–45], its use within a protein engineering environment was only recently demonstrated by [5]. Rabitz, Reetz, and colleagues applied ASRA, combined with an iterative saturation mutagenesis approach (ISM) [47], to successfully identify most (34/45) of the best enantioselective mutants of *Aspergillus niger* epoxide hydrolase (ANEH). The group used a previously evolved mutant LW202 ($E$=115) with $N$=2 substitution positions for a maximum library size of $20^N$=400. It should be noted that the close proximity of the two substitution positions, Phe215 (position 1) and Asn217 (position 2), demonstrates the ability of ASRA to function in the presence of potentially strong epistatic interactions.

*2.4.3 Access/License/Requirements*

No software or webtool for ASRA appears to be available online, nor does there appear to be any explicit licensing requirement for its implementation. For guidelines on the implementation and development of ASRA, the reader is referred to the original papers [5, 43–45].

*2.4.4 Data Preparation*

ASRA only requires a single scaffold sequence and a set of substitution positions to specify the sequence space from which the initial random library sample will be taken. As Feng et al. [5] demonstrated, it may be beneficial to select a mutant that has been evolved previously as the scaffold sequence. However, this ultimately will depend on the requirements of the experiment in question. Additionally, the possible amino acid targets for each substitution position will need to be defined, i.e., whether all 20 possible amino acids are allowed for each position or fewer. The number of substitution positions and possible amino acids allowed at each position will have a direct effect on the size of the library.

*2.4.5 Tool Performance*

By exploiting regularity in structure–property relationships for a library of proteins, i.e., consistent biochemistry, ASRA is able to predict the variation of a particular property across the sampled sequence space. However, it is unable to extrapolate across amino acids that are unsampled across related substitution sites, i.e., the properties of amino acids that remain unobserved at alternative sites cannot be predicted. Additionally, since ASRA is primarily a property prediction tool, it is unable to provide any direct data for property–structure relationships [5, 45].

## 3 Discussion

The computational methods and models discussed within this review provide a powerful aid in mutant variant and library design efforts and are complementary to "wet-lab" methods currently in use. Although most of these tools have been documented over some time and have seen successful experimental use, the fact that many are not generally available as a downloadable or licensable product limits their immediate use for the prospective protein engineer. SCHEMA is the only tool that is provided within a well-developed and up-to-date downloadable toolbox, whereas the others discussed (and many others not mentioned) do not provide such a simple option. This limitation means a lab or research group will usually require an "in-house" programmer to develop, implement, and maintain such tools or need to enter into a collaboration with the developers or license holder.

The choice of which computational approach to use for a project is a decision that is influenced by two fundamental questions: what is the underlying experimental strategy for the project and what data is available? In terms of strategy, is the project attempting to explore a large sample of the possible sequence space (generate more diversity) or is the focus more towards "fine-tuning" an existing variant, i.e., is the strategy a prospective or retrospective one? If exploration, SCHEMA and OPTCOMB are the most appropriate options given their demonstrated uses in facilitating the design of large, mutationally varied and functionally enriched libraries. Alternatively, ASRA and ProSAR would be more appropriate options if the aim were to further evolve an existing variant by performing a narrow, yet efficient, exploration of the local sequence space. In the first round of a directed evolution experiment, ProSAR and ASRA assume very little and at this stage could be used in conjunction with SCHEMA and OPTCOMB to select constraints for the initial mutant population.

Each of the described tools has different requirements in terms of the data needed for effective implementation. For example, finding high-resolution crystal structures that are representative of a given set of proteins can prove difficult. If one does not wish to use a computationally modeled or closely related homologue structure, or there are no known homologues available, SCHEMA will not be applicable. SCHEMA, OPTCOMB, and ProSAR all require MSAs so an initial assessment needs to be made as to which sequence data to use to implement these tools. How best to determine the number of sequences required in the MSA to satisfactorily represent a protein family statistically is a difficult question to answer, given the unique differences between proteins, e.g., size, environmental niche, etc. For example, the *S*2-scoring system of OPTCOMB has similarities to mutual information (MI) analyses—one of the more widely applied methods for identifying residue dependencies involving the analyses of residue probabilities within and between each column of a MSA. Previous work with MI has suggested that the MSA should include more than 125 sequences in order account for background "phylogenetic influences" [48]; however whether or not such an approximation proves useful for a tool such as OPTCOMB is unknown.

Generalizing this discussion, tools such as SCHEMA and OPTCOMB may be more useful when there is no clear evolutionary direction for the library design or when it is desirable to obtain a large sampling "snapshot" of the possible evolutionary sequence space. As such, the data provided by these tools may *initially* prove more appealing to a statistician or bioinformatician rather than a protein engineer. In contrast, ASRA and ProSAR may initially appeal more to the protein engineer, as they can aid in the production of actively improved variants faster than an *en masse* approach.

## References


1. Peisajovich SG, Tawfik DS (2007) Protein engineers turned evolutionists. Nat Methods 4(12):991–994
2. Voigt CA, Martinez C, Wang ZG, Mayo SL, Arnold FH (2002) Protein building blocks preserved by recombination. Nat Struct Biol 9(7):553–558
3. Saraf MC, Gupta A, Maranas CD (2005) Design of combinatorial protein libraries of optimal size. Proteins 60(4):769–777
4. Pantazes RJ, Saraf MC, Maranas CD (2007) Optimal protein library design using recombination or point mutations based on sequence-based scoring functions. Protein Eng Des Sel 20(8):361–373
5. Feng XJ, Sanchis J, Reetz MT, Rabitz H (2012) Enhancing the efficiency of directed evolution in focused enzyme libraries by the adaptive substituent reordering algorithm. Chemistry 18(18):5646–5654
6. Fox RJ, Davis SC, Mundorff EC, Newman LM, Gavrilovic V, Ma SK, Chung LM, Ching C, Tam S, Muley S, Grate J, Gruber J, Whitman JC, Sheldon RA, Huisman GW (2007) Improving catalytic function by ProSAR-driven enzyme evolution. Nat Biotechnol 25(3): 338–344
7. Meyer MM, Silberg JJ, Voigt CA, Endelman JB, Mayo SL, Wang ZG, Arnold FH (2003) Library analysis of SCHEMA-guided protein recombination. Protein Sci 12(8):1686–1693
8. Meyer MM, Hochrein L, Arnold FH (2006) Structure-guided SCHEMA recombination of distantly related beta-lactamases. Protein Eng Des Sel 19(12):563–570
9. Otey CR, Landwehr M, Endelman JB, Hiraga K, Bloom JD, Arnold FH (2006) Structure-guided recombination creates an artificial family of cytochromes P450. PLoS Biol 4(5):e112
10. Romero PA, Stone E, Lamb C, Chantranupong L, Krause A, Miklos AE, Hughes RA, Fechtel B, Ellington AD, Arnold FH, Georgiou G (2012) SCHEMA-designed variants of human arginase I and II reveal sequence elements important to stability and catalysis. ACS Synth Biol 1(6):221–228
11. Endelman JB, Silberg JJ, Wang ZG, Arnold FH (2004) Site-directed protein recombination as a shortest-path problem. Protein Eng Des Sel 17(7):589–594
12. Smith MA, Rentmeister A, Snow CD, Wu T, Farrow MF, Mingardon F, Arnold FH (2012) A diverse set of family 48 bacterial glycoside hydrolase cellulases created by structure-guided recombination. FEBS J 279(24):4453–4465
13. Otey CR, Silberg JJ, Voigt CA, Endelman JB, Bandara G, Arnold FH (2004) Functional evolution and structural conservation in chimeric cytochromes P450: calibrating a structure-guided approach. Chem Biol 11:309–318
14. Heinzelman P, Komor R, Kanaan A, Romero P, Yu XL, Mohler S, Snow C, Arnold F (2010) Efficient screening of fungal cellobiohydrolase class I enzymes for thermostabilizing sequence blocks by SCHEMA structure-guided recombination. Protein Eng Des Sel 23(11):871–880
15. Heinzelman P, Snow CD, Smith MA, Yu XL, Kannan A, Boulware K, Villalobos A, Govindarajan S, Minshull J, Arnold FH (2009) SCHEMA recombination of a fungal cellulase uncovers a single mutation that contributes markedly to stability. J Biol Chem 284(39): 26229–26233
16. Heinzelman P, Snow CD, Wu I, Nguyen C, Villalobos A, Govindarajan S, Minshull J, Arnold FH (2009) A family of thermostable fungal cellulases created by structure-guided recombination. Proc Natl Acad Sci U S A 106(14):5610–5615
17. Li YG, Drummond DA, Sawayama AM, Snow CD, Bloom JD, Arnold FH (2007) A diverse family of thermostable cytochrome P450s created by recombination of stabilizing fragments. Nat Biotechnol 25(9):1051–1056
18. Landwehr M, Carbone M, Otey CR, Li YG, Arnold FH (2007) Diversification of catalytic function in a synthetic family of chimeric cytochrome P450s. Chem Biol 14(3):269–278
19. Larkin MA, Blackshields G, Brown NP, Chenna R, McGettigan PA, McWilliam H, Valentin F, Wallace IM, Wilm A, Lopez R, Thompson JD, Gibson TJ, Higgins DG (2007) Clustal W and clustal X version 2.0. Bioinformatics 23(21): 2947–2948
20. Eswar N, John B, Mirkovic N, Fiser A, Ilyin VA, Pieper U, Stuart AC, Marti-Renom MA, Madhusudhan MS, Yerkovich B, Sali A (2003) Tools for comparative protein structure modeling and analysis. Nucleic Acids Res 31(13): 3375–3380
21. Arnold K, Bordoli L, Kopp J, Schwede T (2006) The SWISS-MODEL workspace: a web-based environment for protein structure homology modelling. Bioinformatics 22(2):195–201
22. Kiefer F, Arnold K, Kunzli M, Bordoli L, Schwede T (2009) The SWISS-MODEL repository and associated resources. Nucleic Acids Res 37:D387–D392
23. Peitsch MC (1995) Protein modeling by e-mail. BioTechnology 13(7):658–660

24. Kelley LA, Sternberg MJE (2009) Protein structure prediction on the Web: a case study using the Phyre server. Nat Protoc 4(3): 363–371
25. Saraf MC, Horswill AR, Benkovic SJ, Maranas CD (2004) FamClash: a method for ranking the activity of engineered enzymes. Proc Natl Acad Sci U S A 101(12):4142–4147
26. Hiraga K, Arnold FH (2003) General method for sequence-independent site-directed chimeragenesis. J Mol Biol 330:287–296
27. Coco WM, Encell LP, Levinson WE, Crist MJ, Loomis AK, Licato LL, Arensdorf JJ, Sica N, Pienkos PT, Monticello DJ (2002) Growth factor engineering by degenerate homoduplex gene family recombination. Nat Biotechnol 20(12):1246–1250
28. Ness JE, Kim S, Gottman A, Pak R, Krebber A, Borchert TV, Govindarajan S, Mundorff EC, Minshull J (2002) Synthetic shuffling expands functional protein diversity by allowing amino acids to recombine independently. Nat Biotechnol 20(12):1251–1255
29. Saraf MC, Maranas CD (2003) Using a residue clash map to functionally characterize protein recombination hybrids. Protein Eng 16(12): 1025–1034
30. Kawashima S, Kanehisa M (2000) AAindex: amino acid index database. Nucleic Acids Res 28(1):374
31. Stemmer WP (1994) Rapid evolution of a protein in vitro by DNA shuffling. Nature 370(6488):389–391
32. Crameri A, Raillard SA, Bermudez E, Stemmer WPC (1998) DNA shuffling of a family of genes from diverse species accelerates directed evolution. Nature 391:288–291
33. Fox R (2005) Directed molecular evolution by machine learning and the influence of nonlinear interactions. J Theor Biol 234(2):187–199
34. Fox R, Roy A, Govindarajan S, Minshull J, Gustafsson C, Jones JT, Emig R (2003) Optimizing the search algorithm for protein engineering by directed evolution. Protein Eng 16(8):589–597
35. Ma SK, Gruber J, Davis C, Newman L, Gray D, Wang A, Grate J, Huisman GW, Sheldon RA (2010) A green-by-design biocatalytic process for atorvastatin intermediate. Green Chem 12(1):81–86
36. Liang J, Mundorff E, Voladri R, Jenne S, Gilson L, Conway A, Krebber A, Wong J, Huisman G, Truesdell S, Lalonde J (2010) Highly enantioselective reduction of a small heterocyclic ketone: biocatalytic reduction of tetrahydrothiophene-3-one to the corresponding (R)-alcohol. Org Process Res Dev 14(1): 188–192
37. Gooding OW, Voladri R, Bautista A, Hopkins T, Huisman G, Jenne S, Ma S, Mundorff EC, Savile MM (2010) Development of a practical biocatalytic process for (R)-2-methylpentanol. Org Process Res Dev 14(1):119–126
38. Savile CK, Janey JM, Mundorff EC, Moore JC, Tam S, Jarvis WR, Colbeck JC, Krebber A, Fleitz FJ, Brands J, Devine PN, Huisman GW, Hughes GJ (2010) Biocatalytic asymmetric synthesis of chiral amines from ketones applied to sitagliptin manufacture. Science 329(5989): 305–309
39. Thayer AM (2006) Competitors want to get a piece of lipitor. Chem Eng News 84(33): 26–27
40. Soskine M, Tawfik DS (2010) Mutational effects and the evolution of new protein functions. Nat Rev Genet 11(8):572–582
41. Tokuriki N, Tawfik DS (2009) Stability effects of mutations and protein evolvability. Curr Opin Struct Biol 19(5):596–604
42. Gumulya Y, Sanchis J, Reetz MT (2012) Many pathways in laboratory evolution can lead to improved enzymes: how to escape from local minima. Chembiochem 13(7):1060–1066
43. Liang F, Feng XJ, Lowry M, Rabitz H (2005) Maximal use of minimal libraries through the adaptive substituent reordering algorithm. J Phys Chem B 109(12):5842–5854
44. Shenvi N, Geremia JM, Rabitz H (2003) Substituent ordering and interpolation in molecular library optimization. J Phys Chem A 107(12):2066–2074
45. McAllister SR, Feng XJ, DiMaggio PA, Floudas CA, Rabinowitz JD, Rabitz H (2008) Descriptor-free molecular discovery in large libraries by adaptive substituent reordering. Bioorg Med Chem Lett 18(22):5967–5970
46. Faber K (2011) Biotransformations in organic chemistry. A textbook, 6th edn. Springer, Berlin
47. Sanchis J, Fernandez L, Carballeira J, Drone J, Gumulya Y, Hobenreich H, Kahakeaw D, Kille S, Lohmer R, Peyralans J, Podtetenieff J, Prasad S, Soni P, Taglieber A, Wu S, Zilly F, Reetz M (2008) Improved PCR method for the creation of saturation mutagenesis libraries in directed evolution: application to difficult-to-amplify templates. Appl Microbiol Biotechnol 81(2):387–397
48. Martin LC, Gloor GB, Dunn SD, Wahl LM (2005) Using information theory to search for co-evolving residues in proteins. Bioinformatics 21(22):4116–4124

# Chapter 22

# Designing Libraries of Chimeric Proteins Using SCHEMA Recombination and RASPP

**Matthew A. Smith and Frances H. Arnold**

## Abstract

SCHEMA is a method for designing libraries of novel proteins by recombination of homologous sequences. The goal is to maximize the number of folded proteins while simultaneously generating significant sequence diversity. Here, we use the RASPP algorithm to identify optimal SCHEMA designs for shuffling contiguous elements of sequence. To exemplify the method, SCHEMA is used to recombine five fungal cellobiohydrolases (CBH1s) to produce a library of more than 390,000 novel CBH1 sequences.

**Key words** Protein engineering, Homologous recombination, SCHEMA, RASPP, Chimeragenesis

## 1 Introduction

SCHEMA recombination shuffles sequence elements (blocks) defined by a set of crossover locations in homologous proteins to generate novel chimeric proteins [1] (*see* Fig. 1). Despite that fact that homologous mutations are more conservative than random mutations, a chimera containing many mutations is less likely to be functional than one closer in sequence to one of its parent proteins. SCHEMA recombination seeks to maximize the probability that a library of chimeric proteins will be functional by using structural information to select crossover locations that minimize disruption of the folded structure. Our metric for disruption is the number of pairs of residues within 4.5 Å of each other that are not present as that same pair in any of the parents. We refer to these nonnative residue-residue contacts as the SCHEMA energy (E) of a chimera. Minimizing the average SCHEMA energy (<E>) of all the chimeras in a library increases the fraction of functional chimeras [2]. For sequence elements that are contiguous along the polypeptide chain, we developed the RASPP [3] computational tool to identify crossovers that minimize <E>.

Elizabeth M.J. Gillam et al. (eds.), *Directed Evolution Library Creation: Methods and Protocols*, Methods in Molecular Biology, vol. 1179, DOI 10.1007/978-1-4939-1053-3_22, © Springer Science+Business Media New York 2014

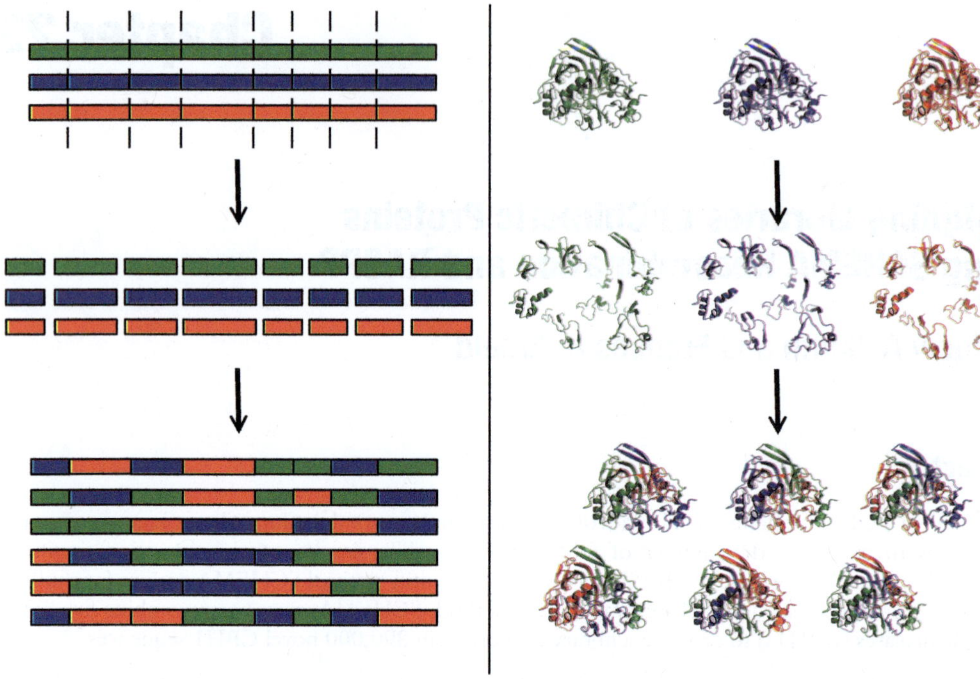

**Fig. 1** SCHEMA recombination. Homologous protein sequences are split into blocks at fixed crossover locations. These blocks are shuffled to generate novel chimeric proteins

Because chimeric proteins retain sequence elements (e.g., catalytic residues) that are shared among the parents, properly folded chimeras usually retain the overall function of the parents. The new combinations of amino acids in other parts of the protein, however, can lead to significant changes in key properties such as stability [4, 5], expression level [6], or substrate specificity [7]. By analyzing a subset of the possible chimera sequences, we can build predictive models and identify the chimeras having useful changes in those properties [8].

In this chapter, we describe the design of a SCHEMA library that recombines five fungal cellobiohydrolases (CBH1s) as an example. We use RASPP to identify optimal libraries having seven crossover sites (eight blocks). Shuffling these blocks among the five homologs generates a recombination library of $5^8 = 390{,}625$ possible sequences. We previously designed a very similar library [6], and the analysis of a subset of chimeras led us to identify chimeric CBH1s that are more stable than any of the five parents.

## 2 Materials

1. A Unix-based computer that can run python scripts (*see* **Note 1**). Python can be downloaded from http://www.python.org/download/.

2. The RASPP toolbox downloaded and unpacked. This is available from https://github.com/mattasmith/SCHEMA-RASPP/archive/master.zip.
3. A multiple sequence alignment of the parental sequences that are to be recombined (*see* **Note 2**). This alignment should be in ALN format (such as that produced by ClustalW), without a header (*see* **Note 3**). As recombination parents in the current example, we picked the CBH1 sequences from *Chaetomium thermophilum*, *Thermoascus aurantiacus*, *Hypocrea jecorina*, *Anaerocellum thermophilum*, and *Talaromyces emersonii*, which share approximately 60 % sequence identity. These CBH1s have a catalytic domain, a linker, and a cellulose-binding domain. The available crystal structures are for the catalytic domain; thus we only considered this domain for recombination (*see* **Note 4**). To eliminate the possibility of generating unpaired disulfide bonds, we mutated two residues in the *T. emersonii* and *T. aurantiacus* CBH1 sequences to cysteine (*see* **Note 5**). We used ClustalW2 [9] to align the parental sequences, and we named our alignment file "CBH1-msa.txt".
4. A PDB structure file of one of the parental sequences (*see* **Note 6**). We used the *T. emersonii* structure, "1Q9H.pdb".
5. A sequence alignment of one of the parental sequences with the sequence from the PDB structure file (*see* **Note 7**). We used ClustalW2 to align the parental sequences, and we named our alignment file "Temer-1Q9H.txt".

## 3 Methods

1. Place the parent sequence alignment file (CBH1-msa.txt), the PDB structure file (1Q9H.pdb), and the PDB alignment file (Temer-1Q9H.txt) in the "schema-tools" folder.
2. Run the following command (*see* **Note 8**) in the "schema-tools" directory:

```
python schemacontacts.py -pdb 1Q9H.pdb -msa CBH1-msa.txt -pdbal Temer-1Q9H.txt -o contacts.txt
```

   This generates a file containing the SCHEMA contacts called "contacts.txt" (*see* **Note 9**).

3. Run the following command (*see* **Note 10**) in the "schema-tools" directory:

```
python rasppcurve.py -msa CBH1-msa.txt -con contacts.txt -xo 7 -o opt.txt -min 15
```

   This RASPP script identifies a set of 8-block candidate libraries with low<E> (*see* **Note 11**). Each block is required to

a

| # <E> | <m> | crossover points |
|---|---|---|
| 16.6800 | 63.5313 | 33 73 105 152 206 374 415 |
| 16.0000 | 64.9914 | 33 73 107 149 204 366 415 |
| 16.2800 | 65.6751 | 33 73 107 149 206 366 415 |
| 16.6400 | 66.6670 | 33 73 107 149 212 366 415 |
| 17.4000 | 67.9348 | 33 73 107 149 204 359 415 |
| 17.6800 | 68.5816 | 33 73 107 149 206 359 415 |
| 18.8000 | 70.0666 | 33 73 107 149 212 356 415 |
| 19.6000 | 70.8609 | 33 73 107 175 221 356 415 |
| 20.4000 | 72.3239 | 77 107 149 212 351 374 415 |
| 20.8800 | 73.1818 | 33 73 107 175 250 366 415 |
| → 20.9600 | 73.8879 | 33 73 107 149 254 366 415 |
| 21.1600 | 74.6592 | 33 73 107 175 264 366 415 |
| 22.2800 | 75.6126 | 33 73 107 175 271 366 415 |
| 22.5600 | 76.6614 | 33 73 107 175 264 359 415 |
| 22.5600 | 78.2912 | 33 73 107 149 264 346 366 |
| 22.6400 | 78.6867 | 33 73 107 149 264 344 366 |
| 23.5600 | 79.6392 | 33 73 107 175 264 338 366 |
| 25.2000 | 80.7276 | 73 107 175 271 329 366 415 |
| 26.6800 | 81.5644 | 43 107 175 264 317 356 406 |

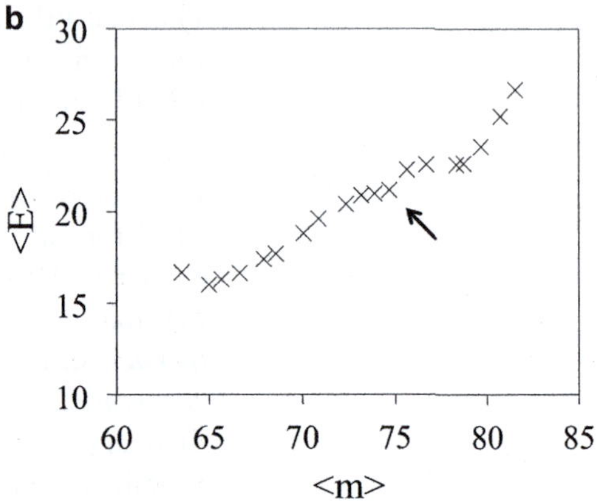


**Fig. 2** Libraries returned by RASPP. (**a**) The contents of "opt.txt", which lists the crossover locations of candidate libraries identified by RASPP. (**b**) A graph of the possible libraries plotting average SCHEMA energy (<E>) of each library against the average number of mutations (<m>). The trade-off between <E> and <m> is apparent. The chosen library is highlighted with an *arrow*. Many other library designs with low average SCHEMA energies and large numbers of mutations are also suitable choices

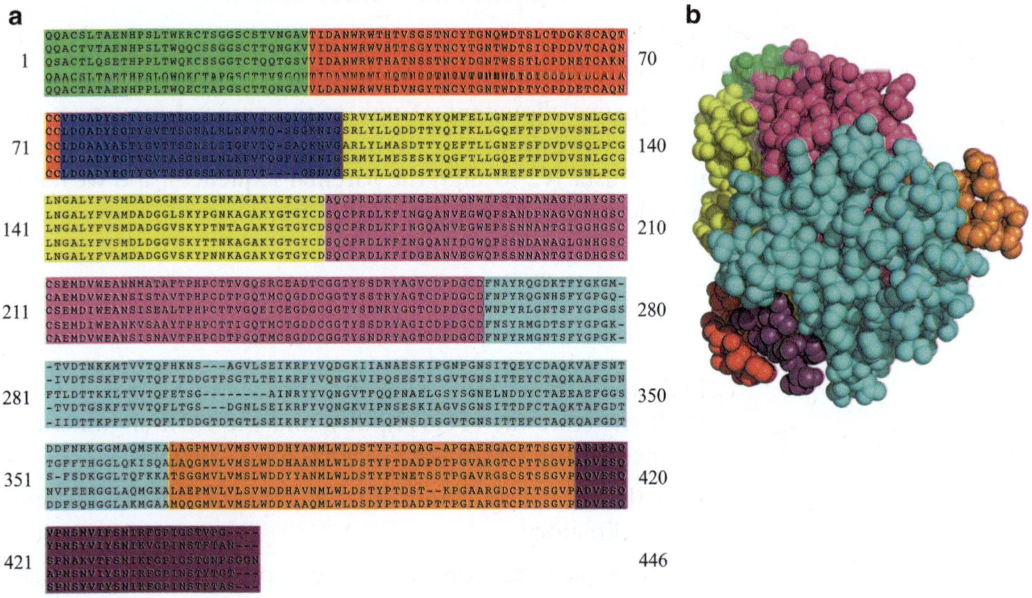


**Fig. 3** Visualizing the chosen RASPP design. (**a**) The multiple sequence alignment of the parent CBH1s with each of the eight blocks highlighted in a *different color*. (**b**) The blocks highlighted on the CBH1 structure "1Q9H.pdb"

have at least 15 mutations. These libraries are saved to the file "opt.txt" (*see* **Note 12**) (Fig. 2).

4. Pick a library from the results file "opt.txt" (*see* **Note 13**). In this case, we pick the library with crossover points [33 73 107 175 264 366 415], <E> = 21.2 and <m> = 74.7 (Fig. 3).

5. Create a text file called "CBH1-xo.txt" that contains the crossover points of the chosen library each separated by a space (*see* **Note 14**). The contents of the text file should be the following:

```
33 73 107 175 264 366 415
```

6. Run the following command (*see* **Note 15**) in the "schema-tools" directory:

```
python schemaenergy.py -msa CBH1-msa.txt -con contacts.txt -xo CBH1-xo.txt -E -m -o energies.txt
```

   This generates a list of all the chimeras in the chosen library along with their SCHEMA energies and number of mutations (*see* **Note 16**). This list is saved to the file "energies.txt".

7. At this point, we constructed a small chimera test set by substituting each block from each parent into the parental sequence from *T. emersonii*; the corresponding genes were synthesized (*see* **Note 17**). We could also have synthesized the genes encoding a different subset of the library (*see* **Note 18**) or even constructed the entire library (*see* **Note 19**). Before expressing the CBH1 chimeras, we add a linker and cellulose-binding domain to the recombined catalytic domains.

## 4 Notes

1. The RASPP toolbox "schema-tools" is written for python 2.6 on a Unix-based system. We recommend using this python release for the RASPP toolbox.
2. As a general rule, when picking sequences for SCHEMA recombination, we try to ensure the sequence identity between the homologs is not lower than ~55 % if individual genes are to be synthesized. In our experience, recombining sequences with much lower identities results in libraries with a high proportion of nonfunctional chimeras, even using SCHEMA. (This may not be a problem if the whole library is constructed and screened for functional chimeras.) The parental sequences are assumed to share the same fold; homologs with >55 % identity are likely to have very similar structures. If a structure is available for multiple parental sequences, we confirm they have the same fold by aligning the parental structures.
3. Lines starting with "#" are ignored in the multiple sequence alignment file. Sequence similarity symbols and trailing numbers are also ignored.
4. SCHEMA library designs require a protein structure. If no structural information is available for a parent sequence, but there are

structures of homologs, we can use MODELLER to build a structure model [10]. An inaccurate homology model hinders SCHEMA library design; an actual structure is preferred.

5. We assumed but did not verify that broken disulfide bonds are destabilizing. In this case, *C. thermophilum*, *H. jecorina*, and *A. thermophilum* CBH1s have ten disulfide bonds, while *T. aurantiacus* and *T. emersonii* have nine disulfide bonds. If the cysteines from the missing disulfide bond are in separate sequence blocks, chimeras with unpaired cysteines can result. We avoided this by modifying the parental sequences of *T. aurantiacus* and *T. emersonii* to include the remaining cysteine pair.
6. A structure is necessary to identify the residue-residue contacts. When possible, we pick a high-resolution structure (<2.0 Å).
7. The sequence of the PDB file can be extracted with the following (run from the "schema-tools" directory):

   ```
   python -c "import pdb; pdb.get('1Q9H.pdb')"
   ```

   We aligned this PDB sequence with the corresponding parent sequence (*T. emersonii* CBH1) from the parental alignment. The parent sequence must have the same identifier in both alignment files ("Temer"), and the identifier of the PDB sequence must be the name of the PDB structure ("1Q9H"). The PDB sequence can be identical to the parent sequence, but this is not always the case; often the PDB sequence will be truncated or contain several point mutations. In our case, we have mutated several of the residues in *T. emersonii* CBH1 to cysteine (*see* **Note 5**).
8. The python script "schemacontacts.py" calculates all of the SCHEMA contacts. Several arguments need to be provided when running this script:
   (a) "-pdb 1Q9H.pdb": name of the PDB structure
   (b) "-msa CBH1-msa.txt": name of the parental sequence alignment
   (c) "-pdbal Temer-1Q9H.txt": name of the PDB sequence alignment
   (d) "-o contacts.txt": name of an output file to store the contacts
9. Each contact is represented as a pair of residue numbers in "contacts.txt". Numbering is given in terms of both the parental sequence alignment and the PDB sequence alignment.
10. The python script "rasppcurve.py" finds crossover points that minimize the average SCHEMA energy for a library. Several arguments need to be provided when running this script:
    (a) "-msa CBH1-msa.txt": name of the parental sequence alignment
    (b) "-con contacts.txt": name of the contacts file

(c) "-xo 7": number of crossovers

(d) "-min 15": minimum number of nonidentical residues in a block (prevents trivial solutions)

(e) "-o opt.txt": name of an output file for the results

This script may take several hours to complete, depending on protein size and computer specifications. Increasing the number of crossovers in a library increases library size and reduces the average number of mutations in a block. The user may want smaller blocks if searching for properties from single point mutations. However, it is harder to find desirable chimeras in larger libraries, and increasing the number of blocks increases the <E> of a library. We chose to split our five parent proteins into eight blocks.

11. There is a trade-off between the average SCHEMA energy of a library (<E>) and the average number of mutations from the closest parent (<m>), which depends on the relative block sizes (*see* Fig. 2b). If all the blocks are evenly sized, <m> is very high, but the solution space of possible libraries is very small and so < E> is large. As block sizes become uneven, the solution space of possible libraries increases. This enables RASPP to find libraries with lower <E>, but these libraries have lower <m>. RASPP is designed to find low <E> libraries for a range of <m>.

12. Each library is defined by seven crossover points. The crossover points are given by the first residue of each new fragment (excluding the first fragment, which is always 1) based on the numbering of the parental sequence alignment. The results file "opt.txt" also gives <E> and the average number of mutations from the closest parent (<m>) for each library.

13. RASPP returns a set of candidate libraries with a range of <m> values. A lower <E> implies more functional chimeras in the library. For moderately sized proteins (250–500 amino acids), we try to pick SCHEMA libraries with <E> less than 30. Protein-specific biochemical and structural knowledge may help users pick from the candidate libraries.

14. Lines starting with "#" are ignored in the crossover file.

15. The python script "schemaenergy.py" lists the chimeras in a library. Several arguments need to be provided when running this script:

    (a) "-msa CBH1-msa.txt": name of the parental sequence alignment

    (b) "-con contacts.txt": name of the contacts file

    (c) "-xo CBH1-xo.txt": name of the crossover file that defines the library

(d) "-E -m": specifies that the chimeras should be listed with their E and m values

(e) "-o energies.txt": name of an output file for the results

16. Chimeras are numbered according to the parental sequence of each block with the numbers ordered from the first block to the last block. Parents are numbered based on the order they appear in the parental sequence alignment. For example, chimera "14221313" has parent 1 as the sequence of its first block, parent 4 as its second block, etc.

17. The fungal CBH1 enzymes have poor heterologous expression in *Saccharomyces cerevisiae*. Because *T. emersonii* CBH1 expresses much better than the other parents, we analyzed the blocks one at a time in the background of *T. emersonii* CBH1. These chimeras tend to have low SCHEMA energies, and they can be easily constructed via overlap extension PCR. Using this "monomera" approach, we identified stable CBH1 chimeras in a SCHEMA library similar to the one presented here [6].

18. We pick a subset of the library to analyze. We ensure every block from every parent is represented independently of one another in this subset. This enables us to model the effect blocks have on biochemical properties such as stability [5].

19. It is possible to construct an entire SCHEMA library in the laboratory by assembling blocks of sequence with specific overhangs [11, 12]. This approach is appropriate for searching for chimeras with specific properties that cannot be predicted from a small library sample.

## Acknowledgments

The authors acknowledge funding from the Institute for Collaborative Biotechnologies through grant W911NF-09-D-0001 from the US Army Research Office and the National Central University, Taiwan, through a Cooperative Agreement for Energy Research Collaboration. M.A.S. is supported by a Resnick Sustainability Institute fellowship.

### References

1. Voigt CA, Martinez C, Wang Z-G, Mayo SL, Arnold FH (2002) Protein building blocks preserved by recombination. Nat Struct Biol 9:553–558
2. Meyer M, Hochrein L, Arnold FH (2006) Structure-guided SCHEMA recombination of distantly related β-lactamases. Protein Eng Des Sel 19:563–570
3. Endelman J, Silberg J, Wang Z, Arnold FH (2004) Site-directed protein recombination as a shortest-path problem. Protein Eng Des Sel 17:589–594
4. Romero P, Stone E, Lamb C, Chantranupong L, Krause A, Miklos A, Hughes R, Fechtel B, Ellington AD, Arnold FH, Georgiou G (2012) SCHEMA-designed variants of human

Arginase I and II reveal sequence elements important to stability and catalysis. ACS Synth Biol 1:221–228

5. Li Y, Drummond DA, Sawayama AM, Snow CD, Bloom JD, Arnold FH (2007) A diverse family of thermostable cytochrome P450s created by recombination of stabilizing fragments. Nat Biotechnol 25:1051–1056
6. Heinzelman P, Komor R, Kanaan A, Romero PA, Yu X, Mohler S, Snow C, Arnold FH (2010) Efficient screening of fungal cellobiohydrolase class I enzymes for thermostabilizing sequence blocks by SCHEMA structure-guided recombination. Protein Eng Des Sel 23: 871–880
7. Otey CR, Landwehr M, Endelman JB, Hiraga K, Bloom JD, Arnold FH (2006) Structure-guided recombination creates an artificial family of cytochromes P450. PLoS Biol 4:e112
8. Heinzelman P, Romero PA, Arnold FH (2013) Efficient sampling of SCHEMA chimera families for identification of useful sequence elements. In: Keasling A (ed) Methods in enzymology: methods in protein design. Elsevier Ltd, Oxford, UK
9. Larkin MA, Blackshields G, Brown NP, Chenna R, McGettigan PA, McWilliam H, Valentin F, Wallace IM, Wilm A, Lopez R, Thompson JD, Gibson TJ, Higgins DG (2007) Clustal W and Clustal X version 2.0. Bioinformatics 23: 2947–2948
10. Eswar N, Webb B, Marti-Renom MA, Madhusudhan MS, Eramian D, Shen M-Y, Pieper U, Sali A (2007) Comparative protein structure modeling using MODELLER. Curr Protoc Protein Sci 50:2.9.1–2.9.31
11. Hiraga K, Arnold FH (2003) General method for sequence-independent site-directed chimeragenesis. J Mol Biol 330:287–296
12. Farrow MF, Arnold FH (2010) Combinatorial recombination of gene fragments to construct a library of chimeras. Curr Protoc Protein Sci 62:26.2.1–26.2.20

# Chapter 23

# Noncontiguous SCHEMA Protein Recombination

**Matthew A. Smith and Frances H. Arnold**

## Abstract

SCHEMA is a method of designing protein recombination libraries that contain a large fraction of functional proteins with a high degree of mutational diversity. In the previous chapter, we illustrated the method for designing libraries by swapping contiguous sequence elements. Here, we introduce the NCR ("noncontiguous recombination") algorithm to identify optimal designs for swapping elements that are contiguous in the 3-D structure but not necessarily in the primary sequence. To exemplify the method, NCR is used to recombine three fungal cellobiohydrolases (CBH1s) to produce a library containing more than 500,000 novel chimeric sequences.

**Key words** Protein engineering, Homologous recombination, SCHEMA, Noncontiguous recombination, NCR, Chimeragenesis

## 1 Introduction

As discussed in the previous chapter, SCHEMA [1] seeks to maximize the probability that a library of chimeric proteins will be functional by using structural information to identify sequence elements ("blocks") that can be swapped. It is advantageous to minimize the average SCHEMA energy (<E>) of all the chimeras in a library, as this increases the fraction of functional chimeras [2]. When recombining sequence elements that are contiguous along the polypeptide chain, RASPP [3] is used to identify optimal crossovers that minimize <E>.

In this chapter, we describe the designing of SCHEMA libraries with even lower <E>s by removing the constraint that blocks be contiguous along the polypeptide chain. These noncontiguous blocks of sequence are still contiguous blocks of structure in the folded protein. We use noncontiguous recombination (NCR) [4] to computationally search for optimal noncontiguous SCHEMA library designs. This approach to chimera design has become feasible now that the genes can be made by total gene synthesis.

Elizabeth M.J. Gillam et al. (eds.), *Directed Evolution Library Creation: Methods and Protocols*, Methods in Molecular Biology, vol. 1179, DOI 10.1007/978-1-4939-1053-3_23, © Springer Science+Business Media New York 2014

Here, we design a SCHEMA library that recombines three fungal cellobiohydrolases (CBH1s) splitting each homolog into 12 blocks. Shuffling these blocks generates a chimera library of $3^{12} = 531{,}441$ possible sequences. Analysis of a library designed in the same manner as that described here [5] allowed the identification of several stabilizing sequence elements. NCR-designed libraries can have significantly lower disruption than RASPP (contiguous) designs from the same parent sequences. Alternatively, NCR enables recombination of parents with lower sequence identity. We recommend analysis of NCR-designed libraries by making an informative sample set of genes and using those to build predictive models, as we have done for RASPP-designed libraries [6].

## 2 Materials

1. A Unix-based computer that can run python scripts (*see* **Note 1**). Python can be downloaded from http://www.python.org/download/.
2. The NCR toolbox downloaded and unpacked. This is available from https://github.com/mattasmith/SCHEMA-RASPP/archive/master.zip.
3. MUSCLE (*see* **Note 2**). This is available for download from

   http://www.drive5.com/muscle/downloads.htm

   Unpack the compressed file and place the executable in the directory "ncr/tools/muscle" (*see* **Note 3**).
4. hmetis (*see* **Note 4**). This is available for download from

   http://glaros.dtc.umn.edu/gkhome/metis/hmetis/download

   Unpack the compressed file and place the hmetis folder in the directory "ncr/tools" (*see* **Note 3**).
5. A multiple sequence alignment of the parental sequences to be recombined (*see* **Note 5**). This alignment should be in FASTA format (*see* **Note 6**), and the file should be named "alignment.fasta". As recombination parents for the example used here, we selected the CBH1 sequences from *Chaetomium thermophilum*, *Hypocrea jecorina*, and *Talaromyces emersonii*, which have about 60 % pairwise amino acid sequence identity. These CBH1s have a catalytic domain, a linker, and a cellulose-binding domain. The available crystal structures are for the catalytic domain; thus we only considered this domain for recombination (*see* **Note 7**). To eliminate the possibility of generating unpaired disulfide bonds, we mutated two residues in the *T. emersonii* CBH1 sequence to cysteine (*see* **Note 8**). We used PROMALS3D [7] to align the parental sequences.

6. A PDB structure file of one of the parental sequences (*see* **Note 9**). We use the *T. emersonii* structure, "1Q9H.pdb". Alternatively, if no structure is provided, the NCR tools can search for suitable structures from the PDB database (*see* **Note 10**).

## 3 Methods

1. Place the parent sequence alignment file (alignment.fasta) in the "ncr" folder. Place the PDB structure file (1Q9H.pdb) in the directory "ncr/structures".
2. Set the "Number of blocks" to 12 and "Find all PDB structures" to 0 in the "init.txt" file (*see* **Note 10**).
3. Run the following command (*see* **Note 11**) in the "ncr" directory:

   ```
   python ncr.py
   ```

   This NCR script identifies a set of candidate libraries with low <E> and sends these results to the terminal window (*see* **Note 12**) (Fig. 1). These libraries are saved in the directory "ncr/output" and listed in the text file "library12_result_list.csv" (*see* **Note 13**).
4. Pick an NCR library (*see* **Note 14**). In this case, we pick the library "library12_2.output", with <E> = 16.8 and <m> = 83.9 (Fig. 2).
5. Certain non-conserved residues still need to be assigned to blocks (*see* **Note 15**). Open "ncr/output/library12_2.output" and assign residues 41, 175, 197, 199, 202, and 442 to blocks G, C, A, A, A, and J, respectively (*see* **Note 16**).

```
Designing libraries...
   library12_1: E = 18.222, m = 84.035, blocks = [17 17 19 15 17 19 17 17 17 17 19 17 ]
-> library12_2: E = 16.778, m = 83.929, blocks = [15 19 16 16 20 20 15 19 17 15 18 18 ]
   library12_3: E = 17.667, m = 83.588, blocks = [15 15 20 16 18 18 17 21 18 19 14 17 ]
   library12_4: E = 16.778, m = 83.779, blocks = [14 16 19 17 20 20 14 17 19 15 19 18 ]
   library12_5: E = 16.222, m = 83.589, blocks = [14 19 17 14 18 20 21 21 18 17 16 13 ]
   library12_6: E = 16.444, m = 83.422, blocks = [14 18 18 15 20 17 21 22 16 13 17 17 ]
   library12_7: E = 16.222, m = 83.412, blocks = [12 17 17 21 22 17 13 17 20 16 18 18 ]
   library12_8: E = 13.556, m = 82.819, blocks = [10 20 17 28 22 19 12 15 18 10 17 20 ]
   library12_9: E = 15.444, m = 82.939, blocks = [11 21 14 26 17 22 12 17 22 11 17 18 ]
  library12_10: E = 13.556, m = 82.682, blocks = [10 19 17 10 12 18 16 17 20 30 17 22 ]
  library12_11: E = 13.778, m = 82.635, blocks = [10 17 20 30 17 22 11 20 14 11 18 18 ]
  library12_12: E = 15.889, m = 82.777, blocks = [10 17 22 11 23 19 17 15 9 26 22 17 ]
  library12_13: E = 13.333, m = 82.507, blocks = [9 13 13 10 22 17 18 20 17 30 22 17 ]
  library12_14: E = 13.778, m = 82.545, blocks = [12 18 18 10 20 19 30 22 17 11 20 11 ]
  library12_15: E = 12.222, m = 81.440, blocks = [11 15 24 10 20 22 12 35 22 6 20 11 ]
```

**Fig. 1** Libraries returned by NCR. The average SCHEMA energy (<E>) and average number of mutations (<m>) for each library are printed to the terminal window. In addition, the output displays the distribution of the mutations among the 12 blocks. Libraries with higher <E> have more evenly sized blocks. The chosen library is highlighted with an *arrow*. We picked this library because we wanted evenly sized blocks to help us efficiently search for stabilizing mutations

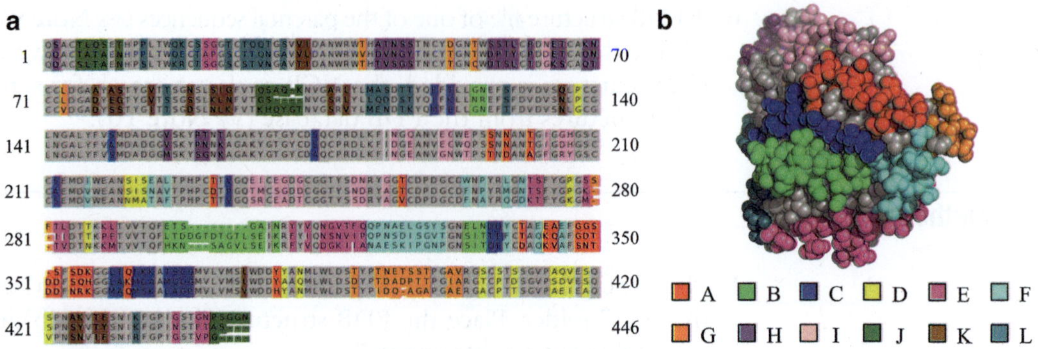


**Fig. 2** Visualizing the chosen NCR design. (**a**) The multiple sequence alignment of the parent CBH1s with each of the 12 blocks highlighted in a *different color*. Conserved residues are colored *gray*. It is clear that the blocks are noncontiguous along the polypeptide chain. (**b**) The blocks highlighted on the CBH1 structure "1Q9H.pdb". Most of the blocks are contiguous structural elements in 3-D

6. Run the following command (*see* **Note 17**) in the "ncr" directory:

```
python picklibrary.py library12_2
```

   This generates a list of all the chimeras in the chosen library along with their SCHEMA energies, number of mutations, and sequences (*see* **Note 18**). This list is saved as a text file "chimeras.output" in the directory "ncr/picked_libraries/library12_2".

7. We synthesize the genes encoding a subset of the chimera library (*see* **Note 19**). Before expressing the CBH1 chimeras, we add a linker and cellulose-binding domain to the recombined catalytic domains.

## 4 Notes

1. The NCR toolbox "ncr" is written for python 2.6 on a Unix-based system. We recommend using this python release for the NCR toolbox.
2. Ensure you download the correct distribution of MUSCLE for your system. For example, on Apple OS X, it might be "muscle3.8.31_i86darwin64.tar.gz". The NCR tools were written for MUSCLE 3.8.
3. The NCR toolbox unpacks as a folder called "ncr". Directories are given relative to this folder. For example, there is a folder in "ncr" called "tools" and the directory would be "ncr/tools".
4. Ensure you download the correct distribution of hmetis for your system. For example, on Apple OS X, it might be "hmetis-1.5-osx-i686.tar.gz". The NCR tools were written for hmetis 1.5.
5. We assume the parental proteins share the same structural fold. If structures are available for more than one parental protein,

we confirm the parents have the same fold by aligning the parental structures. It is important that the sequence alignment is accurate, especially when the parental sequence identities are low.

6. In FASTA format, the name of each sequence begins with “>”, for example, “>Temersonii”. After each name, there should be a return, followed by the corresponding aligned sequence.
7. SCHEMA library designs require a protein structure. If no structural information is available for a parent sequence, but there are structures of homologs, use MODELLER to build a structure model [8]. An inaccurate homology model hinders SCHEMA library design; an actual structure is preferred.
8. We assumed but did not verify that broken disulfide bonds are destabilizing. In this case, *C. thermophilum* and *H. jecorina* CBH1s have ten disulfide bonds, while *T. emersonii* has nine disulfide bonds. If the cysteines from the missing disulfide bond are in separate sequence blocks, chimeras with unpaired cysteines can result. We avoided this by modifying the parental sequence of *T. emersonii* to include the remaining cysteine pair.
9. One or more structures is needed to identify the residue-residue contacts. When possible, we select high-resolution structures (<2.0 Å). If a PDB file contains more than one chain, each chain is automatically split into its own structure file labeled XXXX.A.pdb, XXXX.B.pdb, etc. The NCR tools can handle multiple structures. Residue-residue contacts from multiple structures of the same parent form a parental contact map if these contacts are present in at least 50 % of the structures. If structures from multiple parents are used, each contact is weighted by the fraction of parental contact maps it appears in.
10. The “init.txt” file is in the “ncr” folder. It specifies two parameters for the NCR toolbox:
    (a) “Number of blocks”: The number of blocks in the designed libraries. It can be either a number (e.g., 8) or a range of numbers (e.g., 2–6) for designing a range of libraries with different block sizes.
    (b) “Find all PDB structures”: If 1, the NCR script will search, download, and use all suitable structures from the PDB database. If 0, the user will provide one or more structures.

    Increasing the number of blocks in a library increases library size and reduces the average number of mutations in a block. The user may want smaller blocks if searching for single mutations that cause specific functional changes. However, it is harder to find desirable chimeras in larger libraries, and increasing the number of blocks increases the <E> of a library. We chose to split our three parental proteins into 12 blocks.

11. The python script “ncr.py” generates one or more parental contact maps, calculates the SCHEMA contacts, and searches for low <E> libraries. This script may take several hours to complete, depending on protein size and computer specifications. Progress is displayed in the terminal window. The script uses heuristic algorithms to find near-optimal solutions; thus results will vary slightly each time “ncr.py” is run. However, these variations are very small, and we find we do not identify significantly lower <E> library designs by running the algorithm multiple times.

12. In the terminal window, NCR lists <E> and <m> for each library as well as the distribution of mutations among the 12 blocks. This distribution is given as a list of 12 numbers, each referring to the number of mutations in a block with blocks counting A, B, C, etc. There is a trade-off between the average SCHEMA energy of a library (<E>) and how evenly distributed mutations are among the blocks. If all the blocks are evenly sized, the solution space of possible libraries is small and so <E> is large. As block sizes become uneven, the solution space of possible libraries increases. This enables NCR to find libraries with lower <E>, but libraries with very unevenly sized blocks may not be useful, as it will be more difficult to identify sequence-function relationships in very large blocks, and very small blocks contain few mutations. NCR is designed to find low <E> libraries for a range of block sizes.

13. In noncontiguous recombination, a library is defined by assigning every non-conserved residue to a block. In the library text file “library12_2.output”, a designated block (named “A”, “B”, “C”, etc.) appears beside every non-conserved residue. A dash (“-”) is placed next to every conserved residue. Residues are numbered based on the parental sequence alignment. The results file “library12_result_list.csv” lists <E> and <m> for each library.

14. NCR returns a set of candidate libraries with a range of <m> values. A lower <E> implies more functional chimeras in the library. For moderately sized proteins (250–500 amino acids), we try to pick SCHEMA libraries with <E> less than 30. For noncontiguous recombination of homologs with >55 % sequence identity, often all the candidate libraries have <E> below 30. In our case, we pick a library with evenly sized blocks. This will make it easier to identify stabilizing point mutations within a stabilizing block. Protein-specific biochemical and structural knowledge may also help users select from the candidate libraries. For example, one may wish to conserve a specific region of protein structure, such as an allosteric site or active site, by choosing a library design that has the structural feature in a single block. Note that the <E> value is lower and the <m> value higher in this NCR design than the RASPP design from the previous chapter.

Blocks are not always one contiguous piece of structure. Sometimes, a group of residues will only have SCHEMA contacts with one another and not with the rest of the protein. These "disconnected blocks" can belong to any block without altering <E>. NCR will assign these disconnected blocks to blocks such that <m> is maximized. This can result in a block comprising two separate pieces of structure. These disconnected blocks are apparent when blocks are visualized on the PDB structure. In this case, blocks "A", "G", and "J" each contain a disconnected block.

15. Some non-conserved residues do not have any SCHEMA contacts. These residues often appear on the surface of the protein, in a region that is highly conserved, or in a region where structural information is missing. NCR does not assign these residues to a block, and instead the decision is left to the user. Unassigned residues are printed to the terminal. In this case, residues 41, 175, 197, 199, 202, and 442 have not been assigned a block.
16. Looking at the structure "1Q9H.pdb", we designate each unassigned residue to the same block as one of its neighboring residues. This will slightly alter <m> for the library but leave <E> unaffected. We can alter the block assignments by editing the text file "ncr/output/library12_2.output". In this file, unassigned residues, like conserved residues, have a dash ("-") in place of a block ("A", "B", "C", etc.).
17. The python script "picklibrary.py" generates all the chimeras in a given library. The name of the library "library12_2" needs to be provided as a parameter and appears after "python picklibrary.py". Any non-conserved residues that have not been assigned to a block will be automatically assigned to block A. For a large library such as this one (more than 500,000 chimeras), this script may take several hours to complete.
18. Chimeras are numbered according to the parental sequence of each block with the numbers ordered from the first to the last block. Parents are numbered based on the order they appear in the parental sequence alignment. For example, chimera "132213131322" has parent 1 as the sequence of its first block ("A"), parent 3 as its second block ("B"), etc. The amino acid sequence provided alongside each chimera in "chimeras.output" is built from the parent sequence alignment. It contains dashes ("-") where there are gaps in the alignment. These dashes should be removed when ordering the synthetic genes.
19. These chimeras are very difficult to construct with traditional cloning techniques; each noncontiguous block will usually have mutations distributed throughout the protein sequence. We pick a subset of the library to synthesize and analyze. Typically this subset contains 20–40 chimera sequences and is limited by synthesis cost. We ensure every block from every

parent is represented independently of one another in this subset. This enables us to model the effects of the different blocks on biochemical properties such as stability [9]. We pick a set of chimeras to be most informative using the Submodular Function Optimization Matlab toolbox [10, 11]. Alternatively, we could have selected a set of chimeras that substitute one block at a time into the background of a parent that expresses well, such as *T. emersonii* CBH1 [12].

## Acknowledgments

The authors acknowledge funding from the Institute for Collaborative Biotechnologies through grant W911NF-09-D-0001 from the US Army Research Office and the National Central University, Taiwan, through a Cooperative Agreement for Energy Research Collaboration. M.A.S. is supported by a Resnick Sustainability Institute fellowship.

### References

1. Voigt CA, Martinez C, Wang Z-G, Mayo SL, Arnold FH (2002) Protein building blocks preserved by recombination. Nat Struct Biol 9:553–558
2. Meyer M, Hochrein L, Arnold FH (2006) Structure-guided SCHEMA recombination of distantly related β-lactamases. Protein Eng Des Sel 19:563–570
3. Endelman J, Silberg J, Wang Z, Arnold FH (2004) Site-directed protein recombination as a shortest-path problem. Protein Eng Des Sel 17:589–594
4. Smith MA, Romero PA, Wu T, Brustad EM, Arnold FH (2013) Chimeragenesis of distantly-related proteins by noncontiguous recombination. Protein Sci 22:231–238
5. Smith MA, Bedbrook CN, Wu T, Arnold FH (2013) Hypocrea jecorina cellobiohydrolase I stabilizing mutations identified using noncontiguous recombination. ACS Synth Biol. doi:10.1021/sb400010m
6. Heinzelman P, Romero PA, Arnold FH (2013) Efficient sampling of SCHEMA chimera families for identification of useful sequence elements. In: Keasling A (ed) Methods in enzymology: methods in protein design. Elsevier Ltd, Oxford, UK
7. Pei J, Kim B-H, Grishin NV (2008) PROMALS3D: a tool for multiple protein sequence and structure alignments. Nucleic Acids Res 36:2295–2300
8. Eswar N, Webb B, Marti-Renom MA, Madhusudhan M, Eramian D, Shen MY, Pieper U, Sali A (2007) Comparative protein structure modeling using Modeller. Curr Protoc Protein Sci 2:15–32
9. Li Y, Drummond DA, Sawayama AM, Snow CD, Bloom JD, Arnold FH (2007) A diverse family of thermostable cytochrome P450s created by recombination of stabilizing fragments. Nat Biotechnol 25:1051–1056
10. Krause A (2010) SFO: a toolbox for submodular function optimization. J Mach Learn Res 11:1141–1144
11. Romero P, Stone E, Lamb C, Chantranupong L, Krause A, Miklos A, Hughes R, Fechtel B, Ellington AD, Arnold FH, Georgiou G (2012) SCHEMA-designed variants of human Arginase I and II reveal sequence elements important to stability and catalysis. ACS Synth Biol 1:221–228
12. Heinzelman P, Komor R, Kanaan A, Romero PA, Yu X, Mohler S, Snow C, Arnold FH (2010) Efficient screening of fungal cellobiohydrolase class I enzymes for thermostabilizing sequence blocks by SCHEMA structure-guided recombination. Protein Eng Des Sel 23:871–880

# Chapter 24

# Engineering Proteins by Reconstructing Evolutionary Adaptive Paths

**Vanessa E. Cox and Eric A. Gaucher**

## Abstract

Reconstructing evolutionary adaptive paths (REAP) is a low-throughput technique used to design protein libraries that can be assayed for specific properties such as catalytic function or thermostability. This approach takes advantage of natural selection by using theoretical ancestral proteins as the foundation for library variants. REAP gives rise to smaller libraries but with a higher ratio of viable proteins than other high-throughput techniques. REAP uses analyses of ancestral sequences and signatures of functional divergence to modify extant protein sequences. This allows the experimenter to statistically evaluate which amino acid mutations in which sites within the protein are most likely to produce functional proteins having varied phenotypes.

**Key words** Protein engineering, Ancestral sequence reconstruction, Phylogenetic analysis, Functional divergence, Molecular evolution

## 1 Introduction

With the increase in availability of sequence data, protein engineering can move from creating large, randomly designed libraries to small libraries that utilize highly directed protein engineering. REAP is a low-throughput method that exploits the evolutionary history of a protein family in order to engineer a small, highly functional, phenotypically diverse library of protein variants. The concept behind REAP is to use sequence signatures of functional divergence to identify residues implicated in altered protein phenotype [1]. This, combined with ancestral sequence reconstruction, creates a small library with a high density of functional proteins. These proteins can be assayed for phenotypes including increased acid or base stability and increased thermostability.

Identifying amino acid sites with low replacement rates suggests which residues likely play a role in protein phenotype. Mutating each of these sites with all twenty possible amino acids would quickly create a library too large to be practical. Using the

Elizabeth M.J. Gillam et al. (eds.), *Directed Evolution Library Creation: Methods and Protocols*, Methods in Molecular Biology, vol. 1179, DOI 10.1007/978-1-4939-1053-3_24, © Springer Science+Business Media New York 2014

evolutionary history of a protein and recreating ancestral sequences focuses a library on the site mutations most likely to impart phenotypic diversity while still retaining protein function. By using sequences that theoretically once existed, one can create a higher density of functionally active variants in a library. Incorporating the knowledge gleaned from ancestral sequences allows the researcher to sample sequence space which nature has, in principle, already tested and deemed viable.

Other methods to guide library design do exist and include such techniques as DNA shuffling, consensus sequence approach, ancestral mutation method, and ancestral sequence reconstruction. Due to the unbiased approach, DNA shuffling can be a useful tool; however, it generally results in a large library with many nonfunctional proteins [2]. The expense and difficulty of assaying large libraries, combined with the focused manner in which REAP directs amino acid replacements, makes REAP an appealing alternative approach. A consensus sequence can result in a smaller library, but ignores the context-dependent nature of amino acid replacements and fails to take advantage of natural selection [3]. The ancestral mutation method does take advantage of ancestral sequences but, unlike REAP, does not infer exactly which replacements are most likely to impart a desired phenotype [4]. Ancestral sequence reconstruction also takes advantage of natural selection since ancient sequences theoretically existed and were deemed fit to live. The downside to this approach, however, is that it does not incorporate extant sequence information and is dependent upon the accuracy of computational models. Ultimately, no other method exploits natural selection to create a small library with a high degree of functional proteins as effectively as the REAP method. In one example, a lab used a REAP-designed library of only 93 mutants to engineer a variant of DNA polymerase capable of faithfully incorporating unnatural nucleosides/nucleotides [5].

Generating small libraries has multiple advantages. Large libraries are time-consuming to synthesize, and it is expensive to assay them. This is due in part to the fact that assays for protein function and activity do not often lend themselves to high-throughput screening. Ultimately, REAP was developed to save laboratories time and money to assay variant libraries by directing the engineering of the protein. With the REAP method, the final variant library is small and demonstrates a high degree of functionality as well as phenotypic diversity.

## 2 Materials

Software programs and databases used are cited in Tables 1 and 2.

**Table 1**
**List of databases that contain nucleotide and protein sequence data and other information that may be useful for a REAP analysis**

| | |
|---|---|
| *BRENDA*<br>Comprehensive database of enzymes | http://www.brenda-enzymes.info/ |
| *EBI*<br>European Bioinformatics Institute: DNA, protein, and macromolecular structure databases | http://www.ebi.ac.uk/ |
| *GenBank*<br>Annotated database of publically available DNA sequences | http://www.ncbi.nlm.nih.gov/genbank/genbank/ |
| *NCBI*<br>National Center for Biotechnology Information: resource for biomedical and genomic information | http://www.ncbi.nlm.nih.gov/ |
| *PDB*<br>Protein Data Bank: searchable data bank of protein macromolecular structure | http://www.wwpdb.org/ |
| *PFAM*<br>Database of protein families, searchable by sequence | http://pfam.sanger.ac.uk/ |
| *PIR*<br>Protein Information Resource: nonredundant, annotated database of protein sequences | http://pir.georgetown.edu/ |
| *PRF*<br>Protein Research Foundation: peptide/protein database searchable by sequence | http://www.prf.or.jp/index-e.html |
| *RefSeq*<br>Reference sequences: database of DNA sequences labeled by established annotation | http://www.ncbi.nlm.nih.gov/RefSeq/ |
| *Swiss-Prot*<br>Database of DNA sequences manually annotated and reviewed | http://www.uniprot.org/ |
| *TPA*<br>Third-party annotation: database of DNA sequences annotated by inference and experimentation | http://www.ncbi.nlm.nih.gov/genbank/tpa/ |

## 3 Methods

### 3.1 Assembly of Homologous Sequences

The first step in REAP is to select the extant protein sequences required for all subsequent downstream steps. The sequences selected should represent a diverse set of homologs with different phenotypic properties, generally from different evolutionary lineages or domains of life. Since subsequent analysis of the homologs chosen will be the basis for the diversity represented by a variant library, ensure that the chosen set of extant sequences will be sufficiently diverse to have the properties desired. Paralogs may also be included in this set if there is sufficient sequence similarity for an accurate alignment and if the paralogs offer relevant phenotypic diversity.

**Table 2**
**List of software programs that may be utilized in a REAP analysis**

| | |
|---|---|
| BLAST | http://blast.ncbi.nlm.nih.gov/ |
| Basic local alignment search tool: DNA database searchable by sequence or annotation | |
| *Clustal Omega* | http://www.ebi.ac.uk/Tools/msa/clustalo/ |
| Creates a multiple sequence alignment and phylogenetic tree based on sequence input | |
| *T-Coffee* | http://tcoffee.crg.cat/ |
| Creates a multiple sequence alignment based on DNA, RNA, or protein sequences | |
| *MrBayes* | http://mrbayes.sourceforge.net/ |
| Creates a Bayesian-based phylogenetic tree from a multiple sequence alignment | |
| *DIVERGE* | http://xungulab.com/software/diverge2/document/ |
| Identifies protein functional divergence using site-specific shifted evolutionary rates | |
| *Rate shift analysis server* | http://www.daimi.au.dk/~compbio/rateshift/protein.html |
| Provides rate shift analysis for proteins based on a multiple sequence alignment | |
| *PAML* | http://abacus.gene.ucl.ac.uk/software/paml.html |
| Phylogenetic analysis by maximum likelihood: provides phylogenetic analysis of DNA or protein sequences using maximum likelihood | |

The number of homologous sequences assembled is an important consideration. Enough sequences must be collected so that they represent sufficient functional diversity and the variants designed will have the desired activity. However, if the collection is too large, the subsequent steps may be overly difficult. The goal is to balance the diversity represented by the homologs with the ease of performing REAP with a smaller collection of sequences. Generally, 50–200 sequences will fulfill the needs for this method. In addition to the sequence set, an out-group is also required to "root" the phylogenetic tree. This sequence should be more distantly related to the homologs and is generally selected from a different lineage. Additionally, it may be advisable to include a few extra homologs since some sequences may need to be excluded during the alignment.

There are a number of databases that can be used to find homologous sequences, including BRENDA, EBI, GenBank, NCBI, PFAM, PDB, PIR, PRF, RefSeq, Swiss-Prot, and TPA (Table 1). One can search by protein name or by an Enzyme Commission (EC) number. One also can search by sequence similarity using BLAST (basic local alignment search tool; *see* **Note 1**). However, if one chooses to search using an annotation, name, or EC number, make sure the protein is truly a homolog. This is because some sequences may be mislabeled which will complicate the alignment in the next step [6]. If, while doing the alignment, a sequence is mislabeled or is unusually short in length, it is feasible to exclude that sequence and continue with the protocol.

### 3.2 Generation of a Multiple Sequence Alignment

To create a multiple sequence alignment (MSA), you may begin with one of several software programs such as Clustal Omega [7] or T-Coffee [8]. However, after obtaining the computed alignment, it must be evaluated by eye and adjusted manually. Since the REAP method depends on the assumption that amino acids that align in the MSA share a common ancestor, care must be taken to optimize the MSA and achieve a correct alignment (*see* **Note 2**). This step may be time-consuming, but a correct alignment is crucial to making an accurate phylogenetic tree in the next step. Take the time required to ensure the MSA is correct. Throughout the process of adjusting the MSA, it may be helpful to rerun the alignment frequently to obtain the most updated version.

After using software to create an MSA, analyze the alignment and determine if any sequences do not fit. If any sequence looks like it is not a true homolog, exclude it from the analysis. If sequences were identified by name or EC number, it is possible you may have included a mislabeled sequence. Any sequences which appear not to be legitimate homologs should be removed. Additionally, sequences which exhibit many insertions or deletions may need to be eliminated if there are too many gaps for the sequences to properly align. Ultimately, the goal is to minimize gaps in the alignment. Sequences with an abundance of insertions or deletions can make this step overly complicated and should be excluded.

Once you have decided all the sequences are true homologs, gaps must be removed manually. Trimming the N and C termini to normalize the length of sequences is generally useful (*see* **Note 3**). The N and C termini are commonly subjected to mutations which do not affect overall function. Therefore, these areas may have many gaps and mutations which may complicate the protocol without adding functional diversity to the final protein library. Additionally, if an insertion is only seen in one sequence, it may be safe to simply delete that portion of the sequence; however, generally treatment of insertions is not so straightforward. Insertions or deletions that are responsible for functional diversity must be retained in the MSA as these may lend the desired function to the library's variant proteins. Knowledge of evolutionary relationships can help with refining the MSA and choosing which insertions should be kept since they may impart functional diversity and which ones should be deleted. You may wish to rerun the alignment frequently during these adjustments so you are always working with the most updated version.

It is likely that you will not be able to eliminate all gaps. When you have gaps which will remain in the final MSA, it is important to ensure the amino acids flanking the gap are correctly aligned.

The residues bordering a gap must be examined to determine if the alignment software program put the residues on the correct side of the gap. The alignment should be examined by eye to determine if the residue aligns better with the MSA at the front or back end of the gap.

### 3.3 Creation of a Phylogeny

For each MSA, a phylogenetic tree must be created in order to identify evolutionary relationships and distances. This tree must be created de novo as opposed to using a published phylogeny. A published phylogenetic tree will not necessarily match the tree for a specific protein due to horizontal gene transfer, different mutation rates, and using different sets of extant sequences. A number of methods can be used to create a phylogenetic tree including maximum likelihood, maximum parsimony, Bayesian inference approaches, and distance-based approaches such as neighbor joining [9]. Each method creates a tree using its own set of parameters to determine the best fit. Software programs can be used for this step, such as MrBayes which will create a Bayesian-based tree [10].

Once the evolutionary tree for the protein has been created from the MSA, it should be compared to a published tree that illustrates accepted evolutionary relationships. If discrepancies between the new protein tree and the published tree exist, evaluate the two phylogenies and determine which illustration is most likely correct for the protein. At this point, the MSA should be adjusted until the trees match or any discrepancies are explicable. Errors in the phylogeny of the protein could result from a limited sampling that lacks appropriate representation across relevant phyla. If this seems to be the case, sequences should be added or removed from the MSA and the MSA realigned before reconstructing the phylogeny. Alternatively, you may decide it is likely that a horizontal gene transfer event occurred for a particular protein in which case the phylogenies would not perfectly match. Continue to manipulate the collection of sequences and the MSA until the phylogeny seems consistent and accurate.

### 3.4 Recognition of Signatures of Functional Divergence

The goal of REAP is to identify evolutionary signatures of functional divergence to guide protein engineering and library design [11]. Sites of amino acid conservation suggest that these residues are important for protein function. The MSA and phylogeny allow the comparison of homologs which permits the identification of conserved residues as well as analysis of the rates of conservation over phyla by modeling site-specific rate shifts. Identifying type I and type II functional divergence will allow one to identify which amino acids in the sequences are most likely to be linked to protein function. This in turn can guide the design of the variant library to sample mutations which nature has already tested and theoretically deemed fit to live.

In type I functional divergence, also called heterotachy or covarion-like divergence, a specific site is highly conserved in one portion of a phylogenetic but is variable in another portion of the tree. A heterotachy model models the notion that sites can have different rates of mutation at different points in their evolutionary history. Incorporating a phylogeny enables the comparison of rates of mutation/replacements among branches. The mutation rate, which differs between branches in type I functional divergence, suggests that a residue is functionally relevant in the portion of the phylogeny in which it is conserved. A low mutation rate indicates that an amino acid is important for protein function. A residue that has a high mutation/replacement rate in a different portion of the phylogeny is probably not important for the function of a variant.

Type II functional divergence is defined by low mutation rates across the phylogenetic tree, but the amino acid that is conserved varies between two portions of the tree. This type of pattern suggests that a residue is important for protein activity, but the function of the protein is different between the two portions of the tree. Programs like DIVERGE and rate shift analysis server can calculate the statistical significance of functional divergence using an MSA and phylogeny as input. The information gleaned from these programs is helpful, because it provides a metric to measure the strength of functional divergence and provides a guide by which to decide which particular amino acids should be incorporated in a library of variants. Generally, when designing a library, it is best to focus mutations in sites carrying the highest signature of functional divergence, regardless if it is type I or type II (Fig. 1).

### 3.5 Ancestral Sequence Reconstruction

After identifying sites and phylogenetic branches associated with functional divergence, the next step is to reconstruct and resurrect putative ancestral sequences. In order to perform ancestral sequence reconstruction, first determine how these sequences evolved along the branches of a phylogeny. By using the analyses of the MSA and tree, it is possible to design the library to incorporate sequences associated with functional divergence that have been sampled by nature. These mutations, derived from the ancestral sequences, have, in theory, been tested by natural selection and deemed "fit to survive." By using this method to design a library of variants, the library will incorporate mutations already sampled by nature.

In order to design ancestral sequences, software programs, such as PAML, use an MSA and phylogenetic tree to calculate the probability of a specific residue to occupy a site in an ancestral sequence [12]. PAML uses maximum likelihood as model and calculates the probabilities of amino acids changing along branches based on a model of amino acid replacements. This provides a metric that allows one to determine the likelihood of a specific replacement or a collection of replacements having occurred in

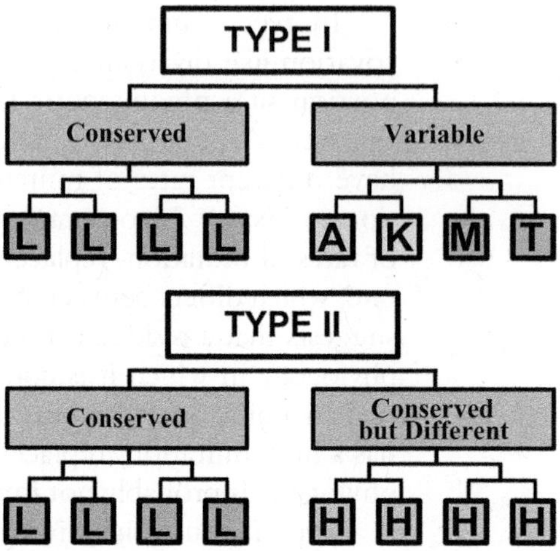


**Fig. 1** Type I and type II functional divergence. *Colored squares* represent the amino acids of extant homologs, represented by their *single-letter abbreviation*, in a specific site in the multiple sequence alignment. Type I functional divergence is depicted by the presence of a conserved leucine in one half of the phylogenetic tree but no sequence conservation in the other half of the tree, as illustrated by the presence of various amino acids at that position. Type II functional divergence is illustrated by the presence of a conserved leucine in one half of the phylogenetic tree and the presence of a conserved histidine at the same site in the other half of the tree. In this example, the amino acid in each lineage is conserved, but the identity of that amino acid differs between lineages. Type I and type II functional divergence can be used to identify sites and amino acids associated with protein phenotype

an ancestral sequence. It is generally most useful to focus on replacements found in the most probable ancestral sequences when designing a variant library.

### 3.6 Creation and Characterization of a Variant Library

A variant library created using REAP will include residues associated with diverse protein function in extant sequences as well as ancestral sequences going back as far as the MSA and tree allow. These sequences highlight replacements that may be tested for a desired protein function. One may synthesize sequences with single or multiple replacements. If one chooses to incorporate multiple replacements into each variant, then each amino acid replacement should be put in multiple variants (i.e., different permutations of mutations) in order to identify possible instances of epistasis. It can also be beneficial to consider protein structure and amino acid properties when designing the variant library and choosing which replacements to incorporate (*see* **Note 4**). Computational models can also help reduce the number of variants and focus the library.

Choosing an appropriate backbone sequence is vital to the success of REAP as it can affect the phenotype imparted by the mutations. Most of the time, the protein exhibiting a phenotype most closely related to the target phenotype is chosen, especially in cases when the goal of the experiment is to enhance function of an extant protein or when it is expected that relatively few independent mutations will be required to produce the desired activity. However, it may be appropriate to choose a different backbone sequence in certain situations, such as when another protein has been better characterized or is known to be easier to purify. Even if the functional characteristics of a candidate backbone protein are more distant from the goal phenotype, it may be a prudent choice if the most closely related protein has been only poorly characterized or if the purification is known to be especially difficult.

Another important factor in the selection of a backbone sequence is protein stability. Since mutations frequently destabilize a protein, it is best to begin with a highly stable protein that will be able to survive some increased instability without losing its three-dimensional structure or function. Either the consensus or ancestral proteins are good choices [13] as these sequences might provide a structure more compatible with some of the likely mutations, thus reducing complications arising from epistasis.

Depending on the synthesis method used to create variants, the library size can be easily controlled. When choosing a synthesis method, consider the sequence space to be assayed against the expense and time involved. The larger the library and the more complex the assay, the more time and resources will be required to characterize the variant library. This creates an incentive to keep the library small. These factors should be balanced to provide a library large enough to offer variants with a desired function, but small enough to be economically feasible.

Select synthesis and assay methods appropriate to the size and design of the library. If one has chosen to use single mutations/replacements, site-directed mutagenesis can be very efficient. However, if one is making a larger library or incorporating multiple changes into each variant, using a commercial company may be an attractive option. However, the cost of using a commercial company may outweigh the convenience if the variant library is expected to be very large. Another commercial option offers the opportunity to specify the frequency of a mutation and the average number of sequence differences per variant, but specific combinations of mutations among the variants will be random. This allows one to test a large number of mutations with a smaller library. However, after the initial assays, more tests would be required to determine the exact replacements responsible for a desired function in all promising variant proteins.

## 4 Notes

1. Performing a BLAST search.

   When using BLAST to collect homologous sequences, you must select a reference sequence. This sequence is commonly the best performing or most studied extant protein; however, two sequences may occasionally need to be used if there is a great deal of divergence among families or domains. For example, two reference sequences may be needed, one for bacteria and another one for eukaryotes.

Usually it is best to use BLAST to search one genome at a time and then choose a sequence from among the best hits. If you BLAST multiple genomes at once, it is easy to introduce bias by selecting sequences most closely related to the reference sequence, which can skew the resulting collection of homologs.

Take care to only collect homologous sequences when using BLAST. It is possible that no homolog may exist within a specific genome or it is possible that sequences can be misidentified. If unsure, consider performing a reciprocal search by using the new sequence as the reference and verifying that the original reference sequence is a best hit. If a true ortholog has multiple hits within a genome, select the best sequence based on alignment, annotation, and gene size.

2. Preference for amino acid sequences.
   Amino acid sequences are often used, but it is certainly possible to use the nucleotide sequences if DNA is appropriate for a particular study. Using amino acids in the alignment allows a more accurate alignment and also eliminates sequences that encode the same amino acid sequence but using different codons. Since an altered protein function is the goal of the REAP method, amino acid replacements, not altered codons, are generally the focus of the analyses. Aligning nucleotides can create a very large collection of sequences to be analyzed as well as complicating the protocol without ultimately adding functional diversity to the variant library.

3. Dealing with the N and C termini.
   Since amino and carboxyl termini are subjected to many insertion and deletion events, they can make it difficult to obtain a proper alignment. It is often most practical to trim them and normalize the length of the collected homologous sequences. This will simplify the sequence reconstruction and prevent these ends from influencing a phylogenetic tree. Often, the ends are involved in species-specific signaling important for gene expression but not necessary for the REAP method. However, in the event that these termini are needed, you may simply use the ends from an extant protein sequence.

4. Use of other information in library design.
   When additional information is available about a specific protein, such information should be incorporated into the design of the library. Information from either experiments or modeling can influence which REAP mutations are given

precedence. For example, a subset of mutations identified by REAP might be used because of their known proximity to the active site of an enzyme. Published structures of the protein or site-directed mutagenesis studies of the protein can be used to guide the final library design. Studies examining the chemical importance of specific sites, such as the conservation of a salt bridge at a particular site, could also be informative. In that case, incorporation of a pair of mutations may be most appropriate to enable retention of the conserved structure in the mutants. Additional information about the tolerance for mutation at conserved sites might already be available from prior studies, which can be used to guide the final design of the variant library.

## Acknowledgments

This work was supported by grants from NASA Exobiology (NNX08AO12G), NASA Astrobiology Institute, and the School of Biology at Georgia Institute of Technology (E.A.G). This work was also supported by the School of Chemistry and Biochemistry and a Cherry Emerson Fellowship from Georgia Institute of Technology (V.E.C.).

### References

1. Cole MF, Gaucher EA (2011) Exploiting models of molecular evolution to efficiently direct protein engineering. J Mol Evol 72:193–203
2. Fan Y, Fang W, Xiao Y, Yang X, Zhang Y, Bidochka MJ, Pei Y (2007) Directed evolution for increased chitinase activity. Appl Microbiol Biotechnol 76:135–139
3. Cole MF, Gaucher EA (2011) Utilizing natural diversity to evolve protein function: applications towards thermostability. Curr Opin Chem Biol 15:399–406
4. Woese CR (1987) Bacterial evolution. Microbiol Rev 51:221–271
5. Chen F, Gaucher EA, Leal NA, Hutter D, Havemann SA, Govindarajan S, Ortlund EA, Benner SA (2010) Reconstructed evolutionary adaptive paths give polymerases accepting reversible terminators for sequencing and SNP detection. Proc Natl Acad Sci U S A 107:1948–1953
6. Benner SA, Gaucher EA (2001) Evolution, language and analogy in functional genomics. Trends Genet 17:414–418
7. Larkin MA, Blackshields G, Brown NP, Chenna R, McGettigan PA, McWilliam H, Valentin F, Wallace IM, Wilm A, Lopez R, Thompson JD, Gibson TJ, Higgins DG (2007) Clustal W and Clustal X version 2.0. Bioinformatics 23:2947–2948
8. Notredame C, Higgins DG, Heringa J (2000) T-Coffee: a novel method for fast and accurate multiple sequence alignment. J Mol Biol 302:205–217
9. Yang Z, Rannala B (2012) Molecular phylogenetics: principles and practice. Nat Rev Genet 13:303–314
10. Huelsenbeck JP, Ronquist F, Nielsen R, Bollback JP (2001) Bayesian inference of phylogeny and its impact on evolutionary biology. Science 294:2310–2314
11. Gaucher EA, Gu X, Miyamoto MM, Benner SA (2002) Predicting functional divergence in protein evolution by site-specific rate shifts. Trends Biochem Sci 27:315–321
12. Yang Z (2007) PAML 4: phylogenetic analysis by maximum likelihood. Mol Biol Evol 24:1586–1591
13. Tawfik DS, Bershtein S, Goldin K (2008) Intense neutral drifts yield robust and evolvable consensus proteins. J Mol Biol 379:1029–1044

# Index

## A



## B



## C



## D



Elizabeth M.J. Gillam et al. (eds.), *Directed Evolution Library Creation: Methods and Protocols*, Methods in Molecular Biology, vol. 1179, DOI 10.1007/978-1-4939-1053-3, © Springer Science+Business Media New York 2014

## E



## F



## G

## S



## T



## U



## W



## Y

9781493910526